Fields, Forces, and Flows in Biological Systems

Fields, Forces, and Flows in Biological Systems

Alan J. Grodzinsky

With the technical and editorial assistance of Dr. Eliot H. Frank

CRC Press
Taylor & Francis Group
Boca Raton London New York

CRC Press is an imprint of the
Taylor & Francis Group, an **informa** business

Garland Science
Vice President: Denise Schanck
Editor: Summers Scholl
Assistant Editor: Alex Engels
Production Editors: Mac Clarke and Georgina Lucas
Cover Design: Andrew Magee
Copyeditor: Mac Clarke
Typesetting: TechSet
Proofreader: Sally Huish

To Gail and Michael

CRC Press
Taylor & Francis Group
6000 Broken Sound Parkway NW, Suite 300
Boca Raton, FL 33487-2742

First issued in paperback 2019

ISBN 13: 978-0-8153-4212-0 (hbk)
ISBN 13: 978-0-367-86435-4 (pbk)

Library of Congress Cataloging-in-Publication Data
Grodzinsky, Alan J.
Fields, forces, and flows in biological systems / Alan J. Grodzinsky; with the technical and editorial assistance of Eliot H. Frank.
    p. cm.
  Includes bibliographical references and index.
  ISBN 978-0-8153-4212-0 (hardback)
1. Biological transport. 2. Biological systems. I. Frank, Eliot H. II. Title.
  QH509.G76 2011
  571.6'4–dc22   2011000915

Preface

SCOPE AND PURPOSE

This textbook describes the fundamental driving forces for mass transport, electric current, and fluid flow as they apply to the biology and biophysics of molecules, cells, tissues, and organs. Basic mathematical and engineering tools are presented in the context of biology and physiology. The chapters are structured in a framework that moves across length scales from molecules to membranes to tissues. Examples throughout the text deal with applications involving specific biological tissues, cells, and macromolecules. In addition, a variety of applications focus on sensors, actuators, diagnostics, and microphysical measurement devices (e.g., bioMEMS/NEMS microfluidic devices) in which transport and electrokinetic interactions are critical.

The book is written for beginning graduate students and advanced undergraduates and is aimed at an audience that has seen basic freshman physics (mechanics, electricity, and magnetism) as well as undergraduate exposure to differential operators and differential equations. In addition, it is hoped that the textbook will be a valuable resource for interdisciplinary researchers, including biophysicists, physical chemists, materials scientists, and chemical, electrical, and mechanical engineers seeking a common language for the subject.

PHILOSOPHY OF THE TEXTBOOK

A primary objective of this text is to integrate the fundamental principles of transductive coupling between chemical, electrical, and mechanical forces and flows that are intrinsic to transport within biological tissues, membranes, macromolecules, and biomaterials. These principles are applied and interpreted in the context of state-of-the-art discoveries and challenges in biology, physiology, and macromolecular science. Thus, a balanced presentation of selected, basic principles from chemical, electrical, mechanical, and materials engineering and science is intended, in order to establish a common language for biological and biomedical engineering students, rather than the disparate languages often used by chemical, electrical, or mechanical engineers alone. However, this text is not intended as simply a compilation of examples in which traditional engineering techniques are applied to problems in physiology. Rather, current problems in biology and biophysics are used to motivate quantitative engineering approaches applicable from the nanometer length scale of biomacromolecules up through the complex structural organization of tissues and organs.

While the global aim of bioengineering curricula is to integrate engineering fundamentals with modern biological and medical science, the underlying interdisciplinary nature of the engineering components themselves can be a blessing and a curse. Some specialized texts by necessity are focused on one or two engineering disciplines connected to physiology. However, there is also a need for foundational bioengineering courses and texts that are cross-disciplinary even within the engineering fundamentals. The topic of transport is an ideal medium in which to achieve this objective.

At the same time, this text is not focused on transport alone. Rather, our objective is to describe more broadly the intra- and intermolecular *fields* and *forces* that affect the biology, physiology, and biophysics of molecules, cells, tissues, and organs. Most biological tissues and macromolecules (e.g., proteins, polysaccharides, and nucleic acids) are electrically charged under physiological conditions. Therefore, it is necessary to describe electrical forces and interactions from first principles, just as fundamentals laws are needed to describe fluid velocity fields and chemical transport. In this way, electrical interactions at multiple length scales can be addressed on an equal footing with forces that derive from local chemical and mechanical gradients. Thus, electrical forces at the nanoscale are fundamental components underlying the integration of molecular structure and biochemistry with tissue-level mechanics, transport, biophysics, and biology.

ORGANIZATION OF THE TEXTBOOK

The organization of the book derives from the order of major topics covered in the MIT Biological Engineering core curriculum subject: chemical transport in electrolyte media (Chapter 1); electrical fields and electrochemically mediated transport (selected sections from Chapters 2 and 3), the concepts of stress and the stress tensor (the early sections of Chapter 4); fluid mechanics and convective transport (Chapter 5); and integrative case studies involving physicochemical interactions at the macromolecular and cellular levels (examples in Chapter 4) and electrokinetic examples fundamental to MEMS and physical chemistry (Chapter 6). At the same time, many sections in Chapters 1, 4, and 7 are also essential components of MIT's undergraduate and graduate courses in molecular, cellular, and tissue biomechanics, including the rheological and deformational behavior of tissues and gels. Thus, the coverage of the textbook is broader than that used solely in a one-term course, and is intended to allow flexibility in choosing the order and content to adapt to the breadth of topics and courses of interest to biological and biomedical engineering students and instructors.

The course at MIT has evolved over many years, and is now typically taken each term by students in biological engineering, mechanical, chemical, and electrical engineering, materials science and engineering, and other departments. Thus, while each student has seen aspects of some of the material, none has seen the breadth of topics covered, and therefore no assumptions are made concerning the students' background, except for exposure to undergraduate-level mathematics and physics. Pedagogically, starting with chemical transport enables the mathematical treatment to focus initially on diffusion of a scalar (solute concentration) before the added complexities of dealing with vector fields (fluid velocity and electric fields). The spirit of the course is such that the instructor focuses each lecture using a current problem from the biological or medical literature, and then uses the text material as the fundamental basis for discussing, modeling, and critically analyzing and interpreting the results. The numerous examples and homework problems in the book are used by the students to gain additional experience and further insight. A solutions manual and figures from the book are available to qualified adopters of the text, and additional homework problems will be available to students on the book web site.

Acknowledgments

It is extremely difficult in a short space to acknowledge the tremendous debt of gratitude that I owe to the many people who have made this book possible. Dr Eliot Frank is the person who is mainly responsible for this book seeing light of day. Eliot is a long-time colleague and member of our research group whose intellectual strengths and technical talents are immeasurable. In addition to all of the figures he has drawn for each chapter and his LaTeX'ing of the text, he has been the go-to person to ask, "what do you think about" this section or that approach to the field.

The inspiration for this book all along has been my own teacher, mentor, and thesis advisor at MIT, Professor James R. Melcher, who tragically passed away at an early age 20 years ago. Jim contributed invaluable insights and sections to the early versions of this book when it was initially organized as a set of course notes for the first graduate subject I taught. In addition, he and the late Professor Hermann Haus, another giant in Electrical Engineering and Computer Science at MIT, taught me how to teach, how to organize a blackboard, and how to best tie together ideas in a logical fashion for presentation in a lecture to students ranging from freshman to advanced graduates.

I have also learned a tremendous amount from faculty who have co-taught this material with me more recently, including Professors William Deen, Doug Lauffenburger, Roger Kamm, and Mark Bathe. Doug has also been instrumental in the gestation and organization of the curriculum in Biological Engineering at MIT, where this material continues within the core. Perhaps the most influential people in my continued education and ongoing efforts for the course are the many students I've had the pleasure to have in class. Some of these extraordinarily gifted students have gone on to be instrumental as teaching assistants for the course before embarking on their own careers. Dr Rachel Miller additionally went on to contribute her invaluable skills in the editing of the final page proofs. Dr Paul Kopesky provided cover art from his own research (marrow-derived progenitor cells in a peptide hydrogel scaffold).

Over the years, many close colleagues have used sections of this book in their own teaching and research, and have thereby contributed invaluable insights and suggestions on the presentation, including Professors Larry Bonassar (Cornell University), Michael Buschmann (École Polytechnique de Montréal), Solomon Eisenberg (Boston University), Moonsoo Jin (Cornell University), Young-Jo Kim (Harvard Medical School), John Kisiday (Colorado State University), Marc Levenston (Stanford University), Tom Quinn (McGill University), Bob Sah (University of California San Diego), Joonil Seog (University of Maryland), Ronald Siegel (University of Minnesota), David Smith and his colleagues (University of Western Australia), and Aryeh Weiss (Jerusalem College of Technology); and Drs Peter Basser (National Institutes of Health), Jonathan Fitzgerald (Merrimack Pharmaceuticals), Jenny Lee (Gates Foundation), Michael DiMicco (Genzyme Corp.), Parth Patwari (Harvard Medical School), and Jeff Sachs (Merck and Co., Inc.).

I would also like to thank the reviewers of this textbook for their excellent suggestions: Michael Buschmann (École Polytechnique de Montréal), Delphine Dean (Clemson University), H. Pirouz Kaveh-pour (University of California at Los Angeles), George Pins (Worcester Polytechnic Institute), and Serafim Rodrigues (Centre National de la Recherche Scientifique). My thanks also go to the reviewers of the initial proposal: Solomon Eisenberg (Boston University), David G. Foster (University of Rochester), Anuradha Godavarty (Florida International University), Evan Morris (Purdue School of Engineering and Technology), and Muhammed H. Zamman (Boston University). Additional thanks go to the Whitaker Foundation for support during the earlier phases of this project, and to Linda Bragman and Han-Hwa Hung for their continued extraordinary efforts in our group at MIT. Finally, the editorial staff at Garland Science have been an absolute pleasure to work with, including initial and current editors for this project, Bob Rogers and Summers Scholl, assistant editor Alex Engels, copyeditor Mac Clarke, and production editor Georgina Lucas.

Contents in Brief

Preface   **v**

Acknowledgments   **vii**

1 Chemical Transport in Electrolyte Media   **1**

2 Electric Fields and Flows in Electrolyte Media   **33**

3 Electrochemical Coupling and Transport   **71**

4 Electrical Interaction Forces: From Intramolecular to Macroscopic   **139**

5 Newtonian Fluid Mechanics   **173**

6 Electrokinetics: MEMS, NEMS, and Nanoporous Biological Tissues   **203**

7 Rheology of Biological Tissues and Polymeric Biomaterials   **239**

Appendix A   Integral Theorems   **287**

Appendix B   Differential Operators in Various Coordinate Systems   **291**

Appendix C   Vector Identities   **299**

Appendix D   System of Units   **301**

Appendix E   Physical Constants   **303**

Index   **305**

Contents in Detail

Preface   v
Acknowledgments   vii

1 Chemical Transport in Electrolyte Media   **1**
  1.1   Introduction   1
  1.2   Diffusive Flux and Continuity   1
  1.3   A Molecular View of Diffusion   5
  1.4   Chemical Reactions: Some Common Examples   10
  1.5   Boundary Conditions and Boundary Value Problems ...   21
  1.6   Diffusion and Chemical Reactions   24
  1.7   Problems   28
  1.8   References   32

2 Electric Fields and Flows in Electrolyte Media   **33**
  2.1   Introduction   33
  2.2   Laws of Electromagnetism   33
  2.3   Maxwell's Equations in Media: Polarization,
       Magnetization, and Conduction   38
  2.4   Electromagnetic Waves   44
  2.5   The Quasistatic Approximations   48
  2.6   EQS and MQS Boundary Value Problems in
       Biological Systems   51
  2.7   Electric Fields and Currents in Conducting
       Biological Media   56
  2.8   Summary   62
  2.9   Problems   63
  2.10 References   69

3 Electrochemical Coupling and Transport   **71**
  3.1   Ion Transport in a Binary Electrolyte   71
  3.2   Coupled Steady State Diffusion Across a Neutral
       Membrane: The Diffusion Potential   77
  3.3   Electrodes and the Measurement of Membrane
       Potentials   84
  3.4   Donnan Equilibrium and the Donnan Potential: Charged
       Membranes and Tissues in Equilibrium   89
  3.5   Fixed Charge Membrane Models: Steady State Diffusion
       Potentials Across Charged Membranes   99
  3.6   Nonsteady Electrolyte Transport Phenomena in Media
       Containing Fixed Charge Groups: Electrodiffusion 107
  3.7   Problems 120
  3.8   References 137

**4 Electrical Interaction Forces: From Intramolecular to
Macroscopic**   **139**
  4.1   Introduction 139
  4.2   Force, Stress, Traction, and the Force Density 139
  4.3   Force Density and the Maxwell Stress Tensor 142
  4.4   Polarization Force Density 144
  4.5   The Diffuse Double Layer: Site of Intra- and Intermolecular
       Electrical Interactions 150
  4.6   The Double Layer in Relation to Native and Synthetic
       Biological Polyelectrolytes: A Historical Perspective ... 156

4.7   Potential Profile for Interacting Plane Parallel
       Double Layers   159
4.8   Force Equilibrium with Interacting Plane Parallel
       Double Layers   161
4.9   Interacting Double Layers: Rate Processes and Electrical
       Terminal Constraints   166
4.10   Problems   168
4.11   References   171

5 Newtonian Fluid Mechanics   **173**
5.1   Introduction   173
5.2   Conservation of Mass   173
5.3   Conservation of Momentum   174
5.4   Inviscid, Incompressible Flow: Bernoulli's Equation   175
5.5   Viscous Forces and Stress–Strain Rate Relations;
       The Navier–Stokes Equation   178
5.6   Plane, Fully Developed Flow of Incompressible,
       Viscous Fluids; Low-Reynolds-Number Flow   182
5.7   Stream Functions   185
5.8   Creep Flow Transfer Relations: Molecular, Cell, and
       Tissue Surfaces   186
5.9   Stokes Drag on a Rigid Sphere   189
5.10   Convective Diffusion: The Role of Convective Mass
       Transport   189
5.11   Diffusion Boundary Layers   191
5.12   Problems   196
5.13   References   201

**6 Electrokinetics: MEMS, NEMS, and Nanoporous
    Biological Tissues**   **203**
6.1   Introduction   203
6.2   Electrocapillary and Electrokinetic Phenomena   203
6.3   Electroosmosis and Streaming Potentials in Charged,
       Porous Membranes and Tissues   208
6.4   Membrane Electrokinetic Coupling: Phenomenological
       and Physical Models   213
6.5   Electrophoresis   217
6.6   Problems   222
6.7   References   237

**7 Rheology of Biological Tissues and Polymeric
    Biomaterials**   **239**
7.1   Introduction   239
7.2   Swelling and Deformational Behavior of Tissues:
       Illustrative Examples   239
7.3   Equilibrium Elastic Behavior   246
7.4   Viscoelastic Behavior   253
7.5   Poroelastic Behavior of Biomaterials: Theories
       and Experiments   259
7.6   Electrokinetic Transduction in Poroelastic Media   268
7.7   Problems   272
7.8   References   285

Appendix A Integral Theorems   **287**
Time Rate of Change of a Surface Integral   287
Time Rate of Change of a Volume Integral   288

Appendix B   Differential Operators in Various Coordinate Systems   **291**

Appendix C   Vector Identities   **299**

Appendix D   System of Units   **301**

Appendix E   Physical Constants   **303**

Index   **305**

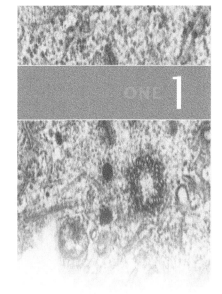

Chemical Transport in Electrolyte Media

1.1 INTRODUCTION

We first consider the diffusive motion of solutes in a fluid electrolyte medium. Low- and high-molecular-weight solutes are of interest, from mobile ions and small electrically neutral solutes to proteins, nucleic acids, glycoproteins, proteoglycans, and biological and chemical pharmaceutical compounds. Concentration gradients of such solutes will cause diffusive solute flow within and across biological tissues, cell membranes, and intracellular and extracellular spaces, and through hydrogels and other porous biomaterials. Fundamentally, diffusion is the transfer of a chemical species from regions of higher concentration to those of lower concentration by the mechanism of random Brownian motion of individual molecules within an ensemble [1–3]. Diffusive processes have also been thoroughly treated regarding carrier motions in semiconductor and solid state materials, gaseous media, and the general description of transport provided by nonequilibrium thermodynamics [4–6].

In this chapter, solute flux results solely from the presence of solute concentration gradients within the electrolyte medium. For the case of charged solutes, the additional electrical migration flux caused by the direct action of an applied electric field will be treated in Chapter 3. Since biological tissues and biomaterials contain fixed-charge groups that induce local, built-in electric fields ("self-fields"), solute migration fluxes will also need to be included in that discussion. The motion of solutes associated with convection of the fluid solvent are introduced in Chapter 5 after a more detailed treatment of Newtonian fluid mechanics.

After considering both a continuum and a molecular view of diffusive motions along with conservation of species (Sections 1.2 and 1.3), the concepts of boundary conditions and the solution of boundary value problems defined by the diffusion equation are introduced (Section 1.5). The importance of solute binding to cell surface receptors, extracellular matrix, and biopolymers in general is included in the context of specific examples in Section 1.4, leading to a discussion of diffusion–reaction rate processes and kinetics in Section 1.6. The ionization of biomolecular charge groups associated with acid–base reactions provides another important set of examples of diffusion–reaction in biological systems. This provides the opportunity to introduce key macromolecular constituents of the extracellular matrix (Section 1.4).

1.2 DIFFUSIVE FLUX AND CONTINUITY

Within an electrolyte medium, empirical evidence has shown that the diffusive flux N_i of solute species i with respect to the solvent is often linearly related to the local gradient in the concentration of that species, c_i, by

$$\boldsymbol{N}_i = -D_i \nabla c_i \tag{1.1}$$

where D_i is the diffusivity and the parameters in (1.1) have the SI units:
$\boldsymbol{N}_i \equiv$ molar flux (mol m^{-2} s^{-1})
$c_i \equiv$ molar concentration (mol m^{-3})
$D_i \equiv$ diffusivity (m^2 s^{-1})
In cartesian coordinates, $\nabla \equiv \boldsymbol{i}_x \partial/\partial x + \boldsymbol{i}_y \partial/\partial y + \boldsymbol{i}_z \partial/\partial z$ ($\boldsymbol{i}_j$ being the unit vector in the jth direction; see Appendix B for definitions of ∇ in other coordinate systems). The flux equation (1.1) can be regarded as a phenomenological constitutive law, often referred to as Fick's first law of diffusion (1855). The linear form of (1.1) corresponds to the case of a dilute solution in an isotropic medium, D_i being independent of concentration. A tensor form of the diffusivity would apply for anisotropic media [3]. We will see in later chapters that additional flux terms are added to (1.1) to account for solute motion caused by convection of the solvent as well as the motion of charged solutes in the presence of an electric field (the electrical migration flux).

1.2.1 Continuity of Solutes with Respect to a Stationary Fluid

Having established the basic point-by-point constitutive relation between the solute flux and the local solute concentration gradient, (1.1), we now use the integral form of continuity (conservation) to describe the global relation between *solute accumulation* in a region of space, the *net flux of solute entering the region*, and the rate at which solutes are *generated or lost by chemical reaction* within that region. The stationary control volume of Figure 1.1 is enclosed by the surface S, and the outward flux of solute species i is denoted by the vector $\boldsymbol{N}_i$. The continuity law then takes the form

$$\frac{d}{dt} \int_V c_i \, dV = -\oint_S \boldsymbol{N}_i \cdot \boldsymbol{n} \, da + \int_V R_i \, dV \tag{1.2}$$

where the left-hand term is the net accumulation of solute in V, and the minus sign in front of the surface integral on the right corresponds to net flux crossing into the control volume. R_i (mol m^{-3} s^{-1}) is the net volume rate of formation of species i by chemical reaction. The volume V and surface S are assumed to be fixed in space.

"Flux" versus "flux density", a word on nomenclature: The "net flux" of species through the surface S in (1.2) corresponds to the closed surface integral of $\boldsymbol{N} \cdot d\boldsymbol{a}$. Therefore, N is strictly the "flux density" of the solute, as used in [7]. However, in much of the literature on mass transport and electrochemical systems, $\boldsymbol{N}$ is simply referred to as the "flux," and we will therefore use that nomenclature throughout. We note this to anticipate any confusion on this point in other subsystems. For example, we will see in Chapter 2 that Gauss' law in the electric field subsystem involves the integral of $\epsilon \boldsymbol{E} \cdot d\boldsymbol{a}$ around a closed surface, which is called the "net electric flux," and $\epsilon \boldsymbol{E}$ is then called the (electric displacement) "flux density."

From (1.2), we can then derive the point-by-point differential form of continuity by using Gauss' theorem,

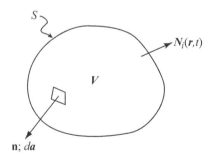

Figure 1.1 A control volume V is enclosed by a surface S. An area element on the surface has unit normal $\boldsymbol{n}$, so that a differential area vector can be defined as $d\boldsymbol{a} = \boldsymbol{n} \, da$. $\boldsymbol{N}_i(\boldsymbol{r}, t)$ is the outward flux across the surface.

$$\oint_S \boldsymbol{N}_i \cdot \boldsymbol{n} \, da = \int_V \nabla \cdot \boldsymbol{N}_i \, dV \tag{1.3}$$

and noting that the time derivative of the left hand term of (1.2) can be brought inside the volume integral since V and S are stationary:

$$\int_V \left(\frac{\partial c_i}{\partial t} + \nabla \cdot \boldsymbol{N}_i - R_i \right) dV = 0 \qquad (1.4)$$

Since the volume element dV is arbitrary, we can set the sum of the integrands in (1.4) to be zero, giving the differential form of continuity in the absence of convective or electrical forces:

$$\boxed{\frac{\partial c_i}{\partial t} = -\nabla \cdot \boldsymbol{N}_i + R_i} \qquad (1.5)$$

Combining the flux constitutive law (1.1) with the continuity law (1.5) in the absence of chemical reactions gives

$$\frac{\partial c_i}{\partial t} = \nabla \cdot (D_i \nabla c_i) \qquad (1.6)$$

For cases in which D_i is a constant independent of position, (1.6) gives the classic form of the diffusion equation (Fick's second law):

$$\frac{\partial c_i}{\partial t} = D_i \nabla^2 c_i \qquad (1.7)$$

where the Laplacian operator in Cartesian coordinates is $\nabla^2 c \equiv \partial^2 c/\partial x^2 + \partial^2 c/\partial y^2 + \partial^2 c/\partial z^2$, and thus the one-dimensional form of the diffusion equation is

$$\frac{\partial c_i}{\partial t} = D_i \frac{\partial^2 c_i}{\partial x^2} \qquad (1.8)$$

1.2.2 Continuity of Solutes with Respect to a Moving Deforming Fluid

We will often have occasion to consider the coupling of electrical, mechanical, and chemical processes occurring in a multicomponent system consisting of a fluid and several distinct chemical species. In formulating the governing laws or equations of change, it is often convenient to focus on a volume of fixed identity that may be moving and deforming. A continuity law must be written for each chemical species that, along with the relevant laws of motion and electromagnetism, serves to uniquely characterize the system of interest. For example, we might be interested in characterizing a small volume of fluid moving in the extracellular space of a deforming tissue and then modeling the diffusion of proteins out of the volume. A simpler case is pictured in Figure 1.2, which depicts a volume of water containing a dilute ionic solution. We wish to write a continuity law relating the time rate of change of electrolyte ions in V to the flux of ions through the surface S.

Defining a volume of fixed identity can be subtle. In a mixture such as that in Figure 1.2, each species has a different velocity—a situation not encountered in viscous fluid flow or heat conduction problems. One must first choose a relevant *local material* velocity for the mixture, a velocity with respect to which the motion of the volume V and surface S can be defined. Typical choices that are used in the membrane filtration literature [8], as well as that of general mixture theory in the study

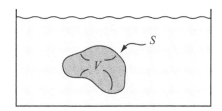

Figure 1.2 A moving, deforming volume of fluid V, with the associated closed surface S.

of transport phenomena [9], are the *mass-averaged velocity* v or the *molar-averaged* velocity v:

$$v \equiv \frac{\sum_i \rho_i v_i}{\sum_i \rho_i} \tag{1.9}$$

$$v \equiv \frac{\sum_i c_i v_i}{\sum_i c_i} \tag{1.10}$$

where ρ_i is the density (kg m^{-3} in SI units) and c_i is the molar concentration (mol m^{-3}) of the ith species.

In aqueous electrolyte media, the velocities (1.9) and (1.10) can be well-approximated by that of the solvent, water. For example, v for a 0.15 M NaCl solution becomes

$$v = \frac{c_{\text{Na}} v_{\text{Na}} + c_{\text{Cl}} v_{\text{Cl}} + c_{\text{H}_2\text{O}} v_{\text{H}_2\text{O}}}{c_{\text{Na}} + c_{\text{Cl}} + c_{\text{H}_2\text{O}}} \tag{1.11}$$

With $c_{\text{H}_2\text{O}} \simeq 55\,\text{M}$ ($= 1000\,\text{g L}^{-1}/18\,\text{g mol}^{-1}$) $\gg 0.15\,\text{M}$, and considering relevant velocity magnitudes, $v \simeq (c_{\text{H}_2\text{O}} v_{\text{H}_2\text{O}})/c_{\text{H}_2\text{O}} = v_{\text{H}_2\text{O}}$ in (1.11) if there is any reasonable fluid convection at all. (In mixtures of gases, the average velocities do not reduce to such a simplified result, and the choice of reference frame (local velocity) is usually one of convenience.)

With the results of (1.11), we can return to Figure 1.2 and define the volume of fixed identity, V. The convecting fluid provides the best reference frame, and V is therefore delineated by always following the same water molecules in the ensemble. If the molecules of interest were labeled, we would always be sure of following the given volume. In actuality, the statistical fluctuation of the water molecules results in a continual exchange of molecules back and forth across S, so that the volume of "fixed" identity must be defined within a statistical context. With this in mind, we write the *continuity* law relating the time rate of change of solute in V to the flux of solute through S using the integral form

$$\frac{d}{dt} \int_{V(t)} c_i \, dV = -\oint_{S(t)} \boldsymbol{N}'_i \cdot \boldsymbol{n} \, da + \int_{V(t)} R_i \, dV \tag{1.12}$$

where $\boldsymbol{N}'_i$ is the flux of the ith species across S and R ($\text{mol m}^{-3}\,\text{s}^{-1}$) is the volume rate of its formation due to chemical reactions. The volume and surface are both time-dependent, as indicated in (1.12); thus, $\boldsymbol{N}'_i$ with respect to the moving, deforming surface S is given by

$$\boldsymbol{N}'_i = -D_i \nabla c_i \tag{1.13}$$

We now use the result of an integral theorem (see Appendix A) that prescribes mathematically how to evaluate the time rate of change of a volume integral when the volume is a function of time. From (A.9) of Appendix A with $\zeta = c_i$,

$$\frac{d}{dt} \int_{V(t)} c_i \, dV = \int_{V(t)} \frac{\partial c_i}{\partial t} \, dV + \oint_{S(t)} c_i v \cdot \boldsymbol{n} \, da \tag{1.14}$$

where v is the fluid velocity, i.e., the velocity of the deforming surface. Equations (1.12) and (1.14) can be combined, Gauss' theorem being used to convert the closed surface integrals to volume integrals, resulting in a differential statement of continuity for the ith species,

$$\frac{\partial c_i}{\partial t} = -\nabla \cdot \boldsymbol{N}'_i - \nabla \cdot c_i v + R_i \tag{1.15}$$

Expansion of the second right-hand term of (1.15) gives

$$\nabla \cdot (c_i \boldsymbol{v}) = \boldsymbol{v} \cdot \nabla c_i + c_i \nabla \cdot \boldsymbol{v} \qquad (1.16)$$

For the most part, we will be dealing with incompressible liquids, for which a physical statement of conservation of mass is (see Chapter 5)

$$\nabla \cdot \boldsymbol{v} = 0 \qquad (1.17)$$

Equations (1.15)–(1.17) taken together give

$$\frac{\partial c_i}{\partial t} + \boldsymbol{v} \cdot \nabla c_i = -\nabla \cdot \boldsymbol{N}_i' + R_i \qquad (1.18)$$

The physical significance of the two terms on the left-hand side of (1.18) can be understood by asking how we might express the time rate of change of $c_i(x, y, z, t)$ for an observer *moving with the fluid*. In general, Taylor expansion of Δc_i gives

$$\Delta c_i = \frac{\partial c_i}{\partial t} \Delta t + \frac{\partial c_i}{\partial x} \Delta x + \frac{\partial c_i}{\partial y} \Delta y + \frac{\partial c_i}{\partial z} \Delta z \qquad (1.19)$$

The last three terms arise since a moving observer would measure a Δc_i if c_i varied in space—even if c_i were independent of time. Dividing by Δt and taking the limit as $\Delta t \to 0$,

$$\lim_{\Delta t \to 0} \frac{\Delta c_i}{\Delta t} \equiv \frac{\mathrm{D} c_i}{\mathrm{D} t} = \frac{\partial c_i}{\partial t} + \frac{\partial c_i}{\partial x} v_x + \frac{\partial c_i}{\partial y} v_y + \frac{\partial c_i}{\partial z} v_z \qquad (1.20)$$

where $\mathrm{D} c_i / \mathrm{D} t$ is the material or *convective derivative*. Equation (1.20) may be written conveniently in vector notation by noting that the last three terms on the right-hand side take the form $\boldsymbol{v} \cdot \nabla c_i$. Thus,

$$\frac{\mathrm{D} c_i}{\mathrm{D} t} = \frac{\partial c_i}{\partial t} + \boldsymbol{v} \cdot \nabla c_i \qquad (1.21)$$

But the right-hand side of (1.21) is identical to the left-hand side of (1.18), giving

$$\boxed{\frac{\mathrm{D} c_i}{\mathrm{D} t} = -\nabla \cdot \boldsymbol{N}_i' + R_i} \qquad (1.22)$$

which equates the total time rate of change of c_i for an observer moving with the fluid to the divergence of the flux of the *i*th species with respect to the moving fluid, accounting for chemical reactions that lead to the generation or recombination of species. This is precisely the continuity law that we were looking for, now written in differential form.

1.3 A MOLECULAR VIEW OF DIFFUSION

We first summarize several key aspects of solute flow by diffusion that have been emphasized in general treatments of this subject. First, there is no net force on any particular solute molecule in the direction of flow. Rather, solute flux is completely determined by random thermal motion in which it is more likely that there is a net flux of solutes flowing from regions of high to low concentration. Second, especially in the limit of dilute solutions, the solute molecules are assumed to undergo collisions primarily with solvent molecules and not with each other.

As a result, the solutes are not directly pushing each other in the direction of flow. Rather, each solute molecule is equally likely to move in any direction after a collision with a solvent molecule. Finally, from a probabilistic perspective, it is extremely unlikely that a given solute molecule will return to the exact position from which it started several collisions earlier.

1.3.1 Diffusivity in the Context of Random Thermal Motion

The concepts summarized above derive from Einstein's series of fundamental papers (1905–1908) on the theory of Brownian motion, which were greatly influenced by experimental observations known at that time (the English translation of the compendium of these papers can be found in [1].) While there have been many derivations of solute diffusivity in terms of random walk models of diffusion, the simplest mathematically is probably that in Einstein's 1908 paper [10], which also provides a heuristic derivation of Fick's first law.

Einstein considered a simple one-dimensional problem pictured in Figure 1.3 in which a dilute solution of molecules has an initial concentration profile $c(x)$ that results in a net diffusive flux of solute, N_x, in the $+x$ direction. The solute molecules undergo random thermal motions in which solutes collide only with the solvent molecules. During a time Δt, the solute molecules in the ensemble are displaced on average by a distance Δx from their previous position. (By analogy to carrier transport in semiconductors and to the kinetic theory of gases, Δx and Δt are often related to the mean free path and intercollision time, respectively, of the collision process.)

We now estimate the number of moles of solute that diffuse from left to right across the midplane (x) in time Δt. Statistically, half the solute molecules in the region between $x - \Delta x$ and x can move to the right across the midplane and the other half to the left, away from the midplane. Similarly, half the molecules between x and $x + \Delta x$ move to the left during the time increment. Therefore, in time Δt, the total

Figure 1.3 Flux across midplane due to a gradient in the concentration of species, $c(x)$.

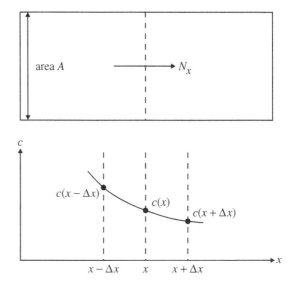

number of moles of solute molecules crossing the midplane to the right (in the $+x$ direction) is

$$N_x(\text{mol m}^{-2}\,\text{s}^{-1})\,A(\text{m}^2)\,\Delta t(\text{s}) = \underbrace{\frac{1}{2}\left[\frac{c(x-\Delta x)+c(x)}{2}\right]A\,\Delta x}_{\text{left-to-right}}$$

$$\underbrace{-\frac{1}{2}\left[\frac{c(x)+c(x+\Delta x)}{2}\right]A\,\Delta x}_{\text{right-to-left}} \qquad (1.23)$$

$$N_x\Delta t = \frac{\Delta x}{4}\left[c(x)-\left(\frac{\partial c}{\partial x}\Big|_x\Delta x\right)-c(x)-\left(\frac{\partial c}{\partial x}\Big|_x\Delta x\right)\right] \qquad (1.24)$$

where the Taylor series has been used to approximate the terms $c(x-\Delta x)$ and $c(x+\Delta x)$ to first order. This leads to an expression relating the flux N_x and the local gradient in concentration at the midplane:

$$N_x = -\left(\frac{(\Delta x)^2}{2\,\Delta t}\right)\frac{\partial c}{\partial x} \qquad (1.25)$$

If we compare this with Fick's first law, (1.1), we arrive at the expression derived by Einstein relating solute diffusivity D to Δx and Δt:

$$\boxed{D = \frac{(\Delta x)^2}{2\,\Delta t}} \qquad (1.26)$$

where Δx is the length of the path taken on average by a solute molecule moving in the x direction during a time Δt. While the mathematical details of this one-dimensional derivation are admittedly simplified compared with those of a fully three-dimensional development, several fundamental physical concepts emerge. Diffusion times are proportional to the square of the distance over which solute diffusion occurs. In addition, the basic units of the diffusivity ($\text{m}^2\,\text{s}^{-1}$) are delineated. Finally, as emphasized in Einstein's paper, "this wandering about of the molecules of the solute in a solution will have as a result that an originally non-uniform distribution of concentration of the solute will gradually give place to a uniform one."

Example 1.3.1 The Stokes–Einstein Relation for Diffusivity of Solutes For larger spherical solutes immersed in a fluid, the relation between solute diffusivity and fluid viscosity was reasoned by Einstein using the kinetic theory of Brownian motion coupled with results from continuum mechanics. The solute was assumed to be large enough that diffusion would involve collisions with many smaller solvent (e.g., water) molecules and, thus, an effective fluid drag would oppose solute motion. From fluid mechanics, the frictional drag force f exerted by a fluid of viscosity μ on a spherical particle of radius a moving at steady velocity U through the fluid was already well known: $f = 6\pi a\mu U$ (Stokes' law, 1851; see Chapter 5, Section 5.9

EXAMPLE

for a detailed derivation). At the same time, the kinetic theory and the law of van't Hoff (1852–1911) were known, the latter relating the concentration of solutes to the osmotic pressure associated with solute collisions with container walls ($P^{os} \simeq RTc_i$, see Chapter 3 and Problem 3.9). The resulting thought model involved a balance between (1) the pressure gradient $\partial P^{os}/\partial x$ (which has the units of a force density ($\mathrm{N\,m^{-3}}$) and is proportional to $RT\,\partial c_i/\partial x$ from van't Hoff's law) and (2) the force density associated with the frictional drag on the total number of solutes per unit fluid volume ($c_i N_{AV}$), in which each solute molecule is subjected to the drag force $6\pi a\mu U$. Referring to Figure 1.3, the "driving force" for diffusive flux in the $+x$ direction is $-\partial P^{os}/\partial x \sim (-\partial c_i/\partial x)$, and, with zero net flux in equilibrium, the balance gives

$$-\frac{\partial P^{os}}{\partial x} = -RT\frac{\partial c_i}{\partial x} = (6\pi a\mu U)(c_i N_{AV}) \qquad (1.27)$$

Combining this relation with the flux equation (1.1) and noting that the product $c_i U$ has the units of flux N_i, we arrive at the Stokes–Einstein relation for the diffusivity:

$$\boxed{D = \frac{k_B T}{6\pi a\mu}} \qquad (1.28)$$

where the Boltzmann constant $k_B = RT/N_{AV}$, R is the universal gas constant, and N_{AV} is Avogadro's number. We also note that for a spherical particle, the radius a can be approximated by (molecular weight)$^{1/3}$. Thus, the diffusivity is weakly dependent on the molecular weight.

1.3.2 The Validity of a Continuum Formulation

In biological systems, descriptions of chemical transport as well as electrical and mechanical fields and flows are most often presented in the context of continuum models. The length scales of the continuum are chosen so that relevant variables of interest can be described by continuous functions of position. These variables include the concentrations of chemical species, the densities of fixed and mobile charges, the mass densities of fluid or solid elements of the biomaterial, and their motions (velocity or deformation). Thus, the continuum concept is based on the ability to choose "infinitesimally small" volumes of interest, small enough to define these variables on a point-by-point basis, but large enough to contain many individual molecules or charges. In electrochemical systems, each charge in the incremental volume of the electrolyte medium will experience the same electric field provided there are many more charges outside that volume; and if there are sufficient numbers of charges in all such volumes, then all field quantities will vary continuously with position in the medium.

The validity of a continuum formulation also reflects the nature of the medium of interest. Chemical species in aqueous media are viewed as undergoing random thermal motion within a "sea" of water molecules. In contrast, Maxwell's macroscopic theory of electromagnetic fields was initially formulated in terms of densities of stationary charges or moving charges (currents) within a vacuum, more like the picture of a gas than a fluid. When reformulated for conducting, polarizable media (e.g., biological media), Maxwell's formulation can distinguish between the microscopic fields associated with the atomic "lattice" structure of a medium and macroscopic averaged fields that ultimately constitute a continuum description.

While charge groups within biological tissues and synthetic polymers are localized to within statistical fluctuations according to molecular structure, there are often sufficient numbers of charge groups present such that meaningful linear, surface, or volume charge densities can be defined in a continuum sense. Superimposed on the distribution of these immobilized charges, there exist in the electrolyte phase space-charge regions having dimensions as small as several tenths of a nanometer. Fluid flows within these porous regions may also occur over characteristic length scales as small as nanometers, such as that in electrokinetic flows within microfluidic channels or pores within the macromolecular matrix of tissues. Thus, we may need to consider simultaneously the length scales associated with chemical concentration gradients, electrical potential gradients, and fluid pressure gradients, all within a complex biological structure.

Of course, the biological system of interest may impose its own natural length scale, e.g., interstitial spacing between macromolecules, distances between channels or receptors along a cell surface, the diameter of cells within a tissue, or the spacing between cells. Thus, we should carefully revisit the validity of a continuum formulation on a case-by-case basis, since there are many examples in which atomic or molecular models can be formulated outside the context of a continuum.

Example 1.3.2 Chemotaxis, Cell Dimensions, and the Continuum Approximation (After a classroom problem from DA Lauffenburger) Gradients in the concentration of chemoattractant factors can cause the migration of cells towards the factor [11]. For example, neutrophils can sense gradients in various factors (e.g., interleukin-8 and leukotriene B4) that are secreted by host cells in response to infection by pathogens such as bacteria, and migrate toward the site of infection as a part of host innate immune defense against pathogens [12]. Bacterial infection thus results in neutrophils migrating from the bloodstream through tissue extracellular matrix towards the chemotactic agents, shown schematically in Figure 1.4. Neutrophils are particularly sensitive to the spatial gradient of the secreted chemotactic factor: they can sense nanomolar levels of chemoattractant concentration and concentration gradients as little as 1% of the nanomolar baseline across the $\sim 10\,\mu$m diameter of a neutrophil.

EXAMPLE

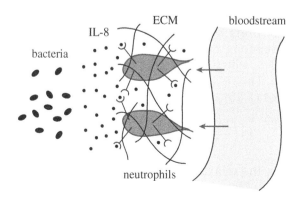

ECM   bloodstream

IL-8

bacteria

neutrophils

Figure 1.4 Schematic of chemoattractant gradient causing neutrophil migration associated with binding of chemotactic factors to cell receptors. ECM, extracellular matrix; IL-8, interleukin-8.

We wish to estimate the minimum critical size of the incremental volume element V_c needed to form the basis of a continuum model so that the concentration of chemoattractant, $c(r)$, can be expressed as a continuous function of position. We assume that the probability distribution of chemoattractant molecules at any position in space can be represented as Gaussian, with an expected value E and an expected standard deviation of $E^{1/2}$. Therefore, the expected relative magnitude of the fluctuation of the distribution is $E^{1/2}/E$. We can relate the expected number of molecules per unit volume at a given position, E, to $c(r)$ and V_c by $E = (V_c)c(N_{AV})$, where N_{AV} is Avogadro's number. Given that the cells can sense a 1% change of a 10^{-9} M baseline concentration of chemoattractant, show that $V_c > 10^{-2}$ nL, which corresponds to a cubic volume element having a side length of about 20 μm. While on some scales, 20 μm may seem small, typical cell diameters are 5–10 μm. Hence, the continuum approximation can be challenged when thinking about continuum equations less than the size of a cell, and care must be taken when thinking about quantitative conclusions versus qualitative estimates using such an approximation. Of course, over long enough time scales, we are greatly helped by the concept of ensemble averages over which we can integrate.

1.4 CHEMICAL REACTIONS: SOME COMMON EXAMPLES

We first consider specific examples of chemical reactions in order to establish working definitions of *reaction rate, reaction order, molecularity*, and the *mechanisms* by which reactions occur. Taken together, these concepts are the focus of the field of chemical kinetics—the subject of many detailed monographs and textbooks within the broader context of physical chemistry [13,14].

The first example is that of an *irreversible, unimolecular first-order* reaction in which the reactant, A, is converted to the product, B:

$$A \xrightarrow{k_1} B \tag{1.29}$$

The rate equation for this first-order reaction states that at constant temperature, the concentration of reactant, [A], decreases at a rate proportional to [A] and the rate constant k_1:

$$-\frac{d[A]}{dt} = k_1[A] \tag{1.30}$$

where k_1 has the units s^{-1}. Thus, the rate of an elementary reaction such as (1.29) depends on the frequency of encounters between the reactant molecules, which is directly proportional to the concentration of the reactants. This is a statement of the Principle of Mass Action. Rearrangement of (1.30) and integration over time gives the form

$$\ln\left(\frac{[A]_0}{[A]}\right) = k_1 t, \quad [A] = [A]_0 e^{-k_1 t} \tag{1.31}$$

where $\tau = 1/k_1$ is the relaxation time of this first-order reaction. Reaction rates are also found to depend on temperature: the frequency of encounters between reactant molecules increases with temperature and, in addition, high-energy encounters are more probable with increasing temperature. Within a restricted range of temperature, Arrhenius found an empirical relation between the rate constant k and temperature, T:

$$k = \mathscr{A} e^{-E_a/RT} \tag{1.32}$$

where E_a is an activation energy found from experiments and $\mathscr{A}$ is a constant.

Detailed treatises on chemical kinetics emphasize that the rate law for a reaction, the order of the reaction, and its rate constant must all be found experimentally, and generally cannot be deduced from a balanced chemical equation. Therefore, in the continuity equation (1.2), dc_i/dt is not necessarily equal to the "reaction rate" R_i except in the special case of a closed system. With Figure 1.5 taken to be such a closed, well-stirred reaction vessel, we can write the integral law of continuity for a reaction within the vessel as:

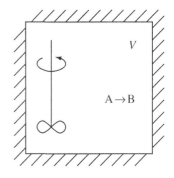

Figure 1.5 Closed, stirred reaction vessel at temperature T.

$$\frac{d}{dt} \int_V c_i \, dV = - \oint_S \mathbf{N}_i \cdot \mathbf{n} \, da \overset{0}{\diagup} + \int_V R_i \, dV \tag{1.33}$$

Since the volume dV in (1.33) is arbitrary, this special case reduces to

$$\frac{dc_i}{dt} = R_i \tag{1.34}$$

assuming that c_i and R_i are spatially uniform within the closed system of Figure 1.5. Under these conditions, the nomenclature of our continuity law (1.33) can be related to that of the rate equation (1.30) with $c_i \equiv [A]$ and $R_i \equiv -k_1[A]$.

A second example is the *reversible, first-order unimolecular* reaction

$$A \underset{k_2}{\overset{k_1}{\rightleftharpoons}} B \tag{1.35}$$

having the rate equation

$$\frac{d[A]}{dt} = -k_1[A] + k_2[B] = R_A \tag{1.36}$$

The Principle of Detailed Balance states that in equilibrium, the forward and reverse rates for each path in the overall reaction must be equal, and a reaction equilibrium constant K_{eq} is thereby defined by setting $d[A]/dt = 0$ in (1.36):

$$k_1[A] = k_2[B] \tag{1.37}$$

$$\frac{[B]}{[A]} = \frac{k_1}{k_2} \equiv K_{eq} \tag{1.38}$$

A *reversible bimolecular* reaction has the form exemplified by the dissociation of acetic acid, CH_3COOH:

$$CH_3COOH \underset{k_2}{\overset{k_1}{\rightleftharpoons}} CH_3COO^- + H^+ \tag{1.39}$$

with an associated rate equation and equilibrium dissociation constant K_d:

$$\frac{d}{dt}[CH_3COOH] = k_2[CH_3COO^-][H^+] - k_1[CH_3COOH] \tag{1.40}$$

$$K_d \equiv \frac{k_1}{k_2} = \frac{[CH_3COO^-][H^+]}{[CH_3COOH]} \tag{1.41}$$

The above examples help to illuminate the definitions of mechanism, molecularity, and order of elementary reactions. A description of the steps through which a reaction proceeds is called the mechanism of the reaction. The individual steps are referred to as unimolecular, bimolecular, termolecular, etc., depending on the number of molecular species that are involved in the reaction. Thus, the reversible reaction (1.39) is bimolecular, since two species are involved in the association to form CH_3COOH, while reactions (1.29) and (1.35) are unimolecular. Finally, the reaction order of such elementary reactions is most often inferred from the power of the concentration of a species in the rate equation for the reaction, i.e., the exponent in the rate law. Thus, reaction (1.40) is first order in $[H^+]$ and first order in $[CH_3COO^-]$, but is said to be second order overall. Other examples of chemical reactions will appear later in our text in the context of biological systems that involve diffusion–reaction, electrical interactions between species, and the effects of motion induced by fluid flow or deformation of biomaterials and tissues.

Example 1.4.1 pH, pK, and the Henderson–Hasselbalch equation In this example, we explore the titration behavior of acetic acid as a model for the ionization behavior of the carboxylic acid residues of proteins and other macromolecules commonly encountered in biological systems. We imagine the closed system of Figure 1.5, but now containing a dilute aqueous solution of acetic acid, a weak organic carboxylic acid having the dissociation reaction (1.39). When the base sodium hydroxide, NaOH, is added to the solution, the titration curves of Figure 1.6 result, giving a graphical relation between the pH of the solution and the amount of NaOH added.

To find an analytical expression for this titration behavior, we use the definition of the equilibrium dissociation constant in (1.41), here renamed K_a. Cross-multiplication of (1.41) yields

$$[H^+] = K_a \frac{[CH_3COOH]}{[CH_3COO^-]} \tag{1.42}$$

EXAMPLE

Figure 1.6 Titration behavior of acetic acid showing the fraction of CH_3COO^- and CH_3COOH groups as a function of the pH of the solution.

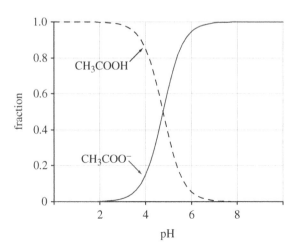

Taking the logarithm of both sides of (1.42), and remembering that pH is defined as the negative of the logarithm of [H$^+$], we write

$$-\log[\text{H}^+] = -\log K_a - \log\left(\frac{[\text{CH}_3\text{COOH}]}{[\text{CH}_3\text{COO}^-]}\right) \quad (1.43)$$

$$\text{pH} = \text{p}K_a + \log\left(\frac{[\text{CH}_3\text{COO}^-]}{[\text{CH}_3\text{COOH}]}\right) \quad (1.44)$$

Equation (1.44) is commonly called the Henderson–Hasselbalch equation. The pH at which 50% of the CH$_3$COOH molecules are dissociated is defined as the pK_a, which is found to be 4.75 (i.e., $K_a = 1.8 \times 10^{-5}$ M). These concepts will come in handy, since fixed-charge groups of biomacromolecules in the intracellular matrix and extracellular matrix and on cell-surface macromolecules are most often acidic or basic groups having titration isotherms similar to that described here. However, since these ionizable charge groups are often closely spaced along individual macromolecules (about 0.5–1 nm apart), additional electrostatic interactions along and between macromolecules complicate the titration behavior of nondilute gels and tissues [15]. Treatment of these electrostatic interactions at the macrocontinuum level (via Donnan equilibrium) is presented in Chapter 3, and at the molecular scale of the electrical double layer surrounding ionized interacting molecules in Chapter 4.

Example 1.4.2 Binding of IGF-1 to IGF-Binding Proteins in the Extracellular Matrix Insulin-like growth factors (IGFs) are peptide hormones secreted by many different cells. IGF-1 is one of the most potent stimulators of cell growth and proliferation, and a potent inhibitor of programmed cell death (apoptosis). IGF-1 also stimulates cells to increase synthesis of extracellular matrix (ECM) macromolecules (e.g., collagens, proteoglycans, and glycoproteins) and to decrease ECM degradation caused by inflammatory cytokines that can upregulate matrix protease activity. IGFs are part of a complex system that includes cell surface receptors (IGF-1R and IGF-2R), two growth factor ligands (IGF-1 and IGF-2), a family of six high-affinity IGF-binding proteins (IGFBP-1,...,-6), and IGFBP-degrading enzymes. Compared with the cell surface receptors, IGFBPs have equal or higher affinity to IGFs. These IGFBPs are soluble cell-secreted proteins that can become enmeshed within the ECM as key players in musculoskeletal, cardiovascular, nerve, and epithelial tissues. They can modulate the roles of IGFs in regulating cell proliferation, differentiation, and cell biosynthesis, and can also act to sequester IGFs within the ECM and thereby regulate delivery of IGFs to cell surface receptors.

In general, intracellular pathway networks and extracellular binding networks can be modeled mathematically, although these models are often complex combinations of coupled ordinary differential equations to represent chemical kinetics and rate laws, along with coupled partial differential equations to account for diffusion–reaction kinetics. In this example, we first explore a simple *reversible, bimolecular first-order*

EXAMPLE

Figure 1.7 (a) Schematics showing insulin-like growth factor 1 (IGF-1) and IGF-2, binding of IGFs to members of the IGF-binding protein (IGFBP) superfamily, IGFBP-1,..., -6, and IGF binding to the IGF type I cell receptor (IGF-1R). (b) Cells and their surrounding extracellular matrix, with IGF diffusing in from upper left. ((a) adapted from Hwa V, Oh Y, and Rosenfeld RG. The insulin-like growth factor-binding protein (IGFBP) superfamily. *Endocr. Rev.* **20**, 761–787 (1999).)

(a)

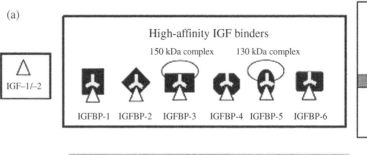

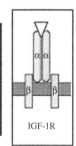

(b)

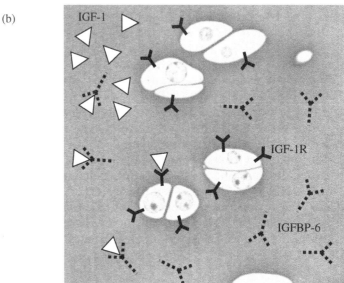

reaction model for the binding between IGF-1 and one of several members of the IGFBP family, IGFBP-6 (Figure 1.7(a)). The resulting binding isotherm is a very important and commonly encountered form in biological systems and in surface physical chemistry. We will then extend this example in Section 1.6 to model diffusion-limited reaction of IGF-1 with IGFBPs in tissues (Figure 1.7(b)), and thereby to describe how such diffusion–reaction kinetics can regulate the transport of ligand growth factors to their intended cell receptor targets.

The reversible, bimolecular first-order reaction for the dissociation of IGF-1 bound to IGFBP-6 can be represented as

$$\underbrace{\text{IGF}}_{\text{bound}} * \text{site} \underset{k_{\text{on}}}{\overset{k_{\text{off}}}{\rightleftharpoons}} \underbrace{\text{IGF}}_{\text{free}} + \text{site} \qquad (1.45)$$

where IGF-1 bound to the IGFBP-6 binding site is represented by the complex, $\text{IGF} * \text{site}$, and dissociation to free IGF proceeds with the rate constant k_{off}. Let us define the concentration of bound IGF as c_B and the concentration of free IGF as c_F. Using this notation, the rate equation for this reaction is

$$\frac{\partial c_B}{\partial t} = k_{\text{on}} c_F [\text{sites}] - k_{\text{off}} c_B \qquad (1.46)$$

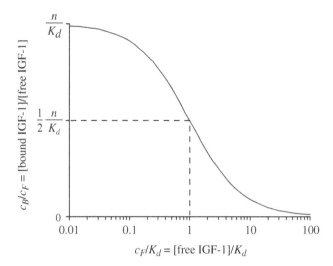

Figure 1.8 Ratio of bound IGF-1 to free IGF-1 as a function of free IGF normalized to K_d.

We can now define the total number density of binding sites, n, as

$$n = \underbrace{[\text{IGF} * \text{site}]}_{\text{occupied}} + \underbrace{[\text{site}]}_{\text{unoccupied}} \tag{1.47}$$

where the concentration of occupied sites is also equal to c_B.

In equilibrium, with $\partial c_B / \partial t = 0$ in the rate equation (1.46), we can define the dissociation constant K_d in terms of c_F, c_B, and n,

$$K_d = \frac{k_{\text{off}}}{k_{\text{on}}} = \frac{c_F[\text{site}]}{c_B} = \frac{c_F(n - c_B)}{c_B} \tag{1.48}$$

and algebraic manipulation can then be used to write the concentration of bound IGF as

$$\boxed{c_B = \frac{n\,c_F}{K_d + c_F}} \tag{1.49}$$

Figure 1.8 shows the ratio of bound to free IGF, c_B/c_F, plotted as a function of c_F, which has a form analogous to the Langmuir binding or adsorption isotherm of physical chemistry. Note that an experimental measurement of this isotherm would enable estimation of the binding constants n and K_d, which has been performed for IGF-1 binding to IGFPB-6 in the dense ECM of articular cartilage [16].

1.4.1 Ionizable Charge Groups in the Extracellular Matrix and on Cell Surfaces

The biological and biophysical properties of tissues are greatly affected by mobile electrolyte ions in the interstitial fluid and by ionized fixed-charge groups attached to the backbones of matrix macromolecules. In order to measure and model the macroscopic density of fixed-charge groups, it is useful to review the location of these groups along the constituent macromolecules. The extracellular matrix (ECM) is a densely assembled network of collagens, elastin, proteoglycans, and glycoproteins, and is a defining feature of all connective tissues. Cellular basement membranes are also composed of ECM macromolecules.

Collagens

The collagens are a family of trimeric ECM molecules secreted by cells and assembled into networks for structural, biomechanical, and other critical functions [17]. Currently, there are 29 reported collagen types encoded by 43 distinct genes (although it has recently been suggested [18] that type XXIX collagen really belongs to the collagen VI family; i.e., the *COL29A1* gene appears to be the *COL6A5* sequence). Collagen is the primary structural protein of the body, comprising about one-third of the total protein content, mainly composed of the fibrillar types I, II, III, and V. All collagens contain triple helical regions formed by three α chains. The exact sequence of amino acids along each polypeptide chain defines collagen's primary structure (Figure 1.9). Secondary structure deals with the spatial arrangement of atoms along a short chain segment, including considerations such as bond angles, bond lengths, and stereochemical features. The tertiary structure accounts for the folding of the entire collagen macromolecule into its triple helical configuration. Aggregations of collagen molecules into various sized fibrils constitute the quarternary structure of the fibrillar collagens (e.g., Figure 1.10). The three α chains of collagens are left-handed helical polypeptide chains wound together in a rope-like fashion as a right-handed superhelix. In fibrillar collagens, each of the three chains contains about 1000 amino acid residues; the entire rod-like procollagen molecule (about 300 kDa) is approximately 300 nm long and 1.4 nm in diameter.

Special properties of the particular amino acid sequences in collagen chains favor interchain hydrogen bonding. This feature greatly affects

Figure 1.9 Diagrammatic representation of several levels of order in the fibrillar collagens: (a) primary structure; (b) secondary structure; (c, d) tertiary structure; (N, amino terminus; C, Carboxyl terminus); (e) quaternary structure. (Adapted from Piez KA. Chemistry of collagen and elastin and biosynthesis of covalent cross-links. In *Aging of Connective and Skeletal Tissue. Thule International Symposium, 3* (Engel A and Larsson T, eds.). Nordiska Bokhandelns Forlag, Stockholm, 1969, pp.15–32.)

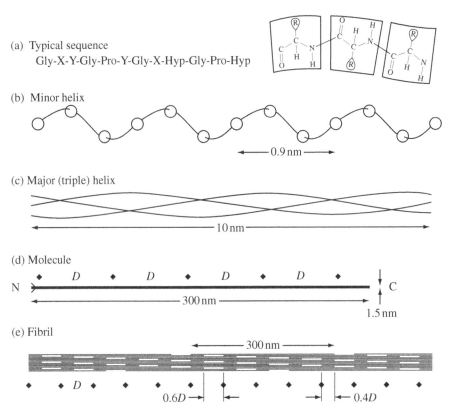

(a) Typical sequence
Gly-X-Y-Gly-Pro-Y-Gly-X-Hyp-Gly-Pro-Hyp

(b) Minor helix
0.9 nm

(c) Major (triple) helix
10 nm

(d) Molecule
N   D   D   D   D   C
300 nm
1.5 nm

(e) Fibril
300 nm
D
0.6D
0.4D

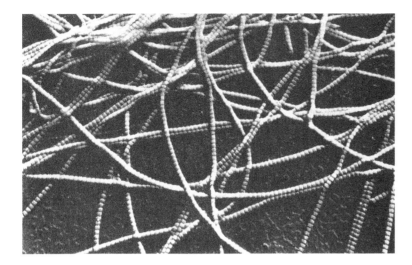

Figure 1.10 Native-type fibrils. (From Hodge AJ. Structure at the electron microscope level. In *Treatise on Collagen*, Volume 1 (Ramachandran GN, ed.). Academic Press, 1967, pp. 185–205.)

the structural rigidity of the resulting molecule, enabling a stable helix to exist at 37°C. In addition, special covalent bonds or crosslinks exist between the three chains (intramolecular crosslinks) and between the molecules of a fibril (intermolecular crosslinks). The swelling of collagen fibrils and collagen-containing tissues *in vivo* and *in vitro* has long been known to be dependent on intramolecular, intermolecular, and interfibrillar collagen crosslinkages. For example, tendon type I collagen fibrils in a neutral pH environment contain 60–65% water by weight. This percentage is even higher at basic or acidic pH owing to electrostatic and osmotic effects. The fibrils imbibe fluid by an amount consistent with crosslink and other ultrastructural constraints. Collagen molecules in solution can lose their helical structure through a process in which the intramolecular hydrogen-bonding structure is broken down. This denaturation process, which is known to be brought about by high temperatures (about 60°C for fibrils) and, in some cases, extremes in pH, results in the randomly coiled polypeptide chains of gelatin. As the mechanical properties of denatured collagen are remarkably different from those of the native helical state, it is essential that the experimenter know the tertiary structure of the collagen contained in a particular fiber, membrane, or tissue specimen.

Collagen charge groups: Collagen is a protein polyampholyte—a polyelectrolyte capable of being either positively or negatively charged. Each molecule has approximately 230 acidic (carboxyl) and 250 basic (amino) groups that can ionize according to the reversible, first-order, bimolecular reactions

$$-COOH \rightleftharpoons COO^- + H^+ \tag{1.50}$$

$$-NH_3^+ \rightleftharpoons NH_2 + H^+ \tag{1.51}$$

Most of these charge groups are ionized at physiological pH and are found primarily as the carboxylic acid side residues of glutamic and aspartic acids (COO^-; $pK_a \simeq 4$ and 3.4, respectively) and the amino residues of lysine, arginine, histidine, and hydroxylysine (NH_3^+; $pK_b \simeq$ 10.5, 12.5, 6, 10.5, respectively) (see Figure 1.11). Since the numbers of acidic and basic groups are almost equal in fibrillar collagens, there is little net charge at pH 7.

The classic titration curve of intact bovine skin corium collagen published by Bowes and Kenten [19] (Figure 1.12) shows a broad isoelectric region (little or no net charge) at $pH \simeq 5–10$. Collagen is one of the few

Figure 1.11 Schematic showing a segment of a polypeptide chain with three amino acids, two of which contain side groups that are ionizable depending on bath pH.

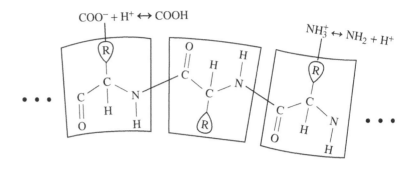

Figure 1.12 Titration curve of intact bovine corium collagen showing the amount of acid or base bound per gram dry collagen versus bath pH. ○, in the absence of added salt; △, in the presence of 0.5 M NaCl. (Adapted from Bowes JH and Kenten RH. The amino-acid composition and titration curve of collagen. *Biochem. J.* **43**, 358–365 (1948).)

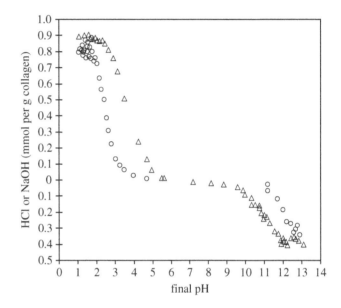

proteins for which the titration behavior has been measured using intact fiber and tissue specimens. Two major conclusions can be drawn from such curves. First, the total amount of strong acid or base that can bind to the specimen is a quantitative measure of the total number of ionizable (titratable) charge groups. Second, the pH at which there is zero net charge, the isoelectric point, is clearly defined. The effect of ionic strength on ionization behavior is also exemplified by the data of Figure 1.12 and is ascribable to electrostatic interactions, which we will discuss further in Chapters 3 and 4.

We note that isolated amino acid molecules in solution have an additional carboxyl and amino group that also ionize (Figure 1.13). However, these latter groups are modified and rendered nonionizable when amino acids bond together to form the polypeptide chains of collagen molecules (or other proteins in general). Thus, factors determining molecular charge configurations are the position of ionizable groups along the polypeptide chains and the pH, ionic strength and other conditions affecting the dissociation reactions (1.50) and (1.51).

Proteoglycans, Glycoproteins, and their Polysaccharide Constituents

The structure and function of the dozens of members of the proteoglycan superfamily have been described in several reviews [20–23]. This superfamily includes the subfamily of large aggregating proteoglycans that can perform structural and biomechanical functions in tissues (e.g., aggrecan, versican, brevican, and neurocan); small leucine-rich

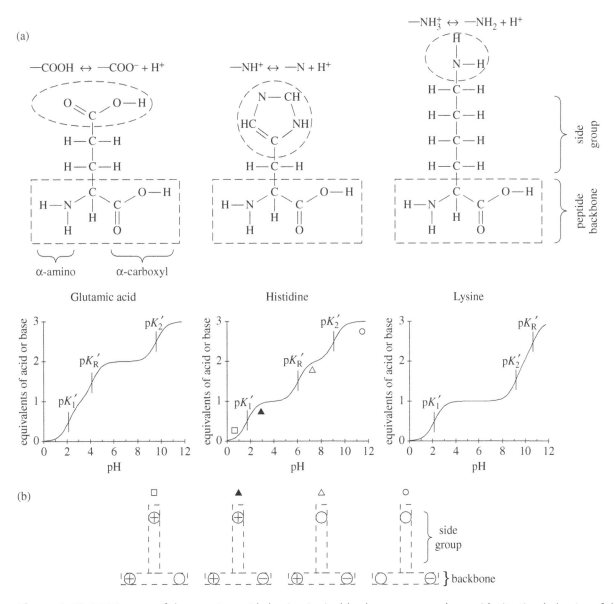

Figure 1.13 (a) Diagram of three amino acids having ionizable charge groups along with titration behavior of dilute solutions of each. (b) Four possible charge configurations of histidine in different pH regions as noted in the titration curve for histidine. ((a) adapted from Lehninger AL. Biochemistry. Worth Publishers, New York, 1970, pp. 67–86.)

proteoglycans that help to regulate collagen fibrillogenesis and act as connecting links between collagens and other proteins of the ECM (e.g., decorin, biglycan, fibromodulin, chondroadherin, asporin, epiphycan, and several others), cell surface proteoglycans that modulate ligand–receptor binding and act as portals to mediate uptake of certain molecules into cells (e.g., syndecans and glypicans), and cellular basement membrane proteoglycans that can bind integrins and other soluble proteins (e.g., perlecan, agrin, and bamacan).

All proteoglycans are composed of a core protein substituted with one or more glycosaminoglycan (GAG) chains (i.e., chains that are covalently bound to the core protein). After intracellular translation of the core protein, GAGs are elongated and sulfated, catalyzed by glycosyltransferases and sulfotransferases in the rough endoplasmic reticulum (rER) and more so in the Golgi apparatus. While the core proteins of some proteoglycans form transmembrane structures that link molecules in the pericellular matrix (PCM) to the cell surface via their GAG chains (e.g., the syndecans), most proteoglycans are secreted into the PCM and ECM.

The biochemical and biophysical properties of the proteoglycans are dominated by their constituent GAG chains, which are linear polymers of disaccharides that include chondroitin sulfate (CS), keratan sulfate (KS), heparin, and heparan sulfate (HS). Each disaccharide may have on the average one KS, two CS, or up to four HS ionized fixed-charge groups, thereby making proteoglycans the most highly charged macromolecules of the ECM. For example, the structure of aggrecan is pictured schematically in Figure 1.14(a); it consists of a $\sim 300\,kDa$ core protein substituted with about 100 CS–GAG chains in the CS domain between the so-called G2 and G3 globular domains of the core, as well as KS GAGs near the G2 domain. Recent breakthroughs have enabled visualization of this complex 3 MDa proteoglycan via atomic force microscopy [24] (Figure 1.14(b)). Aggrecan monomers, in turn, bind noncovalently via their G1 domain to the GAG chain and hyaluronan (hyaluronic acid, HA), and this binding is further stabilized by the binding of link protein situated on HA adjacent to the aggrecan G1 domain. Such aggrecan "aggregates" may contain several aggrecan monomers (e.g., Figure 1.14(c)), up to ~100 aggrecan monomers *in vivo*, the latter giving rise to a huge 300 MDa structure. Cells such as chondrocytes of cartilage synthesize and secrete aggrecan, link protein, and HA, and the process of aggregation occurs extracellularly.

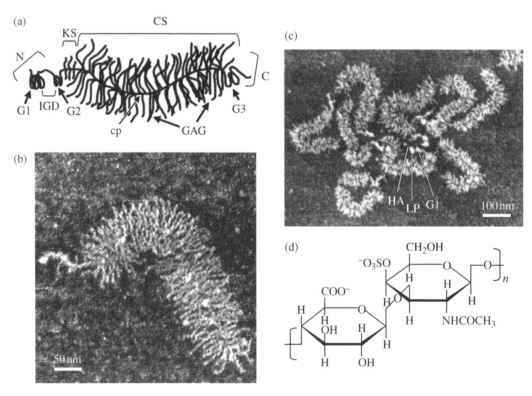

Figure 1.14 (a) Structure of aggrecan. N, amino terminus; G1, G2, G3, globular domains; IGD, interglobular domain between G1 and G2; cp, core protein; KS, keratan sulfate glycosaminoglycan (GAG) region; CS, chondroitin sulfate GAG region; C, carboxyl terminus. (b) AFM height image (tapping mode) of fetal epiphyseal aggrecan monomer. (c) Small aggrecan aggregate with aggrecan G1 and adjacent link protein (LP) bound to short length of hyaluronan (HA). (d) Disaccharide structure of chondroitin-4-sulfate GAG showing ionized carboxyl and sulfate groups. ((a, b) from Ng L, Grodzinsky AJ, Patwari P, et al. Individual cartilage aggrecan macromolecules and their constituent glycosaminoglycans visualized via atomic force microscopy. *J. Struct. Biol.* **143**, 242–257 (2003); (c) from Lee H-Y, Sandy JD, Plaas AHK, et al. Ultrastructure of reconstituted proteoglycan aggregates studied by atomic force microscopy. In *Transactions of 56th Orthopedic Research Society, New Orleans, March 6–9, 2010*, p. 872.)

Since the CS–GAG chains may contain from 20 to 50 disaccharides (e.g., as shown in the chondroitin-4-sulfate structure of Figure 1.14(d)), a single aggrecan aggregate may contain as many as a million ionized negative-charge groups under physiological conditions. These aggregates are packed very densely between the collagen fibrils of cartilaginous and other connective tissues subjected to compressive loading. The presence of such proteoglycans confers a very high equilibrium compressive stiffness to these tissues due to repulsive electrostatic interactions between nearby GAGs along a core protein and between GAGs of neighboring core proteins (see Chapters 4 and 7).

The fact that these GAG charge groups are ionized under physiological conditions can be deduced from chemical titration behavior. The observed pK for carboxyl groups on isolated HA and on CS disaccharides has been reported as approximately 3. The pK of sulfate groups is much lower. Because of variability in the sulfate content of GAG molecules, the average charge per disaccharide unit may vary within a given structure. Thus, in addition to the dissociation reaction (1.50) for carboxyl groups we now include that for sulfate groups:

$$-\text{HSO}_3 \rightleftharpoons \text{SO}_3^- + \text{H}^+ \qquad (1.52)$$

1.5 BOUNDARY CONDITIONS AND BOUNDARY VALUE PROBLEMS

The flux constitutive law (Fick's first law, (1.1)) and species conservation (1.5) were used in Section 1.2 to arrive at the linear diffusion equation, written here to include chemical reactions:

$$\frac{\partial c_i}{\partial t} = D_i \nabla^2 c_i + R_i \qquad (1.53)$$

Solution of the diffusion equation (1.53) for all the practical problems that we will encounter requires use of mathematical methodologies that are well described in detailed texts on differential equations. Linear boundary value problems are defined by the linear differential equation at hand (i.e., having linear differential operators) along with requisite boundary conditions and an initial condition on the concentration $c_i(\boldsymbol{r}, t)$.

1.5.1 Boundary or Interfacial Matching Conditions

Boundary conditions for the second-order partial differential equation (1.53) include conditions on either the concentration of species i, c_i, or its normal derivative $\boldsymbol{n} \cdot \nabla c_i$ (i.e., the normal component of the flux $\boldsymbol{N}_i$). Picturing the volume in Figure 1.1 surrounded by the surface S, we will see that a unique solution to (1.53) exists in the volume V provided that well-posed boundary conditions on c_i, $\boldsymbol{n} \cdot \nabla c_i$, or a combination of the two are specified on the entire surface surrounding the volume. For piecewise-linear problems involving two or more regions in space in which the diffusion equation is to be used to find c_i in each region, the boundary conditions take the form of interfacial or "matching" conditions. (A boundary condition on the scalar c_i is often referred to as a Dirichlet condition and that on the flux $\boldsymbol{n} \cdot \nabla c_i$ is called a Neumann condition.)

Boundary Condition on Flux

Figure 1.15 represents the interface between two regions of space having the normal vector $\boldsymbol{n}$ pointing into region 1. The possibility that a chemical reaction at the interface leads to production of species i is included via the surface generation rate R_{si}. The interfacial matching condition relating fluxes $\boldsymbol{N}_{1i}$ and $\boldsymbol{N}_{2i}$ is derived by applying the integral form of species continuity (1.2) to the pillbox region spanning the interface in Figure 1.15. The integration is carried out in the limit in which the height of the pillbox, h, shrinks to zero. In this limit, there is no contribution from the left-hand side of (1.2); however, the rightmost term will give a contribution since R_{si} is the rate of formation of species i at the interface itself. Carrying out the integration and canceling the areas A on both sides of the equation yields

$$\boldsymbol{n} \cdot (\boldsymbol{N}_{1i} - \boldsymbol{N}_{2i}) = R_{si} \tag{1.54}$$

We will see that this approach is essentially identical to that used in Chapter 2 for deriving the boundary conditions relating the interfacial jumps in the normal components of the electric field $\boldsymbol{E}$ or current density $\boldsymbol{J}$, using the integral forms of Gauss' law and conservation of charge, respectively.

Boundary Condition on Concentration

Suppose, now, that the interface in Figure 1.15 demarcates the boundary between an outer bath (e.g., region 1), and the inside of a tissue or gel in which steric and or electrostatic interactions cause a partitioning of the solute species trying to cross from region 1 to region 2. Alternatively, the interface might be a membrane, e.g., a cell membrane or synthetic membrane. The interfacial matching condition on concentration can then be written as

$$c_{2i} = K_i c_{1i} \tag{1.55}$$

where K_i is the partition coefficient between the outer "bath" region 1 and the "interior" region 2. The condition $K_i = 1$ simply states that the concentration is continuous at the interface.

1.5.2 Uniqueness

Referring again to Figure 1.1, we now argue that there is only one unique concentration distribution $c(x, y, z, t)$ within a specified volume V that obeys the diffusion equation (1.53) and, additionally, satisfies boundary conditions of the type (1.54) and (1.55) that are completely specified

Figure 1.15 Schematic for derivation of the interfacial matching condition on flux.

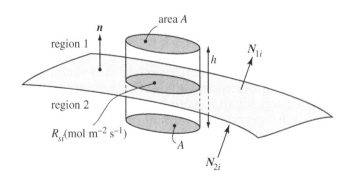

over the surface S surrounding that volume. We start with the simpler case of steady diffusion ($\partial/\partial t \rightarrow 0$) with no volumetric chemical reaction terms ($R_i = 0$ in (1.53)), and the boundary condition that c is specified on the entire surface surrounding the volume. In this case, the diffusion equation reduces to Laplace's equation

$$\nabla^2 c_i = 0 \qquad (1.56)$$

and therefore the flux $\boldsymbol{N}$ has no divergence ($\boldsymbol{N}$ is "solenoidal": $\nabla \cdot \boldsymbol{N} = 0$). To prove that $c_a(x, y, z)$ is the unique solution for solute c, let us assume that there is another Laplacian solution, $c_b(x, y, z)$, that also satisfies the same boundary condition on the surrounding surface. Therefore, defining the difference solution as $c_d = c_a - c_b$, we know that c_d also satisfies (1.56) (by linearity and superposition) and that $c_d = 0$ at the bounding surface, since c_a and c_b each individually satisfy the same boundary condition there. If we can show that the difference solution $c_d = 0$ everywhere within the volume V, then c_a is indeed the unique solution.

First, we argue that c_d cannot possess a maximum or minimum at any point within the volume V. Let us visualize a vector flux line $\boldsymbol{N}_d$ within the volume of Figure 1.1 as it starts on one point on the surrounding surface and reaches another point on the surface. Since $\boldsymbol{N} \sim -\nabla c_d$ by Fick's law and has no divergence, such a flux line cannot start or stop anywhere within V. In addition, since the flux line is in the direction of the negative gradient of c_d, c_d has to continually decrease until the flux line reaches the surface. Similarly, in the opposite direction, c_d has to increase until another point on the surface is reached. Therefore, all maximum and minimum values of c_d have to be located on the surface, and not within V. In summary, c_d at any interior point within V cannot assume a value larger than or smaller than the largest or smallest value of c_d on the surface. But $c_d = 0$ on the surface by definition. Therefore, $c_d = 0$ everywhere within V, and it follows that $c_a = c_b$ is the unique solution within V satisfying the specified boundary condition on S.

The above uniqueness proof can be extended to the case of the full diffusion equation (1.53) and, by analogy, this proof can be also used to show uniqueness of solutions to Laplace's equation and Poisson's equation for the electrical potential in the electro-quasistatic systems of Chapter 2.

1.5.3 Solution Methods for Boundary Value Problems

The problems at the end of this chapter give examples of the use of several methods for solving partial differential equations of the form (1.53); these methods are described in detail in many textbooks [25, 26]. Fourier analysis and the method of separation of variables can provide important physical insight into the spatial distribution and temporal kinetics of diffusion and diffusion–reaction processes [25]. This same method is a powerful tool for solving electrical potential distributions in electroquasistatic problems, in which each variable-separable solution corresponds to a physically realizable charge distribution. The related finite Fourier transform method is also very useful and applicable to such linear boundary value problems where at least one of the spatial dimensions is finite [26].

1.6 DIFFUSION AND CHEMICAL REACTIONS

In this section, we explore rate-limiting phenomena in problems involving solutes that are simultaneously undergoing diffusive transport while participating in chemical reactions within the surrounding medium. Scaling analysis applied to diffusion–reaction laws helps to identify dimensionless parameters that highlight the important characteristic times, rates, and lengths governing solute transport.

1.6.1 Scaling and the Damköhler Number*

To emphasize concepts and simplify the analysis, we first treat the case of steady state one-dimensional diffusion of a solute away from a source of constant solute concentration c_0 and into a medium having sufficient thickness in the x direction to be considered semi-infinite in extent (Figure 1.16(a)). Within the medium ($x > 0$), the solute binds to sites distributed uniformly in space in a manner described by a first-order, irreversible reaction. This system could represent delivery of a drug into a tissue from a patch placed at $x = 0$ having an elution mechanism that maintains constant drug concentration at the surface.

Steady state transport of the drug can be modeled as

$$\frac{\partial c}{\partial t} = D\nabla^2 c \underbrace{-kc}_{+R} \tag{1.57}$$

$$\frac{d^2 c}{dx^2} = \frac{k}{D}c \tag{1.58}$$

where D is the effective diffusivity of the drug in the tissue, taken to be a constant independent of concentration. The one-dimensional diffusion equation (1.58) is subject to the boundary conditions $c = c_0$ at $x = 0$ and $c \to 0$ at $x \to \infty$. Solution of (1.58) involves superposition of terms $Ae^{-\lambda x} + Be^{+\lambda x}$, where $\lambda = \sqrt{k/D}$. Physical constraints (i.e., the boundary condition at $x \to \infty$) lead to the solution

$$c = c_0 e^{-\lambda x} \tag{1.59}$$

which has the spatial distribution shown in Figure 1.16(b). Thus, the disappearance of drug due to binding is characterized by the spatial

Figure 1.16 (a) A drug patch releases drug solute at the surface ($x = 0$) while maintaining constant solute concentration c_0. The drug diffuses in the x direction into a medium modeled as semi-infinite, and binds to sites within the medium in a manner described by (1.58) in the steady state. (b) Schematic of drug solute concentration profile $c(x)$ resulting from diffusion combined with a first-order, irreversible reaction; comparison with profiles corresponding to large and small values of the Damköhler number D_a.

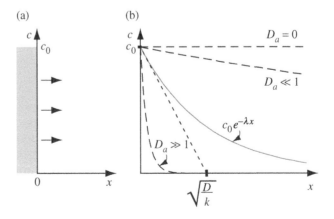

---

*After a classroom problem from WM Deen.

decay constant $\sqrt{D/k}$, which is shorter for smaller D and for higher reaction rate k.

Additional physical insight can be gained by scaling analysis of the diffusion–reaction equation (1.58). Defining normalized parameters $\hat{c} = c/c_0$ and $\hat{x} = x/L$, (1.58) becomes

$$\frac{d^2\hat{c}}{d\hat{x}^2} - \frac{L^2 k}{D}\hat{c} = 0 \tag{1.60}$$

The dimensionless factor $L^2 k/D$ in (1.60) is one form of the Damköhler number D_a. In general, the Damköhler number represents the ratio of a mass transport time to a chemical reaction time (or the equivalent inverse ratio of a chemical reaction rate to a mass transport rate). Here, mass transport is dominated by diffusion, which is represented by the diffusion time over a characteristic length L as L^2/D (remembering the Einstein relation for diffusivity, (1.26)). The characteristic reaction time is k^{-1}, thereby giving

$$D_a = \frac{\text{characteristic diffusion time}}{\text{characteristic reaction time}} = \frac{L^2/D}{k^{-1}} \tag{1.61}$$

Thus, for cases of rapid diffusion (short L^2/D) and slow reaction times (relatively small k, or large k^{-1}), $D_a \ll 1$ (Figure 1.16(b)) and the drug solute is transported well into the tissue. Conversely, for slow diffusion and fast reaction, $D_a \gg 1$ and the drug quickly binds to sites close to the patch ($x = 0$). (Note that in many mass transport texts, the form of the Damköhler number (1.61) is also referred to as the square of the Thiele modulus.)

Example 1.6.1 Diffusion–Reaction in a Finite System This example is an extension of the steady state drug delivery problem of Figure 1.16(a), but more realistically treats a system of finite thickness L (Figure 1.17). The drug concentration at $x = 0$ is c_0 as before, but the boundary at $x = L$ is modeled as impermeable to solutes so that the flux is zero there ($N = 0$ at $x = L$). Show that the steady state concentration profile satisfying these boundary conditions is

$$c(x) = c_0 \frac{\cosh \lambda(L - x)}{\cosh \lambda L} \tag{1.62}$$

Note that for large enough L such that $L \gg \lambda^{-1}$, this solution reduces to $c(x) = c_0 e^{-\lambda x}$, (1.59), consistent with physical reasoning.

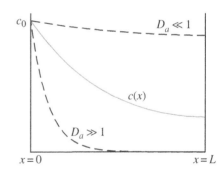

Figure 1.17 Solute release into a finite-thickness medium.

1.6.2 Diffusion–Reaction with Fast Reaction Kinetics

In this section, we focus on a very common and important category of diffusion–reaction problems in which chemical reaction times are very fast, essentially instantaneous, compared with the diffusion times of interest. For characteristic lengths as small as 10–$100\,\mu\text{m}$, diffusion-limited reaction processes can lead to dramatically slower solute transport compared with diffusion with no reaction, even when the reaction is instantaneous. For example, acid–base reactions such as the charge-group ionization reactions of Section 1.4 are very fast; however, changes in the ionization state of proteins at cell surfaces or in tissue ECM caused by changes in the pH or ionic strength of the medium can be relatively slow because of diffusion-limited reactions during equilibration. Another example is that of the binding of growth factors delivered to cell receptors within cultured tissue specimens, which we now consider in more detail.

Example 1.6.2 IGF-1 Binding to IGFBPs in Extracellular Matrix

In Example 1.4.2, we derived an equilibrium binding isotherm (1.49) characterizing the bimolecular, first-order reversible binding of the growth factor IGF-1 to IGFBP-6, one of a family of IGF-binding proteins often found in the ECM of tissues. Here, we use that result to address the important question of transport and delivery of such growth factors to their ultimate cell receptor targets. We will see that the presence of these IGFBPs in the matrix can dominate the kinetics of transport. Of specific interest is the delivery of IGF-1 for the repair and regeneration of musculoskeletal tissues (e.g., cartilage, tendon, and ligament) as well as muscle tissues themselves.

For quantitative measurements of one-dimensional protein transport within and across dense tissues, transport chambers such as that shown schematically in Figure 1.18 can be used. We assume a uniform distribution of IGFBP-6 binding sites fixed within the tissue matrix. From the literature on cartilaginous tissues, it is known that the density of IGFBP-6 sites is approximately 100-fold higher than the density of cell receptors normalized to tissue volume (i.e., about 50 nM versus 0.5 nM, respectively, even though there are about 10,000 IGF receptors per chondrocyte). In addition, IGFBP-6 has an equal or greater affinity for IGF-1 than the cell receptor. Taken together, we start by neglecting the presence of cells, and assume that there is one dominant binding partner (IGFBP-6) of all the binding protein family members in the tissue. We now proceed to model the effects of IGF-1 binding to IGFBP-6 on the transient transport of IGF-1 into the tissue slice of thickness L in Figure 1.18. In particular, we wish to derive an expression for the effective diffusivity of IGF-1 inside the tissue in terms of the dilute solution diffusivity D_{IGF}, the equilibrium dissociation constant K_d, and the density of IGF-binding sites, n.

Initially, there is no IGF-1 in the system. At $t = 0^+$, the concentration of IGF-1 is raised to the value c_1 in the external left-hand chamber bath, while the concentration in the right-hand chamber is negligible compared with c_1 during the period of the experiment ($c_2(x = L) = 0$). Both chambers are well stirred and we also assume that the partition coefficient at the

EXAMPLE

Figure 1.18 Schematic of transport chamber for measurement of diffusion of IGF-1 into tissue containing a uniform distribution of IGFBP-6 along with cells and their IGF-1 cell receptors. At $t = 0^+$, the IGF-1 concentration in the left chamber is raised to c_1 while the concentration on the right can be modeled as remaining near zero for the duration of the experiment ($c_2 = 0$). Experimentally, this can be achieved by making the right-hand chamber volume large enough based on the time for diffusion of IGF into the tissue width L (see Problem 1.7). The final steady state (linear) concentration profile within the tissue, as well as the concentration profile at $t = \tau_{dr}$ is shown, where τ_{dr} is the characteristic diffusion–reaction time constant derived in this example.

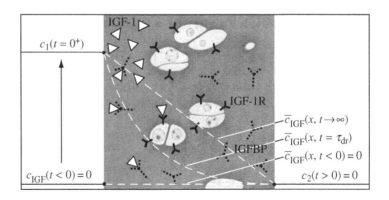

boundaries is $K = 1$; therefore, the concentrations just inside the tissue–bath interfaces are the same as just outside the tissue. (In reality, IGF-1 is a slightly basic protein, and, because of the net negative charge of the ECM of cartilaginous tissues, $K \approx 1.4$ owing to Donnan electrostatic effects, a phenomenon discussed in detail in Chapter 3). With the concentration of free and bound IGF-1 denoted by c_F and c_B, respectively, the overall diffusion reaction equation is

$$\frac{\partial \bar{c}_F}{\partial t} = D_{\text{IGF}} \frac{\partial^2 \bar{c}_F}{\partial x^2} + R_{\text{IGF}} \tag{1.63}$$

$$R_{\text{IGF}} = -\frac{\partial \bar{c}_B}{\partial t} \tag{1.64}$$

$$\frac{\partial \bar{c}_F}{\partial t} + \frac{\partial \bar{c}_B}{\partial t} = D_{\text{IGF}} \frac{\partial^2 \bar{c}_F}{\partial x^2} \tag{1.65}$$

Equation (1.65) states that molecular diffusion of IGF-1 into the tissue can lead to an increase in the concentration of both free and bound IGF-1 inside. (We use the overbar notation $\bar{c}$ to distinguish intratissue from external bath concentration.) From (1.46), the rate equation for binding of IGF-1 to IGFBP-6 is

$$\frac{\partial \bar{c}_B}{\partial t} = k_{\text{on}} \bar{c}_F (n - \bar{c}_B) - k_{\text{off}} \bar{c}_B \tag{1.66}$$

In general, the diffusion–reaction equation (1.65) and the rate equation (1.66) together constitute a coupled nonlinear set of equations. However, it is known that the binding reaction is very fast compared with diffusion times of interest, i.e., $D_a \gg 1$. Therefore, to simplify the analysis and obtain an analytical expression for the effective diffusivity, we assume that at any instant in time and at any position in space, the binding reaction reaches equilibrium very quickly and the local concentration of bound IGF-1 is related to the local concentration of free IGF-1 by the quasi-equilibrium form of the rate equation (1.66) with $\partial/\partial t = 0$:

$$\bar{c}_B(x, t) \sim \frac{n \bar{c}_F(x, t)}{K_d + \bar{c}_F(x, t)} \tag{1.67}$$

The diffusion equation (1.65) then takes the form

$$\frac{\partial \bar{c}_F(x, t)}{\partial t} + \frac{\partial}{\partial t} \left(\frac{n \bar{c}_F(x, t)}{K_D + \bar{c}_F(x, t)} \right) = D_{\text{IGF}} \frac{\partial^2 \bar{c}_F}{\partial x^2} \tag{1.68}$$

By using the chain rule for differentiation, (1.68) can then be written as

$$\frac{\partial \bar{c}_F}{\partial t} \left(1 + \frac{n K_d}{(K_d + \bar{c}_F)^2} \right) = D_{\text{IGF}} \frac{\partial^2 \bar{c}_F}{\partial x^2} \tag{1.69}$$

The diffusion–reaction equation (1.69) is nonlinear in c_F; however, for small enough concentrations $c_F \ll K_d$, it can be written in the form

$$\frac{\partial \bar{c}_F}{\partial t} = D_{\text{eff}} \frac{\partial^2 \bar{c}_F}{\partial x^2} \tag{1.70}$$

$$D_{\text{eff}} = \frac{D_{\text{IGF}}}{1 + n/K_d} \tag{1.71}$$

Equation (1.70) now has the form of a standard diffusion equation, but with an effective diffusivity D_{eff} smaller than D_{IGF} by an amount $1 + n/K_d$ that depends on the values of the IGFBP-6 binding site density n and the equilibrium dissociation constant K_d. For the case of articular cartilage [16], experiments using ^{125}I-IGF-1 in small radiotracer concentrations (10^{-9} M) satisfying the linearization limit revealed that $n \simeq 50$ nM and $K_d \simeq 5$ nM, so that D_{eff} was an order of magnitude smaller than D_{IGF}. Equation (1.70) can be solved for $\bar{c}_F(x, t)$ using separation of variables (as in Problem 1.3) giving the result,

$$\bar{c}_F(x, t) = c_1 \left(1 - \frac{x}{L}\right) - \frac{2Kc_1}{n\pi} \sum_{n=1}^{\infty} \sin\left(\frac{n\pi x}{L}\right) e^{-t/\tau_n} \qquad (1.72)$$

$$\tau_n = \frac{L^2}{n^2\pi^2 D_{eff}} \qquad (1.73)$$

which includes the partition coefficient K. We see that the characteristic diffusion–reaction time constant τ_{dr} can be derived from the $n = 1$ term of the Fourier series solution, which, for constant concentration boundary conditions, takes the form

$$\boxed{\tau_{dr} = \frac{L^2}{\pi^2 D_{eff}}} \qquad (1.74)$$

1.7 PROBLEMS

Problem 1.1 Continuity: Fixed and Moving Frame Representations Two inertial reference frames in relative motion, defined by the Galilean coordinate transformations

$$\boldsymbol{r}' = \boldsymbol{r} - \boldsymbol{v}t \qquad (1.75)$$

$$t' = t \qquad (1.76)$$

are shown in Figure 1.19.

The equation of continuity (1.18) is written with respect to fixed (laboratory) frame coordinates. That is, $c = c(x, y, z, t)$ and time and space derivatives are expressed with respect to (x, y, z, t) coordinates. Note that $\boldsymbol{N}'$ is the flux of species with respect to the moving frame, i.e., the frame defined by the mass- or molar-averaged velocity of the fluid-particle system. $\boldsymbol{N}'$ in (1.18) is expressed in terms of laboratory coordinates, $\boldsymbol{N}' = \boldsymbol{N}'(x, y, z, t)$, and so are its derivatives.

Show that (1.18) can be derived by first writing the continuity equation in the moving frame and then using (1.75) and (1.76) to transform back to the laboratory frame; i.e., start with

$$\frac{\partial c_i'(x', y', z', t')}{\partial t'} = -\nabla \cdot \boldsymbol{N}_i'(\boldsymbol{r}') + R_i' \qquad (1.77)$$

where the primed frame moves at velocity $\boldsymbol{v}$ relative to the laboratory frame, i.e., the mass-central velocity of the fluid.

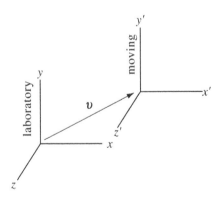

Figure 1.19 Laboratory and moving reference frames.

Problem 1.2 Derive the Einstein Relation Between Diffusivity and Electrical Mobility We will see in Chapter 2 that application of a steady electric field to an ionic (electrolyte) solution gives rise to an electrical migration flux for each mobile ion. The average (ensemble) velocity v imparted to each ionic species in solution by the electric field is linearly proportional to the electric field, with a proportionality constant equal to the electrical mobility $u_i = (q \Delta t)/m$, the product of the ionic charge q and the inter-collision time Δt divided by the ion mass, m. Show that the relation between ionic mobility and diffusivity, called the "Einstein relation," takes the form $D_i/u_i = k_B T/q = RT/|z|F$, where k_B is the Boltzmann constant, T the temperature, R the universal gas constant, F the Faraday constant (10^5 C mol^{-1}), and $|z|$ the magnitude of the valence of the ion. This can be accomplished by relating the kinetic energy of the ion ($\frac{1}{2}mv^2$) to the thermal energy ($k_B T$), and invoking the molecular theory of Brownian motion, also originally proposed by Einstein (1907) [27].

Problem 1.3 Transient Diffusion–Reaction Across Tissue The spatial–temporal evolution of IGF-1 transport into the tissue of Figure 1.18 is described by the diffusion equation (1.70). The solution for $\bar{c}_F(x, t)$ is given by (1.72). Show that this is the correct form of the solution by using separation of variables to derive the answer.

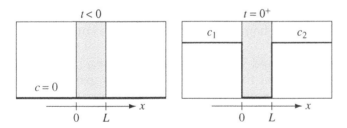

Figure 1.20 Diffusion of a binary electrolyte into a membrane after the concentration in the bath is raised from zero to $c_1 = c_2 = c_0$ at $t = 0^+$.

Problem 1.4 Membrane Completely Immersed in the Same Electrolyte Bath (Figure 1.20)

(a) At $t = 0^+$, a binary electrolyte is added on both sides of a neutral (uncharged) membrane such that $c_1 = c_2 = c_0$. Using the method of separation of variables, find the concentration profiles inside the membrane, $\bar{c}(x, t)$, for $t > 0$. Compare the form of your answer with that of Problem 1.3.

(b) Find the total amount of electrolyte that has entered the membrane (per unit membrane area) at a given time t.

Problem 1.5 Diffusion Out of a Cylindrical Vessel An infinitely long cylindrical rod (vessel) containing electrolyte at concentration c_0 is immersed in an infinitely large bath of distilled water with $c = 0$ for $t < 0$, as shown in Figure 1.21. The bath is so large and *well stirred* that we may assume $c = 0$ at $r = R$ for all time $t > 0$, even as electrolyte diffuses out of the rod. Find $c(r, t)$ inside the rod ($r \leq R$) for $t \geq 0$.

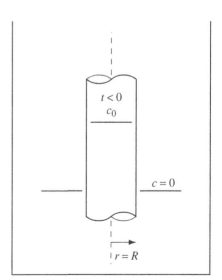

Figure 1.21 Diffusion out of a cylindrical vessel.

Problem 1.6 Time Lag Measurement of the Diffusion Coefficient: An Extremely Important and Commonly Used Experimental Technique

(**a**) Based on your answer to Problem 1.3, find an analytical expression for the flux of IGF evaluated at the right-hand edge of the tissue, $x = L$, at any time t throughout the transient diffusion process across the tissue. By integrating the flux with respect to time t, find the total amount of IGF-1 that has moved across the tissue into the right-hand chamber in a time t. (If you divide by the volume V of the right-hand chamber, you'll have an expression for the concentration $c_2(t)$. While $c_2(t)$ may be small enough to approximate it as zero for the purposes of solving Problem 1.3, $c_2(t)$ can still be quantitatively measured.)

(**b**) As $t \to \infty$, the graph of $c_2(t)$ versus time approaches a straight line as a function of t. Find the extrapolated intercept of the line on the t axis, and show that this time intercept is related to the solute diffusivity D_{eff} by the value $L^2/(6D_{\text{eff}})$.

Problem 1.7 The Time Constant for Equilibration of $c_1(t)$ and $c_2(t)$ Across the Tissue of Figure 1.18 In Problems 1.3 and 1.6, we have assumed that c_1 and c_2 are constants independent of time. In reality, if we wait long enough, these concentrations will equilibrate such that $c_1 = c_2 = c_1/2$, assuming that $c_2 = 0$ at $t = 0$. This problem asks you to find an expression for the "chamber equilibration" time constant.

(**a**) Assume that c_1 and c_2 are uniform within each respective bath compartment for all time (i.e., the compartments are well stirred). Assume that sufficient time has passed that a linear concentration profile of c exists within the tissue for all times of interest during the overall slow equilibration process. Find an expression for the solute flux across the tissue in terms of c_1, c_2, L, and D. (Assume that the partition coefficient $K = 1$; assume that right-hand and left-hand chambers both have volume V.)

(**b**) Using conservation of mass ($Vc_1(t) + Vc_2(t) = \text{constant} = V\beta$), derive a differential equation for the time rate of change of the uniform bath concentration $c_1(t)$ as solute moves from left to right across the tissue.

(**c**) Obtain the solution to the differential equation (this should be a simple exponential decay), and show that your solution satisfies the expected infinite-time response, $c_1(t = \infty) = \beta/2$. Compare your exponential decay time constant with the characteristic diffusion time for a solute moving across a length L, $\tau_{\text{diff}} = L^2/(\pi^2 D)$, and clarify the validity of the assumption of the "steady state" linear concentration profile in Problem 1.3.

Problem 1.8 Dynamics of pH Changes in Charged Membranes and Tissues Derive the diffusion–reaction law for transport of H^+ ions through a protein matrix in which H^+ also binds to amino acid side groups. (Refer to Figure 1.22 and the titration curve of Figure 1.12.)

We are given that the net fixed charge density of the protein matrix at pH 7 is

$$\frac{\rho_{m0}}{F} = -[COO^-]_0 + [NH_3^+]_0 \qquad (1.78)$$

and we will consider only the range of pH < 6, for which $[NH_3^+]_0 \simeq$ constant, but $[COO^-](x, t)$ will change as H^+ changes, i.e., $\bar{\rho}_m(x, t) \geq 0$ as carboxyl groups are neutralized at lower pH. We use the following definitions:

$[H^+] \equiv$ concentration of unbound (free) H^+ ions
$[COO^-] \equiv$ concentration of free carboxyl groups
$[COOH] \equiv$ concentration of neutralized carboxyl groups
$H_B = [COOH] \equiv$ concentration of bound H^+
$C_T = [COO^-] + [COOH] \equiv$ concentration of binding sites

At time $t = 0^+$, a step jump in H^+ concentration is imposed in electrolyte baths on both sides of the protein membrane. The electrolyte baths contain NaCl at a concentration c_0 at an initial pH such that $\overline{[H^+]} \ll \overline{[Cl^-]}$ inside the membrane. To find $\bar{c}_{H^+}(x, t)$ in the membrane, answer the following:

(**a**) Write continuity equations for $\overline{[Na^+]}$, $\overline{[Cl^-]}$, and $\overline{[H^+]}$ inside the membrane. Include a binding term $\partial H_B / \partial t$ in the $\overline{[H^+]}$ continuity law. In general, we need to include electrical migration as well as diffusion terms in these continuity equations, along with Gauss' law and Faraday's law (Chapter 2) to fully account for electrical interactions between multiple ion species and fixed-charge groups. However, we will see in Chapter 3 that as long as the species of interest (H^+ here) is present at a concentration low compared with another ionic species (e.g., Na^+, Cl^-), then we can neglect such electrical coupling for the "minority" H^+ species and focus on diffusion of H^+ alone.

(**b**) Assume that binding of H^+ ions to the COO^- groups of the protein can be described by the first-order, reversible reaction

$$COOH \underset{k_2}{\overset{k_1}{\rightleftharpoons}} COO^- + H^+ \qquad (1.79)$$

$$K_{eq} \equiv \frac{\overline{[COO^-]}\,\overline{[H^+]}}{\overline{[COOH]}} \qquad (1.80)$$

Assume that the reaction (1.79) is so fast compared with diffusion times that the equilibrium (1.80) is achieved instantaneously. With this assumption, replace the $\partial H_B / \partial t$ term in the H^+ continuity law by a term involving $\overline{[H^+]}$, C_T, and K_{eq}.

(**c**) When $\bar{\rho}_m \gg F[H^+]_{bath}$ is satisfied, find the appropriate limiting nonlinear diffusion equation that can be used to solve for $\bar{c}_{H^+}(x, t)$.

(**d**) For incremental changes in $\overline{[H^+]}$, the nonlinear diffusion equation (the answer to part (c)) can be linearized. Find the effective diffusion coefficient in terms of $\overline{[H]}_0$, K_{eq}, and C_T

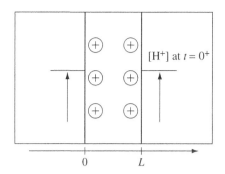

Figure 1.22 Charged membrane subjected to a step change in external H^+ concentration at $t = 0^+$. (For further details on the kinetics of proton diffusion–reaction and the associated kinetics of polyelectrolyte matrix swelling forces see [28].)

and evaluate in the limits $\overline{[\text{H}]}_0 \ll K_{\text{eq}}$, $\overline{[\text{H}]}_0 \gg K_{\text{eq}}$ (i.e., $\text{pH}_0 \gg \text{p}K$ and $\text{pH}_0 \ll \text{p}K$, where $\text{p}K \simeq 3.5$ from Figure 1.12).

(**e**) Find $\bar{c}_{\text{H}^+}(x, t)$ due to a step jump in external bath pH on both sides. Assume that $\bar{c}_{\text{H}^+} \equiv \bar{c}_\alpha$ at $x = 0^+$ and that δ^- is known and is constant for $t \geq 0^+$. (Hint: express your answer in terms of a Fourier series and $\bar{c}_\alpha$.)

What is the effective diffusion–reaction time? Compare with τ_{diff} of Problem 3.3 in Chapter 3. (Don't be confused by the fact that only the left-side bath concentration is raised in Problem 3.3, while both sides are raised here.)

1.8 REFERENCES

[1] Einstein A (1956) *Investigations on the Theory of the Brownian Movement*. Dover Publications, New York.

[2] Glasstone S (1940) *Textbook of Physical Chemistry*. Van Nostrand, New York, p. 255.

[3] Crank J (1975) *The Mathematics of Diffusion*, 2nd ed. Oxford University Press, Oxford.

[4] Reif F (1965) *Fundamentals of Statistical and Thermal Physics*. McGraw-Hill, New York, Chapter 12.

[5] Adler RB, Smith AC, and Longini RL (1964) *Introduction to Semiconductor Physics*. Wiley, New York, p. 62.

[6] de Groot SR and Mazur P (1962) *Non-Equilibrium Thermodynamics*. North-Holland, Amsterdam (reprinted 1984 by Dover, New York).

[7] Newman JS and Thomas-Alyea KE (2004) *Electrochemical Systems*, 3rd ed. Wiley-Interscience, New York.

[8] Kobatake Y and Kamo N (1973) Transport processes in charged membranes. In *Progress in Polymer Science Japan* (Murahashi I, ed.). Wiley, New York, pp. 257–302.

[9] Bird RB, Stewart WE, and Lightfoot EN (2007) *Transport Phenomena*, Revised 2nd ed. Wiley, New York, Chapter 17.

[10] Einstein A (1908) Elementare Theorie der Brownschen Bewegung. *Z. Elektrochem. angew. physik. Chem.* **14**, 235–239. An English translation, "The elementary theory of the Brownian motion," can be found in [1].

[11] Tranquillo RT, Lauffenburger DA, and Zigmond SH (1988) A stochastic model for leukocyte random motility and chemotaxis based on receptor binding fluctuations. *J. Cell Biol.* **106**, 303–309.

[12] Lin F, Nguyen CM-C, Wang S-J, et al. (2004) Effective neutrophil chemotaxis is strongly influenced by mean IL-8 concentration. *Biochem. Biophys. Res. Commun.* **319**, 576–581.

[13] Daniels F and Alberty RA (1987) *Physical Chemistry*, 7th ed. Wiley, New York.

[14] Gardiner WC Jr (1969) *Rates and Mechanisms of Chemical Reactions*. Benjamin, Menlo Park, CA.

[15] Tanford C (1961) *Physical Chemistry of Macromolecules*. Wiley, New York.

[16] Garcia AM, Szasz N, Trippel SB, et al. (2003) Transport and binding of insulin-like growth factor I through articular cartilage. *Arch. Biochem. Biophys.* **415**, 69–79.

[17] Gordon MK and Hahn RA (2010) Collagens. *Cell Tissue Res.* **339**: 247–257.

[18] Fitzgerald J, Rich C, Zhou FH, and Hansen U (2008) Three novel collagen chains, $\alpha4(\text{VI})$, $\alpha5(\text{VI})$, and $\alpha6(\text{VI})$. *J. Biol. Chem.* **283**, 20170–20180.

[19] Bowes JH and Kenten RH (1948) The amino-acid composition and titration curve of collagen. *Biochem. J.* **43**, 358–365.

[20] Iozzo RV (1998) Matrix proteoglycans: from molecular design to cellular function. *Annu. Rev. Biochem.* **67**, 609–652.

[21] Lander AD and Selleck SB (2000) The elusive functions of proteoglycans: in vivo veritas. *J. Cell Biol.* **148**, 227–232.

[22] Schaefer L and Iozzo RV (2008) Biological functions of the small leucine-rich proteoglycans: from genetics to signal transduction. *J. Biol. Chem.* **283**, 21305–21309.

[23] Heinegård D (2009) Proteoglycans and more—from molecules to biology. *Int. J. Exp. Pathol.* **90**, 575–586.

[24] Ng L, Grodzinsky AJ, Patwari P, et al. (2003) Individual cartilage aggrecan macromolecules and their constituent glycosaminoglycans visualized via atomic force microscopy. *J. Struct. Biol.* **143**, 242–257.

[25] Lin CC and Segel LA (1988) *Mathematics Applied to Deterministic Problems in the Natural Sciences*. SIAM, Philadelphia.

[26] Deen WM (1998) *Analysis of Transport Phenomena*. Oxford University Press, New York.

[27] Einstein A (1907) Theoretische Bemerkungen über die Brownsche Bewegung. *Z. Elektrochem. angew. physik. Chem.* **13**, 41–42. An English translation, "Theoretical observations on the Brownian motion," can be found in [1].

[28] Nussbaum JH and Grodzinsky AJ (1981) Proton diffusion–reaction in a protein polyelectrolyte membrane and the kinetics of electromechanical forces. *J. Membr. Sci.* **8**, 193–219.

PROBLEM

Electric Fields and Flows in Electrolyte Media

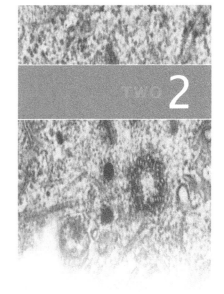

TWO 2

2.1 INTRODUCTION

The interactions of electric and magnetic fields with biological media are of fundamental importance in the study of transport, physiology, microfluidics (microelectromechanical systems (MEMS) and nanoelectromechanical systems (NEMS)) for biomolecule separation, purification, and identification, as well as a host of applications at the molecular, cell, and tissue levels. Illustrative examples are presented in this chapter to summarize (1) the laws of electromagnetism that form the basis of electrochemical and electromechanical transductive coupling in biological systems, and (2) the rate-limiting processes that govern the kinetics associated with electrical interactions. Together, these laws and rate processes are fundamental to the understanding of ionic flux, transport of charged macromolecules through porous tissues and biomaterials, electrokinetic interactions in tissues and microfabricated devices, and many other related topics. These laws also provide the foundation for quantifying *forces* of electrical origin, which are critical to the understanding of intra- and intermolecular motions, cell membrane–ligand interactions, and forces that govern the rheology of tissues and gels.

The laws of electricity and magnetism are embodied in Maxwell's equations, which relate electric (E) and magnetic (H) fields to their sources, electric charge (ρ_e) and current (J) densities. Maxwell's equations are first written in the context of free space (Section 2.2), and then recast to include the effects of linear dielectric, magnetic, and conducting media (Section 2.3). Time-varying electric and magnetic fields can also act as sources of magnetic and electric fields, respectively; this fundamental coupling is embedded in the concept of electromagnetic waves and electrodynamics (Section 2.4). We then restrict our attention to "quasistatic" subsets of Maxwell's equations, which are appropriate at low enough frequencies or, equivalently, when electromagnetic wavelengths are long compared with characteristic dimensions of biological tissues, cells, and molecules (Section 2.5). In this text, we will focus primarily on electric field systems (electroquasistatic systems, Section 2.6), since electrical interactions are most fundamental to native biological tissues, to molecules, and to biotransport in electrolyte media.

2.2 LAWS OF ELECTROMAGNETISM

2.2.1 Integral Form of Maxwell's Equations in Free Space

Maxwell's equations constitute a concise mathematical representation of almost two hundred years of experiments on electromagnetism, its sources, and the interactions between electromagnetic fields and materials. Along with the Lorentz force law, they summarize all of classical electrodynamics [1].

Gauss' Law

In the study of electrostatics, the electric field is often defined in terms of the measured force f on a charge q given by the Lorentz law: $f = qE$. Coulomb's law for the force between two charges and the experimentally observed linear superposition of forces can then be used to find the electric field produced by many individual charges [2]. Gauss' law extends this approach to relate the electric field intensity, $E(r, t)$, to the local density of electric charges. Thus, if $\rho_e(r, t)$ is the volume charge density within any arbitrary volume V enclosed by a surface S (Figure 2.1), then the net *electric flux* through the surface is equal to the total charge Q_{net} in V:

$$\oint_S \epsilon_0 E \cdot da = \int_V \rho_e \, dV = Q_{net} \tag{2.1}$$

where ϵ_0 is the *permittivity* of free space $= 8.85 \times 10^{-12} \, \text{F m}^{-1}$. The charge Q_{net} has units of coulombs (C), ρ_e has units of C m^{-3}, and $\epsilon_0 E$ is called the *electric displacement flux density* and has units of surface charge density, C m^{-2}. We note that ρ_e is defined as $\Delta q / \Delta V$. Since we are concerned with macroscopic fields and a continuum formulation of Maxwell's equations, we specify that the volume element ΔV is large enough to contain many individual charges, but small enough compared with the size of the system. This enables ρ_e to be defined as a continuous function of position.

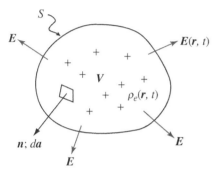

Figure 2.1 A control volume V is enclosed by surface S. An area element on the surface has a unit normal n so that a differential area vector can be defined as $da = n \, da$. A charge distribution ρ_e within V results in an electric field E.

Example 2.2.1 Electric Field Due to a Spherically Symmetric Space Charge Consider a spherical distribution of charge $\rho_e(r)$ in the region $r < R$ such that

$$\rho_e(r) = \begin{cases} \rho_0 & (r < R) \\ 0 & (r > R) \end{cases} \tag{2.2}$$

where ρ_0 and R are given constants. Symmetry demands that the resulting electric field E must be radially directed and dependent only on r:

$$E = i_r E_r(r) \tag{2.3}$$

By Gauss' law, the integral of $\epsilon_0 E$ on a spherical surface at radius r is equal to the charge contained in the spherical volume:

$$\oint_S \epsilon_0 E \cdot da = \epsilon_0 E_r 4\pi r^2 = \begin{cases} \rho_0 \dfrac{4\pi r^3}{3} & (r < R) \\ \rho_0 \dfrac{4\pi R^3}{3} = q & (r > R) \end{cases} \tag{2.4}$$

where q is the total charge. This gives the r dependence of the electric field:

$$E_r(r) = \begin{cases} \dfrac{\rho_0 r}{3\epsilon_0} & (r < R) \\ \dfrac{q}{4\pi\epsilon_0 r^2} & (r > R) \end{cases} \tag{2.5}$$

The solution for $r > R$ gives the familiar $1/r^2$ Coulomb's law dependence of a point charge.

EXAMPLE

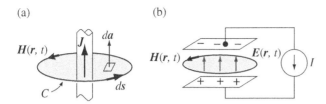

Figure 2.2 (a) The magnetic field $H(r, t)$ is evaluated along a closed contour C that encloses a current density J. (b) A current source I supplies charge to a parallel plate capacitor, resulting in a time-varying electric field $E(r, t)$ and magnetic field $H(r, t)$ in the free space between the plates.

Ampère's Law

Ampère's law relates the magnetic field intensity H to its source, the *current density J*, and the time rate of change of the electric field:

$$\oint_C H \cdot ds = \int_S J \cdot da + \frac{d}{dt} \int_S \epsilon_0 E \cdot da \qquad (2.6)$$

The current density J, having units of $A\,m^{-2} (= C\,s^{-1}\,m^{-2})$, represents the rate of charge transport per unit area, $J = \rho_e v$, where v is the velocity of the moving charge density. Thus, the line integral of H around a closed contour equals the net current passing through the surface spanned by the contour plus the time rate of change of the displacement flux density $\epsilon_0 E$ through the surface. This last term on the right-hand side of (2.6) is classically referred to as the *displacement current*. Examples of contours for the evaluation of the magnetic field arising from the two right-hand terms of (2.6) are illustrated in Figure 2.2. The actual experiments performed by Ampère focused on the magnetic fields produced by currents in physical wires (i.e., the first right-hand term of (2.6)). It was Maxwell's theoretical work that led to the inclusion of the displacement current term, which was reasoned, in part, by analogy to Faraday's law of magnetic induction, below. Maxwell publicly credited this theoretical insight to the fundamental experimental studies of Faraday [3].

Faraday's Law

By analogy with the displacement current term in Ampère's law, Faraday's law, also known as the law of electromagnetic induction, states that an electric field may be induced by a time-varying magnetic field:

$$\oint_C E \cdot ds = -\frac{d}{dt} \int_S \mu_0 H \cdot da \qquad (2.7)$$

where μ_0 is the *permeability* of free space $= 4\pi \times 10^{-7}\,H\,m^{-1}$; see Figure 2.3. The minus sign in (2.7) is associated with Lenz's law, which states that magnetic fields cause electric fields (currents) to be induced in coils in such a direction as to *oppose* a change in the magnetic flux linking the coil. Faraday's law is the governing basis of electric generators, motors, transformers, and the means for distribution of electric power. These and countless other contributions of Faraday in the fields of chemical and physical science and engineering have had immeasurable effects on modern society [4].

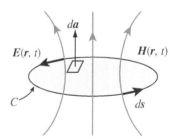

Figure 2.3 The integral of the electric field $E(r, t)$ evaluated along a closed contour C is determined by the time rate of change of the magnetic flux density $\mu_0 H(r, t)$ linking the surface spanned by the contour C.

Gauss' Law for Magnetic Fields

It is an empirical fact that there is no appreciable density of magnetic monopoles (analogous to ρ_e) known to exist in nature. Therefore, by analogy to Gauss' integral law for electric fields, there can be no net magnetic flux out of any region enclosed by a surface:

$$\oint_S \mu_0 \boldsymbol{H} \cdot d\boldsymbol{a} = 0 \tag{2.8}$$

2.2.2 Differential Form of Maxwell's Equations

The integral forms of Maxwell's equations apply to any arbitrary volume and surface (written above in free space). These integral forms are especially useful for calculating electric and magnetic fields when the distributions of the charge and current sources have a high degree of symmetry, which simplifies evaluation of the various contour, surface, or volume integrals. In most practical applications, however, the distribution of sources is spatially complex. Such cases involve solution of boundary value problems, which require the use of differential forms of Maxwell's equations. We can convert the integral forms to the equivalent differential forms by using Gauss' and Stokes' integral theorems:

$$\text{Gauss' theorem} \qquad \oint_S \boldsymbol{A} \cdot d\boldsymbol{a} = \int_V (\nabla \cdot \boldsymbol{A}) \, dV \tag{2.9}$$

$$\text{Stokes' theorem} \qquad \oint_C \boldsymbol{A} \cdot d\boldsymbol{s} = \int_S (\nabla \times \boldsymbol{A}) \cdot d\boldsymbol{a} \tag{2.10}$$

Applying Gauss' integral theorem to Gauss' electric field law, (2.1),

$$\oint_S \epsilon_0 \boldsymbol{E} \cdot d\boldsymbol{a} = \int_V \nabla \cdot \epsilon_0 \boldsymbol{E} \, dV = \int_V \rho \, dV \tag{2.11}$$

Since the volume dV is arbitrary, the integrands inside the volume integrals must be equal, and we obtain Gauss's law in differential form:

$$\nabla \cdot \epsilon_0 \boldsymbol{E} = \rho_e \tag{2.12}$$

Similarly, we can apply Stokes' theorem to Ampère's integral law, (2.6):

$$\int_S (\nabla \times \boldsymbol{H}) \cdot d\boldsymbol{a} = \int_S \boldsymbol{J} \cdot d\boldsymbol{a} + \frac{d}{dt} \int_S \epsilon_0 \boldsymbol{E} \cdot d\boldsymbol{a} \tag{2.13}$$

We assume, here, that the surface S is fixed in time, so that the time derivative on the right-hand side can be taken inside the integral. Since the surface S is arbitrary, the integrands must be related by

$$\nabla \times \boldsymbol{H} = \boldsymbol{J} + \frac{\partial \epsilon_0 \boldsymbol{E}}{\partial t} \tag{2.14}$$

Finally, Faraday's law (2.7) and Gauss' law for the magnetic field (2.8) become, respectively,

$$\nabla \times \boldsymbol{E} = -\frac{\partial \mu_0 \boldsymbol{H}}{\partial t} \tag{2.15}$$

$$\nabla \cdot \mu_0 \boldsymbol{H} = 0 \tag{2.16}$$

2.2.3 Conservation of Charge

Charge conservation, a fundamental physical law, is implicit to Maxwell's equations. We can formulate an integral form of conservation of charge inside a control volume V such as that pictured in Figure 2.1. The time rate of decrease of the charge within the control volume must equal the net loss of charge flowing out of this volume (i.e., that due to the flux of current leaving the surface of the control volume):

$$\oint_S \mathbf{J} \cdot d\mathbf{A} = -\frac{d}{dt}\int_V \rho_e \, dV \tag{2.17}$$

Using Gauss' integral theorem, we can convert the integral form into the differential form of conservation of charge (again taking the volume V to be fixed):

$$\oint \mathbf{J} \cdot d\mathbf{A} = \int_V \nabla \cdot \mathbf{J} \, dV = -\int_V \frac{\partial \rho_e}{\partial t} \, dV \tag{2.18}$$

For an arbitrary volume element dV, the second equality in (2.18) gives:

$$\nabla \cdot \mathbf{J} = -\frac{\partial \rho_e}{\partial t} \tag{2.19}$$

The above derivation was based on physical reasoning. It is interesting and important to note, however, that charge conservation is consistent with Maxwell's equations, and is therefore not an independent equation. If we take the divergence of Ampère's law, we get

$$\nabla \cdot (\nabla \times \mathbf{H}) = \nabla \cdot \mathbf{J} + \frac{\partial}{\partial t}(\nabla \cdot \epsilon_0 \mathbf{E}) \tag{2.20}$$

Since the divergence of the curl is zero, the left-hand side of (2.20) is zero. We can substitute Gauss' law for electric charge into (2.20) and again obtain the law of conservation of charge:

$$\nabla \cdot \mathbf{J} + \frac{\partial \rho_e}{\partial t} = 0 \tag{2.21}$$

2.2.4 The Lorentz Force Law

The Lorentz force law describes the force exerted on a moving charge in the presence of electric and magnetic fields. This empirical law serves to define the field quantities $\mathbf{E}$ and $\mathbf{H}$ in terms of the force on a test charge q. In free space, the Lorentz force law is

$$\mathbf{f} = q[\mathbf{E} + (\mathbf{v} \times \mu_0 \mathbf{H})] \tag{2.22}$$

where $\mathbf{f}$ is the force (with units of $\mathrm{N} = \mathrm{kg\,m\,s^{-2}}$). We can also write the Lorentz force law in terms of the force per unit volume, $\mathbf{F}$ (with units of $\mathrm{N\,m^{-3}}$):

$$\mathbf{F} = \rho_e \mathbf{E} + \mathbf{J} \times \mu_0 \mathbf{H} \tag{2.23}$$

with the implication that forces acting on charge carriers with the medium are converted by momentum transfer into a force density acting on the medium itself.

In summary, Maxwell's equations describe the effects of charges and currents on electric and magnetic fields, and the Lorentz law, in turn, describes the effects of fields on moving charges. These laws, together with Newton's law (which relates the Lorentz force to charged particle velocity v) constitute a complete description of the temporal and spatial evolution of electric and magnetic fields in free space, and, with extensions, to media [1].

2.3 MAXWELL'S EQUATIONS IN MEDIA: POLARIZATION, MAGNETIZATION, AND CONDUCTION

When electric or magnetic fields interact with materials, the applied fields can induce *polarization*, *magnetization*, and/or *electrical conduction* within the media, which, in turn, can perturb the field inside and outside the material. The form of Maxwell's equations written for free space in the previous section must now be modified to include the presence of such phenomena. In addition, macroscopic material constitutive laws that describe polarization, magnetization, and conduction must be developed. We focus initially on linear, isotropic materials and then ultimately consider boundary value problems that include interfaces between media having different material properties.

2.3.1 Polarization

Biological materials, macromolecules, and electrolyte media are made of molecules and atoms that may have permanent dipole moments, such as water [5]. In addition, application of an electric field can induce atomic-level dipole moments even in the absence of a permanent moment. For example, positive nuclei and negative electron clouds can be displaced from each other by the applied field, creating an internal "back-field" at the atomic level. Although net neutral, one region of a dipolar molecule appears positive while another region is negative. Although the distribution of dipoles in a biomaterial is typically random, an applied $\boldsymbol{E}$ field can also lead to orientation of fixed and induced dipoles such that they become partially aligned with the field, as depicted in Figure 2.4. The degree of alignment will depend on how freely such dipoles can rotate, given the stiffness of molecular bonding within the material. The challenge is to incorporate the effects of these atomic and molecular polarization phenomena into averaged macroscopic sources, and thereby obtain the resultant averaged macroscopic fields (i.e., the sum of the applied field and the local atomic level back-fields). This leads to a reformulation of the macroscopic Maxwell's equations incorporating the effects of media [6].

Dipoles are modeled as pairs of opposite charges, $\pm q$, separated by a vector distance $\boldsymbol{d}$ directed from the $-$ to the $+$ charge, so that the *dipole moment* $p \equiv q\boldsymbol{d}$. If N is the number density of dipoles, we define the *polarization density* $\boldsymbol{P}$ by

$$\boldsymbol{P} = Nq\boldsymbol{d} \tag{2.24}$$

Gauss' law, (2.12), was previously written for free space, in which ρ_e represented freely mobile charge in the absence of media. We now distinguish and account for the presence of both polarization charge ρ_p (sometimes called paired or bound charge) and freely mobile or

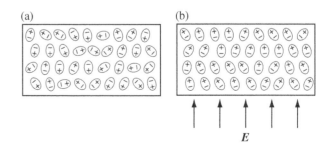

Figure 2.4 (a) A material containing randomly oriented dipoles. (b) In an electric field, the dipoles tend to align with the field.

unpaired charge ρ_e (e.g., ions in saline), which can enter, leave, or be conducted through a medium:

$$\nabla \cdot \epsilon_0 \boldsymbol{E} = \rho_e + \rho_p \tag{2.25}$$

Consider a medium containing dipoles (but no unpaired charge) as depicted in Figure 2.5, and a volume V within the medium. Some dipoles are seen to straddle the surface, leaving behind a charge $-q\boldsymbol{d} \cdot d\boldsymbol{a}$ within the volume. The net charge Q within this volume is determined by the dipole density at the surface. However, this must be equal to the charge ρ_p contained within the volume (since $\rho_e = 0$ in this case),

$$Q = -\oint_S Nq\boldsymbol{d} \cdot d\boldsymbol{a} = -\oint_S \boldsymbol{P} \cdot d\boldsymbol{a} = \int_V \rho_p \, dV \tag{2.26}$$

and by Gauss' theorem we conclude that

$$-\nabla \cdot \boldsymbol{P} = \rho_p \tag{2.27}$$

Gauss' law can now be rewritten as

$$\nabla \cdot \boldsymbol{D} = \nabla \cdot (\epsilon_0 \boldsymbol{E} + \boldsymbol{P}) = \rho_e \tag{2.28}$$

where the *displacement flux density* is defined as

$$\boldsymbol{D} = \epsilon_0 \boldsymbol{E} + \boldsymbol{P} \tag{2.29}$$

In accordance with charge conservation (Section 2.2.3), Ampère's law must be modified appropriately:

$$\nabla \times \boldsymbol{H} = \boldsymbol{J}_e + \frac{\partial \boldsymbol{D}}{\partial t} \tag{2.30}$$

where we have explicitly used the *unpaired charge* current $\boldsymbol{J}_e$ as the current density (e.g., an ionic current density in biological media). The paired charge current density is implicitly included in the $\partial \boldsymbol{D}/\partial t$ term.

2.3.2 Magnetization

Magnetization is associated with the magnetic dipole moments of individual electrons and electron orbital motions in a material. These magnetic dipole moments can be modeled in terms of macroscopic fields and parameters in a manner similar to that of the electric dipole discussion. We first rewrite the free space Gauss' law for the magnetic field (2.16) as

$$\nabla \cdot \mu_0 \boldsymbol{H} = -\nabla \cdot \mu_0 \boldsymbol{M} \tag{2.31}$$

where $\boldsymbol{M}$ is the *magnetization density*, and $\mu_0 \boldsymbol{M}$ is in direct analogy to the *polarization density* $\boldsymbol{P}$ (where μ_0 is included by convention). Gauss' law for the magnetic field then takes the form

$$\nabla \cdot \boldsymbol{B} = \mu_0 \nabla \cdot (\boldsymbol{H} + \boldsymbol{M}) = 0 \tag{2.32}$$

where the *magnetic flux density* $\boldsymbol{B}$ is defined as

$$\boldsymbol{B} = \mu_0(\boldsymbol{H} + \boldsymbol{M}) \tag{2.33}$$

At the same time, we rewrite Faraday's law to include the presence of magnetic materials as

$$\nabla \times \boldsymbol{E} = -\frac{\partial \boldsymbol{B}}{\partial t} \tag{2.34}$$

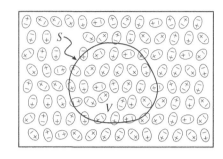

Figure 2.5 Control volume V within a material containing polarization charge. Some dipoles straddle the surface S, resulting in a residual net charge within the volume.

2.3.3 Linear Isotropic Media

The polarization density P is generally related to E through a material constitutive relation. For linear, isotropic materials,

$$P = \chi_e \epsilon_0 E, \qquad D = \epsilon_0 (1 + \chi_e) E \equiv \epsilon E \qquad (2.35)$$

where the second equation follows from (2.29). χ_e is the *electric susceptibility* and $\epsilon = \epsilon_0(1 + \chi_e)$ is the intrinsic *electric permittivity* of the material.

In a similar manner,

$$M = \chi_m H, \qquad B = \mu_0(1 + \chi_m) H \equiv \mu H \qquad (2.36)$$

where χ_m is the *magnetic susceptibility* and $\mu = \mu_0(1 + \chi_m)$ is the intrinsic *magnetic permeability*. (For all materials, $\chi_e > 0$. However, a material can be diamagnetic ($\chi_m < 0$) or paramagnetic, ferrimagnetic, or ferromagnetic ($\chi_m > 0$). In fact, all matter exhibits diamagnetism, but the effect is negligible ($-\chi_m < 0.001$) and is overwhelmed if one of the other forms of magnetism is present.)

For water, ϵ is about $80\epsilon_0$. Values of ϵ for many biological and synthetic materials have been measured and tabulated [6]. Interestingly, the magnetic permeability μ for most biological tissues and cells is essentially μ_0.

It is important to note that in linear, isotropic media, the form of Maxwell's equations remains the same as that in free space, with ϵ_0 and μ_0 replaced by ϵ and μ. Extensions for the case of anisotropic media are enabled by the use of permittivity and permeability tensors ϵ_{ij} and μ_{ij} [7].

While we will generally restrict ourselves to *piecewise* linear, isotropic materials in the rest of this text, important broad classes of nonlinear materials must be acknowledged. For instance, *piezoelectric* materials are characterized by a polarization P proportional to an externally applied mechanical stress. Another example is the common magnet, which exhibits a permanent magnetization M independent of external magnetic fields.

2.3.4 Ohmic Conduction and Generalized Ionic Transport

For most electrolyte media of interest, the molar flux N of ions of the ith species can be represented by the empirical relation (measured with respect to a fixed coordinate system)

$$N_i = +\frac{z_i}{|z_i|} u_i c_i E - D_i \nabla c_i + c_i v \qquad (2.37)$$

where

$N_i \equiv$ flux $(\text{mol m}^{-2}\,\text{s}^{-1})$
$z_i \equiv$ valence of the ith species
$c_i \equiv$ concentration (mol m^{-3})
$u_i \equiv$ mobility $(\text{m}^2\,\text{V}^{-1}\,\text{s}^{-1})$
$D_i \equiv$ diffusion coefficient $(\text{m}^2\,\text{s}^{-1})$
$E \equiv$ electrical field (V m^{-1})
$v \equiv$ velocity (m s^{-1}) of medium

and

$$\frac{D_i}{u_i} = \frac{RT}{|z_i|\,F} \simeq \frac{25.7}{|z_i|} \text{ mV at } 25°C \qquad (2.38)$$

Species	Mobility u_i (m^2 V^{-1} s^{-1})	Diffusion coefficient D_i (m^2 s^{-1})
Cations in H$_2$O at 25°C		
H$^+$	36.30×10^{-8}	9.33×10^{-9}
K$^+$	7.62×10^{-8}	1.96×10^{-9}
Ba^{2+}	6.59×10^{-8}	0.84×10^{-9}
Na$^+$	5.19×10^{-8}	1.33×10^{-9}
Li$^+$	4.01×10^{-8}	1.03×10^{-9}
Anions in H$_2$O at 25°C		
OH$^-$	20.52×10^{-8}	5.27×10^{-9}
SO$_4^{2-}$	8.27×10^{-8}	1.06×10^{-9}
Cl$^-$	7.91×10^{-8}	2.03×10^{-9}
NO$_3^-$	7.40×10^{-8}	1.90×10^{-9}
HCO$_3^-$	4.61×10^{-8}	1.18×10^{-9}
Electrons in Ge at 25°C	0.39	0.010
Electrons in Si at 25°C	0.15	0.004
Holes in Ge at 25°C	0.19	0.005
Holes in Si at 25°C	0.06	0.002

Table 2.1 Some mobilities and diffusion coefficients at 25°C (infinite dilution).

where $F \equiv$ Faraday constant ($\simeq 96{,}500$ C mol^{-1}). Equation (2.38) is the Einstein relationship between ionic mobility and diffusivity, a function of temperature alone. This relationship is based on the assumption that ions are distributed in equilibrium according to Boltzmann statistics, valid in the limit of dilute solution (see Sections 1.2 and 3.1 for a more detailed discussion).

The first two terms on the right-hand side of (2.37), the *migration* and *diffusion* fluxes, respectively, reflect the relative motion of ions with respect to the medium. The last term, or *convection* flux, reflects motion of the medium itself. Equation (2.37) does not include the effect of magnetic fields, which would serve to add a term consistent with the Hall effect. In Section 1.2, we considered continuity and fluxes of species with respect to a moving frame of reference, where the moving frame is defined with respect to a molar-averaged or mass-averaged velocity. (The average is taken over all solute and solvent species.) We will also have occasion to use a particle flux rather than a molar flux, which can be obtained by replacing the molar concentration c_i by the number density n_i (m^{-3}). Thus, c_i and n_i are related by Avogadro's number, $n_i = N_0 c_i$. For example, a 0.1 molar (0.1 M) solution of a given ion has the equivalent number density,

$$n_i \simeq \left(6 \times 10^{23} \, \frac{1}{\text{mole}}\right)\left(0.1 \, \frac{\text{moles}}{\text{liter}}\right)\left(10^3 \, \frac{\text{liter}}{\text{m}^3}\right) = 6 \times 10^{25} \, \text{m}^{-3}$$

Typical values of diffusion coefficients and number densities are shown in Tables 2.1 and 2.2.

The current density $\boldsymbol{J}$ (A m^{-2}) associated with the molar flux (2.37) can be found by recognizing that

$$\boldsymbol{J}_i = z_i F \boldsymbol{N}_i \tag{2.39}$$

$$\boldsymbol{J} = \sum_i \boldsymbol{J}_i = \sum_i z_i F \boldsymbol{N}_i \tag{2.40}$$

$$\boldsymbol{J} = \sum_i |z_i| \, F u_i c_i \boldsymbol{E} + \sum_i (-z_i F D_i \nabla c_i) + \sum_i z_i F c_i \boldsymbol{v} \tag{2.41}$$

Table 2.2 Comparative number densities and conductivities.

Material	$n_i(\mathrm{cm}^{-3})$	$\sigma\,(\Omega^{-1}\,\mathrm{m}^{-1})$
0.1 M NaCl	6×10^{19}	1.07
Copper	$n_e \simeq 10^{22}$	5.8×10^7
Si (intrinsic)	$n = p \simeq 10^{10}$	3.36×10^{-4}
Si (doped) $(N_d = 10^{16})$	$n_e = 10^{16}, n_p = 10^4$	2.4
Quartz		10^{-18}

For a wide variety of materials having charge carrier concentrations and mobilities that are independent of E (i.e., Ohmic conductors), we define the electrical conductivity σ for a linear, isotropic medium from the first term of (2.41):

$$\sigma = \sum_i |z_i|\, F u_i c_i \tag{2.42}$$

The third term on the right-hand side in (2.41) contains the *net* charge ρ_e $(\mathrm{C\,m}^{-3})$

$$\rho_e = \sum_i z_i F c_i \tag{2.43}$$

Then, (2.41) can be written (in the limit of *negligible molecular diffusion*) as

$$\boldsymbol{J} = \sigma \boldsymbol{E} + \rho_e \boldsymbol{v} \tag{2.44}$$

The full constitutive law (2.41) can be regarded as an empirical relation (written in the limit of dilute solution) which is obeyed for a wide variety of electrolyte media and solute species. Equation (2.37) or (2.41) is the Nernst–Planck equation with material convection added. It can be "derived" thermodynamically by considering gradients in the electrochemical potential μ, but this representation itself is essentially empirical. Various limiting cases of the constitutive law can be found in which only a certain subset of terms on the right-hand side of (2.41) are important. For example, when diffusion can be neglected, the law has the form (2.44). If there is no material motion or deformation, so that $v = 0$, or if there is zero *net* charge ρ_e, then the electric current is determined by a combination of electrical migration (drift) and diffusive flows. This is often the case for biological membranes (see, e.g., [8]), tissues, and hydrogel scaffolds having very low hydraulic permeability, and, interestingly, the case of a semiconductor.

In many types of homogeneous conductors in which the concentrations of all species are uniform (e.g., metals and well-mixed electrolyte solutions), we will find that diffusion can be neglected in favor of migration except for small enough length scales on the order of an electrical Debye length or less (we will derive this result formally in Section 3.1, and show that the Debye length is a measure of the decay of an electric field away from a fixed charge boundary). However, even in the presence of concentration gradients, the migration term in (2.41) can still dominate the diffusion term for a certain regime of applied electric field E. We examine this regime by evaluating the ratio of these two terms, with reference to Figure 2.6.

$$\left|\frac{\text{migration}}{\text{diffusion}}\right| \simeq \left|\frac{u\,c\,\boldsymbol{E}}{D\nabla c}\right| \tag{2.45}$$

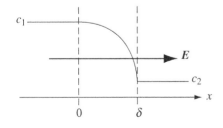

Figure 2.6 Applied field E in a region of nonuniform concentration c of an ionic species.

For the case $c_1 \gg c_2$, we have $c \approx c_1/2$ and $\nabla c \approx c_1/\delta$ for the region of thickness δ in Figure 2.6. Incorporating the above approximation and

NaCl at 25°C	
Concentration (Eq. L^{-1})	$\Lambda(\Omega^{-1}\,m^2\,Eq^{-1})$
	$\equiv \sigma/cz = F(u_+ + u_-)$
Infinite dilution	0.012645
0.0005	0.012450
0.001	0.012374
0.005	0.012065
0.01	0.011851
0.02	0.011551
0.05	0.011106
0.1	0.010674

Table 2.3 Equivalent conductance varies with concentration.

the Einstein relation into (2.45), we find that migration will dominate diffusion when

$$|E\delta| \gg \frac{2RT}{|z|F} \qquad (2.46)$$

For the case of a univalent ion and a thickness $\delta = 10\,\text{nm}$, we find that the electric field must be greater than $5 \times 10^6\,\text{V m}^{-1}$. Therefore, an applied potential drop of magnitude $\gg 50\,\text{mV}$ will lead to a situation in which the flow of univalent ions is governed primarily by the first term of (2.41). If (2.46) is not satisfied, then diffusion must be taken into account. We can also arrive at the inequality (2.46) by comparing the average time it takes for a carrier to travel the distance δ by the mechanisms of migration and diffusion. For migration to dominate, the characteristic migration time $\tau_{\text{mig}} = \delta/v = \delta/uE$ must be much less than the characteristic diffusion time $\tau_{\text{diff}} \sim \delta^2/2D$. (This diffusion time may be derived from the diffusion equation, Fick's second law (Section 1.2), or from Einstein's random walk analysis of diffusion (Section 1.3).)

Two additional issues regarding the conductivity (2.42) merit consideration. First, in evaluating the conductivity, care must be used, since the mobility u_i is in general a function of concentration owing to ion–ion interactions. This can be seen from a comparison of experimentally measured equivalent conductances for the same electrolyte at different concentrations, where the equivalent conductance $\Lambda(\Omega^{-1}\,m^2\,Eq^{-1}) \equiv \sigma/cz = F(u_+ + u_-)$ for a $z:z$ electrolyte. Table 2.3 shows values of Λ for NaCl at 25°C. Note that the D_i and u_i of Table 2.1 refer to the limit of infinite dilution. In view of the Einstein relation (2.38), it should not be surprising that D_i also varies with concentration owing to ion–ion interactions.

Another important issue regarding (2.41) and (2.42) is the presence of chemical reactions (e.g., electrolysis) at the electrode–electrolyte interface. As an example, we consider the electrolyte cell pictured in Figure 2.7, consisting of a homogeneous bath of 0.1 M NaCl between inert (e.g., platinum) electrodes. We might attempt to model the bath

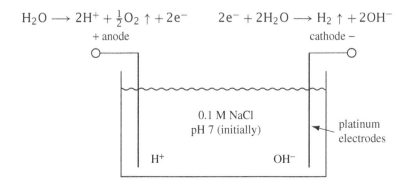

$H_2O \longrightarrow 2H^+ + \tfrac{1}{2}O_2 \uparrow + 2e^-$ \qquad $2e^- + 2H_2O \longrightarrow H_2 \uparrow + 2OH^-$

+ anode \qquad\qquad cathode −

0.1 M NaCl
pH 7 (initially)

platinum electrodes

H$^+$ \qquad OH$^-$

Figure 2.7 Applied current can alter $c_i = c_i(\mathbf{r}, t)$ in general. Classroom demonstration: Cell length $= 10\,\text{cm}$; cell and electrode cross-sectional area $A = 10\,\text{cm}^2$. (The reaction $2e^- + 2H^+ \to H_2 \uparrow$ can occur at low pH.)

as an ohmic conductor according to (2.44), and calculate σ from (2.42). However, if a steady current is applied to the cell, *electrochemical reactions must take place at each electrode/electrolyte interface*. This follows from the requirement of steady state continuity of current at the interfaces. Since electrons carry current in the external metal circuit, while ions do so in the solution, charge transfer reactions (oxidation or reduction) are necessary at the phase boundaries to provide continuity of current. The dominant electrode reactions consistent with platinum electrodes in an initially pH-neutral solution are shown in Figure 2.7. It is seen that, in the face of finite electric current, the electrode reaction products will lead to new contributions to σ in (2.44). Thus, the Ohmic model breaks down to some extent. Further, all species present in the cell will have time- and space-varying concentrations $c_i(\boldsymbol{r}, t)$, which continually evolve so as to maintain charge neutrality even as new species are generated. Diffusion and electrically induced fluid convection may also be important and may affect $c_i(\boldsymbol{r}, t)$.

A quick calculation (referring to Figure 2.7) will show that the initial concentrations $c_{Na^+} = c_{Cl^-} = 0.1$ M can be perturbed significantly by the electrode reactions. From Table 2.2, $\sigma \simeq 1\,\Omega^{-1}\,m^{-1}$ initially; application of 6 V across the electrodes leads to an initial current I_0 (based on the dimensions of Figure 2.7),

$$I_0 = V\frac{A\sigma}{\ell} \simeq 0.06\,\text{A}$$

At the left electrode, this current is consistent with the generation of H^+ ions at a rate

$$\frac{I(\text{C s}^{-1})}{F(\text{C mol}^{-1})} = \frac{6 \times 10^{-2}}{9.65 \times 10^4} = 6.2 \times 10^{-7}\,\text{mol s}^{-1}$$

Over a period of minutes, the total ionic strength can be changed by several percent and the pH by over 5 units in the vicinity of the electrode! Note that rapid stirring of the bath can negate completely the changes in pH and ionic strength. (Why?) Thus, convection is extremely important in the study of electrode kinetics, and in physiological experiments employing static or slowly varying applied currents at implanted metal electrodes.

Therefore, the constitutive law (2.41) most often assumes the role of one more equation in the unknown concentrations $c_i(\boldsymbol{r}, t)$. This law must be added to the list of other physical laws describing the electrical, chemical, and mechanical behavior of systems under consideration, in order to solve self-consistently for all the parameters of interest.

Tables 2.4 and 2.5 list resistivities and dielectric constants for various biological tissues [9].

2.4 ELECTROMAGNETIC WAVES

Even in the absence of net charge and current, time-varying magnetic fields can induce electric fields (Faraday's law), and time-varying electric fields can induce magnetic fields (Ampère's law). This coupling is the basis of electromagnetic wave propagation embodied in the wave equation. For the case of a medium having no free charges and currents ($\boldsymbol{J} = 0$, $\rho_e = 0$), we can take the curl of Faraday's law and substitute in $\nabla \times \boldsymbol{H}$ from Ampère's law (in source-free form):

$$\nabla \times (\nabla \times \boldsymbol{E}) = -\mu\epsilon\frac{\partial^2 \boldsymbol{E}}{\partial t^2} \tag{2.47}$$

corresponding to a linear, isotropic medium (μ, ϵ).

Table 2.4 Resistivities of animal and human tissues, ρ (Ω·m).

Frequency	Muscle	Heart muscle	Liver	Lungs	Spleen	Kidneys	Brain	Fatty tissue	Bone	Bone marrow	Blood	Plasma	0.9% NaCl solution
10 Hz	a 9.6	a 9.6	a 8.4, 12.2	a 11	—	—	—	—	—	—	—	—	—
100 Hz	a 8.9	a 9.2	a 8, 10.6	a 11.1	—	—	—	—	—	—	b 1.66	—	—
1 kHz	a 8	a 7.5, 9.3	a 7.7, 8, 9.7	a 10	a 10	—	—	a 15–50 c 17–25	—	—	b 1.66 d 1.47	b 0.6	—
10 kHz	9.8	e 8.3–9 g 7–13	e 10–16	e 14–19	g 2.6–4.3	—	e 5–8 g 4.5–5.5	—	—	—	e 1.2–1.35 h 1.3–1.8	—	—
100 kHz	i 1.7–2.5 f 5.2	i 1.9–2.4	a 4.6 i 2.2–5.5 g 5.5–8 j 4.2	i 1.65–2	i 2.5–5	i 1.5–2.7	i 4.6–8.8	—	—	—	d 1.47	—	—
1 MHz	i 1.6–2.1 f 2.5	i 1.8–2.3 g 4–5.5 j 4	i 2.1–4.2	i 1.5–2.8	i 2.3–3.8	i 1.4–2.5	i 4.3–7	—	—	—	d 1.4	—	—
10 MHz	i 1.5–1.7	1.4–1.8	i 1.8–2.6	i 1.1–1.5	i 1.5–1.7	i 1.2–1.7	i 3–4.5	—	—	—	d 0.9	—	—
100 MHz	k 1–1.3 i 1.2–1.6 m 1.4–2 m 1.2–1.5	i 1.3–1.7	k 1.2–1.45 i 1.5–2 m 1.8–2.1 m 1.5–1.8	k 0.95–1.3 i 1–1.4	k 0.85–1.05 1.1–1.5 m 1.5 m 1.2	k 1–1.2 1–1.5 m 1.3–1.6 m 0.9–1.4	k 1.6–2.3 i 2–3 m 2.2–2.6 i 1.8–2	k 11.7–12.5 i 15 m 22–43 m 17–25	—	m 41–53 m 30–50	l 0.82 m 1.2–1.5 m 0.8–1	k 0.61 l 0.7 m 0.8 m 0.6	0.6
1 GHz	n 0.75–0.79 o 0.81–0.84 p 0.77	o 0.83–1	n 0.98–1.06 0.92–1.03 p 1	o 1.37	—	o 0.81–0.82	—	7–14 11–35 p 25	20	n 10–23	n 0.64–0.72 o 0.80	n 0.54	0.49 0.56 0.53
10 GHz	n 0.12 q 0.13	—	n 0.15–0.17	—	—	—	—	n 2.4–3.7 q 2.1	n 1.5 q 1.3	n 0.60–2 q 1	q 0.11 q 0.095 r 0.093	n 0.09	0.09
24 GHz	—	—	—	—	—	—	—	t 0.71	t 0.71	—	t 0.038	—	—

Notes: (a) dog, *in situ*, body temperature; (b) sheep, 18°C (plasma, 37°C); (c) dog, *in situ*, body temperature; (d) rabbit, 20°C; (e) rabbit, fresh tissue, 23°C; (f) rabbit, fresh tissue, 23°C; (g) human and various animals, fresh tissues, 18°C; (h) sheep, 18°C; (i) human, minced tissue, 23°C; (j) rabbit, minced tissue, 18°C; (k) human, minced tissue, 37°C; (l) sheep, 20°C; (m) ox and pig, fresh tissue, 20°C; (n) dog and horse, fresh tissue and blood, 38°C; bone and bone marrow, 25°C; (o) human, fresh tissue, 23°C; (p) ox, minced tissue, 22°C; (q) human, fresh tissue, 37°C; (r) human, fresh tissue, 35°C; (s) frog, fresh tissue, 25°C; (t) human, fresh tissue, 37°C. (From [9], pp. 40–43.)

Table 2.5 Dielectric constants, ϵ/ϵ_0.

Frequency	Muscle	Heart muscle	Liver	Lungs	Spleen	Kidneys	Brain	Fatty tissue	Bone	Bone marrow	Blood	Plasma	0.9% NaCl solution
10 Hz	a 10^7	a 7×10^6	a 16×10^6	a 8×10^6	—	—	—	—	—	—	—	—	—
100 Hz	a 800×10^3 s 10^6	a 800×10^3 a 820×10^3	a 900×10^3 a 850×10^3	a 450×10^3	—	—	—	a 150×10^3	—	—	—	— —	— —
1 kHz	a 130×10^3 f 100×10^3 s 170×10^3	a 300×10^3	a 150×10^3	a 90×10^3	—	—	—	a 50×10^3	—	—	—	—	—
10 kHz	a 50×10^3 a 60×10^3 90×10^3 f 50×10^3	100×10^3	a 50×10^3 a 60×10^3	a 30×10^3	—	—	—	a 20×10^3	—	—	d 2903 d 2800	— —	— —
100 kHz	f 20×10^3 s 30×10^3	—	g 7000–12000	—	—	—	—	—	—	—	d 2740	—	—
1 MHz	f 2000	—	g 1200–2000	—	—	—	—	—	—	—	d 2040	—	—
10 MHz	—	—	—	—	—	—	—	—	—	—	d 200	—	—
100 MHz	m 69–73 m 71–76	—	k 65–75 m 72–74 m 70–79	—	m 88–90 m 100–101	m 83–84 m 87–92	m 70–75 m 81–83	m 8–13	—	m 7–8	m 72–74 m 73–76	m 82 m 76	78
1 GHz	n 49–52 o 53–55 p 61	o 53–57	n 46–47 o 44–52 p 50	0.35	—	o 53–56	—	n 4.3–7.5 o 3.2–6 p 9.5	n 8	n 4.3–7.8	n 58–62 63	n 69	72 78
10 GHz	n 40–42 q 29	—	n 34–38	—	—	—	—	n 3.5–3.9 q 3.6	n 8 q 6.6	n 4.4–6.6 5.8	n 50–52 45 48	n 61	66
24 GHz	—	—	—	—	—	—	—	t 3.4	t 6.3	—	t 3.2	—	—

Notes: **(a)** dog, *in situ*, body temperature; **(b)** sheep, 18°C (plasma, 37°C); **(c)** dog, *in situ*, body temperature; **(d)** rabbit, 20°C; **(e)** rabbit, fresh tissue, 23°C; **(f)** rabbit, fresh tissue, 20°C; **(g)** human and various animals, fresh tissues, 23°C; **(h)** sheep, 18°C; **(i)** human, minced tissue, 18°C; **(j)** rabbit, minced tissue, 23°C; **(k)** human, minced tissue, 37°C; **(l)** sheep, 20°C; **(m)** ox and pig, fresh tissue, 20°C; **(n)** dog and horse, fresh tissue and blood, 38°C; **(o)** human, fresh tissue, 25°C; **(p)** ox, minced tissue, 22°C; **(q)** human, fresh tissue, 37°C; **(r)** human, fresh tissue, 35°C; **(s)** frog, fresh tissue, 25°C; **(t)** human, fresh tissue, 37°C. (From [9], pp. 40–43.)

We expand the left-hand side using the vector identity for the curl of the curl, and using Gauss' law in the absence of net charge:

$$\nabla \times (\nabla \times \boldsymbol{E}) = \nabla(\nabla \cdot \boldsymbol{E}) - \nabla^2 \boldsymbol{E} \qquad (2.48)$$

Thus, the wave equation for the electric field is

$$\nabla^2 \boldsymbol{E} = \mu\epsilon \frac{\partial^2 \boldsymbol{E}}{\partial t^2} \qquad (2.49)$$

and a corresponding wave equation can be written for the magnetic field:

$$\nabla^2 \boldsymbol{H} = \mu\epsilon \frac{\partial^2 \boldsymbol{H}}{\partial t^2} \qquad (2.50)$$

The wave propagates with the velocity $c = 1/\sqrt{\mu\epsilon}$, which in free space is equal to $c = 3 \times 10^8 \, \mathrm{m\,s^{-1}}$.

Example 2.4.1 Solutions to the Wave Equations   Consider an electric field intensity of the form $\boldsymbol{E} = \hat{\boldsymbol{x}} E_0 \cos(\omega t - kz + \theta)$. This solution describes a plane wave having an electric field component in the x direction and propagating in the z direction with a radian frequency ω and a propagation velocity $v = \omega/k$. The phase of the electric field is described by θ, and k is the wavenumber. Inserting this equation back into the wave equation in free space, we find that

$$k^2 E_0 \cos(\omega t - kz + \theta) = \mu_0\epsilon_0 \omega^2 E_0 \cos(\omega t - kz + \theta) \qquad (2.51)$$

Simplifying, we find that

$$v = \frac{\omega}{k} = \frac{1}{\sqrt{\mu_0\epsilon_0}} = c \qquad (2.52)$$

and, hence, we see that the wave propagates at the speed of light in free space.

   Exercise: Use Ampère's law to find the magnetic field corresponding to this plane wave solution.

Example 2.4.2 Vector Representation of Plane Wave Solution   Here, we generalize the electromagnetic plane wave solution in the previous example to a wave traveling in the $\boldsymbol{k}$ direction using complex notation.

   Let the electric and magnetic field intensities have the following forms:

$$\boldsymbol{E}(\boldsymbol{r}, t) = \boldsymbol{E}_0 e^{j(\omega t - \boldsymbol{k} \cdot \boldsymbol{r})} \qquad (2.53)$$

$$\boldsymbol{H}(\boldsymbol{r}, t) = \boldsymbol{H}_0 e^{j(\omega t - \boldsymbol{k} \cdot \boldsymbol{r})} \qquad (2.54)$$

where the imaginary unit j is defined by $j^2 = -1$ (the notation j rather than i is used in certain fields to avoid confusion with the electrical current i). Using either wave function we find that $v = \omega/k = 1/\sqrt{\mu\epsilon}$, which in free space is equal to the speed of light. A substitution into Gauss' law for the electric and magnetic fields

EXAMPLE

EXAMPLE

gives

$$\boldsymbol{k} \cdot \boldsymbol{E}_0 = 0, \quad \boldsymbol{k} \cdot \boldsymbol{H}_0 = 0 \tag{2.55}$$

which indicates that both the electric and magnetic field intensities are *perpendicular to the direction of propagation*. Substitution into Faraday's law and Ampère's law gives

$$\boldsymbol{k} \times \boldsymbol{E}_0 = \omega\mu\boldsymbol{H}_0, \quad \boldsymbol{k} \times \mu\boldsymbol{H}_0 = -\frac{\omega}{v^2}\boldsymbol{E}_0 \tag{2.56}$$

which indicates that the direction of propagation, the electric field intensity, and the magnetic field intensity are orthogonal. Finally, from either of the last two equations, we can derive that the electric and magnetic field intensities are related by $\boldsymbol{E} = \boldsymbol{v} \times \mu\boldsymbol{H}$.

Note that the solution presented in Example 2.4.1 can be found by the use of Euler's equation ($e^{jx} = \cos x + j \sin x$). If we expand the solutions for the electric field intensity presented in this problem, we get that

$$\boldsymbol{E}(\boldsymbol{r}, t) = \boldsymbol{E}_0 e^{j(\omega t - \boldsymbol{k}\cdot\boldsymbol{r})} = \boldsymbol{E}_0[\cos(\omega t - \boldsymbol{k}\cdot\boldsymbol{r}) + j\sin(\omega t - \boldsymbol{k}\cdot\boldsymbol{r})] \tag{2.57}$$

where the real part of this equation differs only by a phase term and the direction of propagation from the solution seen before.

Finally, there exist many important problems concerning the beneficial or harmful effects of high-frequency electromagnetic radiation of various types (for good reviews, see [10, 11]). In this regard, the complete set of Maxwell's equations must be used when the quasistatic approximations are no longer valid. Table 2.6 shows the various regions of the electromagnetic spectrum.

2.5 THE QUASISTATIC APPROXIMATIONS

2.5.1 Electroquasistatic Field Systems

In the study of biological systems at the tissue, cell, and molecular levels, including protein, nucleic acid, and polysaccharide

Table 2.6 Regions of the electromagnetic spectrum from infralow to superhigh frequencies in which $h\nu < kT$ [9].

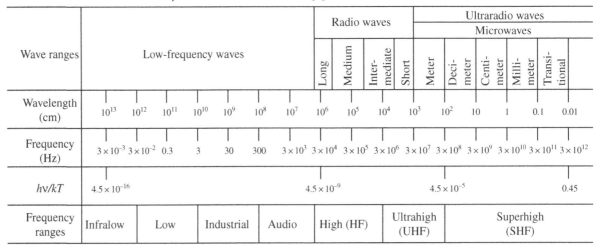

structure–function, together with a wide range of device applications (e.g., MEMS, NEMS, microfluidics, and hydrogel scaffolds), electric field interactions are of primary importance, while magnetic and electromagnetic phenomena play a secondary role. Therefore, in our formulation of electrochemical and electromechanical coupling interactions, we are often justified in using an *electroquasistatic* (EQS) subset of Maxwell's equations. If the local magnetic field is negligibly small, or if the time rate of change of the magnetic field is small enough, then Faraday's law takes the limiting form

$$\nabla \times \boldsymbol{E}(\boldsymbol{r}, t) = -\frac{\partial \boldsymbol{B}(\boldsymbol{r}, t)}{\partial t} \approx 0 \qquad (2.58)$$

while Gauss' law remains,

$$\nabla \cdot [\epsilon \boldsymbol{E}(\boldsymbol{r}, t)] = \rho_e(\boldsymbol{r}, t) \qquad (2.59)$$

The quasistatic form of Faraday's law, (2.58), together with Gauss' law, fully specifies the vector electric field $\boldsymbol{E}$, and the electric and magnetic fields are then essentially decoupled from each other when this limit is valid. Thus, the EQS approximation neglects magnetic induction in Faraday's law, (2.58). The electric field can be solved given charge sources $\rho_e(\boldsymbol{r}, t)$ in the volume as well as charges on surfaces in the region of interest. If the magnetic field produced by the resultant time-varying electric field is of interest, it can then be calculated from Ampère's law,

$$\nabla \times \boldsymbol{H}(\boldsymbol{r}, t) = \boldsymbol{J}(\boldsymbol{r}, t) + \epsilon \frac{\partial \boldsymbol{E}(\boldsymbol{r}, t)}{\partial t} \qquad (2.60)$$

(The divergence equation for the magnetic field is not included in the EQS formulation, since it usually is of no interest in EQS systems.)

In an EQS system, the curl of the electric field is zero. Since the curl of the gradient of any scalar is zero, we can represent the electric field in terms of the gradient of a new scalar potential, the *electric potential* $\Phi(\boldsymbol{r}, t)$:

$$\boldsymbol{E}(\boldsymbol{r}, t) = -\nabla \Phi(\boldsymbol{r}, t) \qquad (2.61)$$

The potential $\Phi(\boldsymbol{r}, t)$ has units of volts (V). The negative sign corresponds to the convention that the electric field points from high to low potential. We note from (2.58) and (2.61) that the electric potential $\Phi(\boldsymbol{r}, t)$ in biological or electrochemical systems is defined uniquely only in regions where the magnetic flux density $\boldsymbol{B}$ or its time derivative is negligibly small.

Before proceeding, it is important to understand the fundamental assumption leading to the above EQS approximation. In short, if the propagation time ℓ/c for an electromagnetic wave traveling at velocity c over a characteristic length ℓ is much less than the excitation period T of interest ($T = 1/f$, radian frequency $\omega = 2\pi f$) so that

$$\frac{\omega \ell}{c} \ll 1 \qquad (2.62)$$

then propagating wave effects are unimportant. This is equivalent to saying that the characteristic length of interest in the system is much smaller than the wavelength of an electromagnetic wave in the medium of interest. For example, biological cells typically have diameters on the order of $10\,\mu$m. *A glance at Table 2.6 suggests that electric fields at microwave frequencies are still essentially quasistatic over the length scale of the living cell!* For any given situation, the error made by neglecting the right-hand side of (2.58) can be calculated by finding the magnetic field from (2.60) and then evaluating the additional component of $\nabla \times \boldsymbol{E}$ that was initially discarded. (For a capacitive structure of characteristic length ℓ, such a calculation will result in the inequality $(\omega\ell/c)^2 \ll 1$, which must be satisfied in order to neglect the additional component of $\boldsymbol{E}$ produced by magnetic induction) [12].

To complete our description of the EQS field system, we add the conservation-of-charge law

$$\nabla \cdot \boldsymbol{J}(\boldsymbol{r}, t) = -\frac{\partial \rho_e(\boldsymbol{r}, t)}{\partial t} \tag{2.63}$$

which is a differential statement of the conservation of charge. Note that (2.63) can be found by taking the divergence of both sides of (2.60). Further, (2.63) provides the only information concerning time rates of change of field and source quantities—information not found in (2.58) and (2.59). Given information about ρ_e and material polarization, (2.58) and (2.59) are sufficient to determine the dominant electric field, and (2.60) serves to determine the magnetic field of the EQS system if needed.

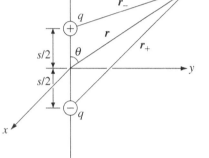

Figure 2.8 Schematic of the quantities used to find the electric potential at a point p far away from an electric dipole.

Example 2.5.1 Electric Dipole   Find the potential at a point p far away from an electric dipole using superposition.

From Gauss' law (2.1), the electric field of a point charge q in free space is

$$\boldsymbol{E}(\boldsymbol{r}, t) = \frac{1}{4\pi\epsilon_0} \frac{q(t)}{r^2} \hat{\boldsymbol{r}} \tag{2.64}$$

Integration of (2.64) gives the electric potential

$$\Phi(r, t) = \frac{1}{4\pi\epsilon_0} \frac{q(t)}{r} \tag{2.65}$$

Figure 2.8 shows a charge dipole having charges $\pm q$, with a separation of s. Let the distance to the point p from the midpoint of the dipole be r, and let the angle made with the axis of the dipole be θ. Using superposition, we can find the electric potential at point p situated a distance r_+ and r_- from the charges $\pm q$:

$$\Phi = \frac{1}{4\pi\epsilon_0} \left(\frac{q}{r_+} - \frac{q}{r_-} \right) \tag{2.66}$$

Using the law of cosines, we find that

$$r_\pm^2 = r^2 + \left(\frac{1}{2}s\right)^2 \mp rs \cos\theta \simeq r^2 \mp rs \cos\theta \tag{2.67}$$

for s small compared with r, which, when substituted back into the formula for the electric potential, gives

$$\Phi(r, \theta) \simeq \frac{1}{4\pi\epsilon_0} \frac{qs \cos\theta}{r^2} \tag{2.68}$$

Note that the electric potential is proportional to $1/r^2$ for the dipole, but $1/r$ for the monopole. Finally, we emphasize that this solution is not restricted to static charge distributions; if the dipole charges are time-varying, the resulting potential has the same time-variation, in keeping with the EQS assumption.

2.5.2 Magnetoquasistatic Field Systems

Though electric field interactions are of primary importance in physiological processes, we will find important examples in which externally applied electromagnetic or magnetic fields have important consequences (often due to the electric fields and currents that they induce inside organisms). Magnetic resonance imaging (MRI) is an important example. Another interesting case study is the present-day use of applied time-varying magnetic fields for deep brain stimulation [13], and in an attempt to induce and enhance the healing of bone fractures [14]. Research on such devices began decades ago, and the devices are still used clinically today. For the study of such magnetic field systems, the *magnetoquasistatic* (MQS) subset of Maxwell's equations is

$$\nabla \times \boldsymbol{H}(\boldsymbol{r}, t) = \boldsymbol{J}(\boldsymbol{r}, t) + \epsilon \frac{\partial \boldsymbol{E}(\boldsymbol{r}, t)}{\partial t} \qquad (2.69)$$

$$\nabla \cdot \mu \boldsymbol{H}(\boldsymbol{r}, t) = 0 \qquad (2.70)$$

$$\nabla \times \boldsymbol{E}(\boldsymbol{r}, t) = -\frac{\partial \boldsymbol{B}(\boldsymbol{r}, t)}{\partial t} \qquad (2.71)$$

where μ is the macroscopic permeability describing a linear, isotropic, magnetizable material. The continuity law for the MQS case is

$$\nabla \cdot \boldsymbol{J}(\boldsymbol{r}, t) = 0 \qquad (2.72)$$

to be consistent with the divergence of (2.69), neglecting the displacement current term. Once again, we note that (2.69) and (2.72) are relevant for inductive-like systems in a frequency range where propagation effects can be ignored.

2.6 EQS AND MQS BOUNDARY VALUE PROBLEMS IN BIOLOGICAL SYSTEMS

2.6.1 EQS: Poisson's and Laplace's Equations

From the EQS field subsystem of Section 2.5, we can rewrite Maxwell's equations for the EQS case in terms of the potential. Combining Gauss' law (2.59) with the quasistatic form of Faraday's law, (2.61) gives *Poisson's equation*:

$$\nabla^2 \Phi(\boldsymbol{r}, t) = -\frac{\rho_e(\boldsymbol{r}, t)}{\epsilon} \qquad (2.73)$$

In a region where there are no free charges, the right-hand side of Poisson's equation reduces to zero and gives *Laplace's equation*:

$$\nabla^2 \Phi(\boldsymbol{r}, t) = 0 \qquad (2.74)$$

Once the scalar potential in a region of space is found by solving Poisson's or Laplace's equation, the electric field $\boldsymbol{E}$ can then be found directly using (2.61). It is therefore usually simpler to solve, first, for the scalar Φ than the three components of the vector $\boldsymbol{E}$.

2.6.2 EQS: Boundary Conditions

To complete the formulation of the EQS problem, we refer to Figure 2.9, which pictures a region of space containing linear, homogeneous, conducting dielectric materials. A current dipole is located within the region, and we wish to find the field $\boldsymbol{E}$ and potential Φ everywhere in space. (The heart is often approximated as a current dipole in cardiac physiology models (see Problem 2.4), and the configuration of Figure 2.9 might then represent various organs within the body and their effects on the distribution of the electric potential at the outer surface corresponding to that measured in an electrocardiogram. Alternatively, the figure might represent various intracellular organelles and the distribution of the potential within a cell.)

At the boundaries between regions 1 and 2, for example, the relevant boundary conditions associated with Faraday's law (2.58), Gauss' law (2.59), and conservation of charge (2.63) are, respectively,

$$\boldsymbol{n} \times (\boldsymbol{E}_1 - \boldsymbol{E}_2) = 0 \qquad \Longleftrightarrow \qquad \Phi_1 = \Phi_2 \tag{2.75}$$

$$\boldsymbol{n} \cdot (\epsilon_1 \boldsymbol{E}_1 - \epsilon_2 \boldsymbol{E}_2) = \sigma_s \tag{2.76}$$

$$\boldsymbol{n} \cdot (\boldsymbol{J}_1 - \boldsymbol{J}_2) + \nabla_\Sigma \cdot \boldsymbol{K} = -\frac{\partial \sigma_s}{\partial t} \tag{2.77}$$

where σ_s is the surface charge density ($\mathrm{C\,m^{-2}}$), and we have allowed for the possibility of a surface current $\boldsymbol{K}$ ($\mathrm{A\,m^{-1}}$) at the interface between the two regions. In (2.77), $\nabla_\Sigma \cdot \boldsymbol{K}$ denotes the divergence of $\boldsymbol{K}$ in the plane of the surface. Thus, each EQS law has an associated boundary condition written in terms of the potential Φ or its normal derivative (or, equivalently, the tangential or normal components of $\boldsymbol{E}$). It can be shown that the condition on continuity of tangential $\boldsymbol{E}$ in (2.75) is equivalent to the condition that the potential Φ is continuous at the interface.

Uniqueness Theorem: It can also be shown (see, e.g., Problem 2.5) that a potential distribution $\Phi(\boldsymbol{r}, t)$ that obeys Poisson's equation or Laplace's equation is uniquely specified within a volume V provided that it is also equal to the specified Φ or its normal derivative $\boldsymbol{n} \cdot \boldsymbol{E}$ at the surfaces S surrounding that volume [1]. Therefore, the boundary conditions above help to specify conditions at the surface S and relate Φ or $\boldsymbol{E}$ to any sources that might exist at that surface (e.g., σ_s). *This Uniqueness Theorem is extremely important: it guarantees that if we have found a solution to Poisson's or Laplace's equation that also satisfies the boundary conditions, either by detailed mathematical analysis or simply by guessing (using physical reasoning), then we have found the only solution!*

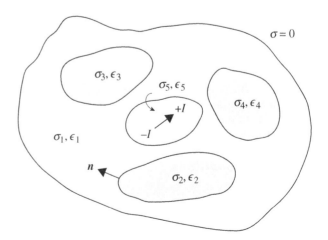

Figure 2.9 Region containing linear, homogeneous lossy dielectric media and a current dipole source.

2.6.3 **Charge Relaxation in EQS**

In the bulk of a homogeneous, isotropic, conducting dielectric medium, assuming no diffusion or convection for now, there will be no steady state free charge $\rho_e(\mathbf{r}, t)$. This can be seen by combining conservation of charge (2.63) with the Ohmic constitutive law $\mathbf{J} = \sigma \mathbf{E}$ and Gauss' law (2.59) to obtain

$$\frac{\partial \rho_e}{\partial t} + \nabla \cdot \sigma \mathbf{E} = \frac{\partial \rho_e}{\partial t} + \frac{\sigma}{\epsilon} \nabla \cdot \epsilon \mathbf{E} = 0 \tag{2.78}$$

$$\frac{\partial \rho_e}{\partial t} + \frac{\sigma}{\epsilon} \rho_e = 0 \tag{2.79}$$

$$\rho_e(\mathbf{r}, t) = \rho_e(\mathbf{r}, 0) \, e^{-t/\tau} \tag{2.80}$$

$$\tau = \epsilon/\sigma \equiv \text{charge relaxation time} \tag{2.81}$$

The result (2.80) implies, physically, that unless there is a steady source of bulk free charge, ρ_e will decay to zero at a rate characterized by the charge relaxation time ϵ/σ. For a 0.1 M NaCl solution, $\epsilon/\sigma \simeq 10^{-9}$ s. During the relaxation process, charge will tend to accumulate at surfaces or interfaces between materials having differing material properties. Of course, free charge can accumulate in the bulk of a nonhomogeneous material; (2.78) would then become

$$\frac{\partial \rho_u}{\partial t} + \nabla \cdot [\sigma(\mathbf{r}) \mathbf{E}] = \frac{\partial \rho_u}{\partial t} + \sigma(\mathbf{r}) \nabla \cdot \mathbf{E} + \mathbf{E} \cdot \nabla \sigma(\mathbf{r}) = 0 \tag{2.82}$$

Equation (2.82) together with the appropriate boundary conditions determines the distribution of the electric field inside the medium. Convection can be included simply by combining the full equation, (2.44), with (2.63). In the steady state, we would then be faced with solution of the problem

$$\sigma(\mathbf{r}) \nabla^2 \Phi + \nabla \Phi \cdot \nabla \sigma(\mathbf{r}) = 0 \tag{2.83}$$

Thus, in a homogeneous material ($\nabla \sigma(\mathbf{r}) \equiv 0$), the problem reduces to a solution of Laplace's equation

$$\nabla^2 \Phi = 0 \tag{2.84}$$

subject to the appropriate boundary conditions. We can see that free charge can accumulate at boundaries between homogeneous media having different values of σ and ϵ by combining (2.76), (2.77), and (2.44) with $v = 0$, which yields a modified "boundary condition" in the sinusoidal steady state:

$$\mathbf{n} \cdot (\sigma_1 + j\omega\epsilon_1) \hat{\mathbf{E}}_1 = \mathbf{n} \cdot (\sigma_2 + j\omega\epsilon_2) \hat{\mathbf{E}}_2 \tag{2.85}$$

where we use the notation $\mathbf{E}(t) = \mathrm{Re}[\hat{\mathbf{E}} \exp(j\omega t)]$, $\hat{\mathbf{E}}$ is the complex amplitude, and we have omitted the surface current term. Thus, discontinuities in σ and/or ϵ can lead to a discontinuity of normal $\mathbf{E}$, the relative importance of which depends on the frequency ω. This can be seen by rewriting (2.85) as

$$\mathbf{n} \cdot \sigma_1 \left(1 + \frac{j\omega\epsilon_1}{\sigma_1}\right) \hat{\mathbf{E}}_1 = \mathbf{n} \cdot \sigma_2 \left(1 + \frac{j\omega\epsilon_2}{\sigma_2}\right) \hat{\mathbf{E}}_2 \tag{2.86}$$

If the relaxation time ϵ/σ is much shorter than the excitation period $(1/\omega)$, so that $(\omega\epsilon/\sigma) \ll 1$, then (2.86) simply reduces to the steady-state limit of the conduction boundary condition (2.77):

$$\mathbf{n} \cdot \sigma_1 \hat{\mathbf{E}}_1 = \mathbf{n} \cdot \sigma_2 \hat{\mathbf{E}}_2 \tag{2.87}$$

Figure 2.10 Region containing conducting, magnetizable media in an imposed magnetic field.

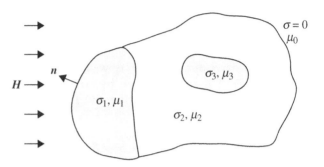

In this limit, free charge "instantaneously relaxes" to the boundary relative to excitation times of interest. The boundary value problem is then seen to be a solution of Laplace's equation (2.84), with (2.87) providing the boundary condition on $\boldsymbol{n} \cdot \boldsymbol{E}$. Once E_n is found from the relevant discontinuity in J, the surface charge σ_s can finally be found from (2.76). In the opposite limit, $\omega\epsilon/\sigma \gg 1$, capacitive effects are seen to play the dominant role in determining the time rate of accumulation of free charge at material boundaries. As an example, consider a situation where the material properties are such that $\omega\epsilon_1/\sigma_1 \gg 1$ and $\omega\epsilon_2/\sigma_2 \ll 1$. Then (2.86) reduces to

$$\boldsymbol{n} \cdot j\omega\epsilon_1 \hat{\boldsymbol{E}}_1 = \boldsymbol{n} \cdot \sigma_2 \hat{\boldsymbol{E}}_2 \tag{2.88}$$

If $\omega\epsilon_1/\sigma_2 \ll 1$, we conclude that $\boldsymbol{n} \cdot \hat{\boldsymbol{E}}_2 \simeq 0$ is the appropriate boundary condition for the problem at hand.

In the above discussion of boundary conditions that are consistent with EQS field systems, the highest-order charge singularity considered thus far has been a surface charge density σ_s. We will find that higher-order singularities, such as the charge dipole layer or "electrical double layer," will also be important in many physiological problems. The double layer and corresponding boundary conditions will be discussed in Problem 2.3 and in Chapter 6.

2.6.4 MQS and Magnetic Diffusion

To formulate the boundary value problem for the distribution of magnetic fields and current densities in conducting, magnetizable materials, we refer to Figure 2.10. At the boundary between regions 0 and 1, for example, the boundary conditions appropriate to Ampère's law (2.69), Gauss' law, (2.70) and conservation of current (MQS) (2.72), respectively, are

$$\boldsymbol{n} \times (\boldsymbol{H}_0 - \boldsymbol{H}_1) = \boldsymbol{K} \tag{2.89}$$

$$\boldsymbol{n} \cdot (\mu_0 \boldsymbol{H}_0 - \mu_1 \boldsymbol{H}_1) = 0 \tag{2.90}$$

$$\boldsymbol{n} \cdot (\boldsymbol{J}_1 - \boldsymbol{J}_2) + \nabla_\Sigma \cdot \boldsymbol{K} = 0 \tag{2.91}$$

where again we have allowed for the presence of a surface current $\boldsymbol{K}$ at the interface between the two regions.

To derive a relation for the distribution of $\boldsymbol{H}$ analogous to (2.78) or (2.82) for the electric field system, we combine (2.69) and (2.71) with a constitutive law for electric current relevant to the *MQS* case. The constitutive law may be postulated as an empirical observation,

$$\boldsymbol{J} = \sigma \left(\boldsymbol{E} + \boldsymbol{v} \times \boldsymbol{B} \right) \tag{2.92}$$

where the second term on the right-hand side accounts for motion of the conducting medium with velocity v in a magnetic field. (Equation (2.92) can also be derived by considering the invariance of Ohm's law, written in the moving frame as $\boldsymbol{J}' = \sigma \boldsymbol{E}'$ and using the transformation laws for $\boldsymbol{J}'$ and $\boldsymbol{E}'$ between inertial reference frames [15, 16].) Restricting the discussion to linear, homogeneous, isotropic materials characterized by conductivity σ and permeability μ, the curl of (2.69) using (2.92) to substitute for $\boldsymbol{J}$ yields

$$\nabla \times (\nabla \times \boldsymbol{H}) = \nabla \times \boldsymbol{J} = \sigma \nabla \times \boldsymbol{E} + \sigma \nabla \times (\boldsymbol{v} \times \boldsymbol{B}) \tag{2.93}$$

$\boldsymbol{E}$ in (2.93) can be eliminated in favor of $\boldsymbol{H}$ by using Faraday's law (2.71):

$$\frac{1}{\mu \sigma} \nabla \times (\nabla \times \boldsymbol{B}) = -\frac{\partial \boldsymbol{B}}{\partial t} + \nabla \times (\boldsymbol{v} \times \boldsymbol{B}) \tag{2.94}$$

$$\frac{\partial \boldsymbol{B}}{\partial t} = \frac{1}{\mu \sigma} \nabla^2 \boldsymbol{B} + \nabla \times (\boldsymbol{v} \times \boldsymbol{B}) \tag{2.95}$$

where (2.95) follows from (2.94) using the vector identity $\nabla \times (\nabla \times \boldsymbol{B}) = -\nabla^2 \boldsymbol{B} + \nabla(\nabla \cdot \boldsymbol{B})$. In the absence of motion, (2.95) takes the form of a diffusion equation with an effective magnetic field "diffusivity" of $1/\mu\sigma$; hence (2.95) is often referred to as the magnetic diffusion equation with the addition of a convective term. Thus, time rates of change of the magnetic flux ($\partial \boldsymbol{B}/\partial t$) can be accounted for by an "oozing" of flux through the dissipative, conducting material, and by a rate of change of flux due to material motion, respectively. In the absence of motion, the depth of penetration of a magnetic field into a conducting material that is infinite in extent can be conceptualized easily in the sinusoidal steady state using scaling analysis ($\partial/\partial t \rightarrow j\omega \sim 1/T$; $\nabla^2 \rightarrow 1/L^2$):

$$\frac{1}{T}|\boldsymbol{B}| \simeq \frac{2}{\mu\sigma L^2}|\boldsymbol{B}| \tag{2.96}$$

The quantity $2/\mu\sigma L^2$ has dimensions of inverse time, and is the inverse of the characteristic time for the field to diffuse across a characteristic length L. (This can be seen by analogy with molecular diffusion, where the Einstein random walk diffusion time $\tau_{\text{diff}} \sim L^2/2D$ is analogous to $\tau_m = L^2/2(1/\mu\sigma)$; the "2" in the denominator here would actually come from $\sqrt{j}$ in an exact derivation of the *skin effect*). When the period of the excitation T is on the order of the magnetic diffusion time τ_m, where

$$\tau_m = \frac{L^2}{2(1/\mu\sigma)} \tag{2.97}$$

the characteristic length L corresponds to the *skin depth* δ of the conducting material at the corresponding frequency ω:

$$\delta \simeq \sqrt{\frac{2}{\omega\mu\sigma}} \tag{2.98}$$

In summary, the distribution of a magnetic field in a moving medium is characterized by (2.95) subject to the boundary conditions (2.89) and (2.90). The corresponding current density can be found by solving for $\nabla \times \boldsymbol{H}$, i.e., (2.69). The analogy between magnetic diffusion and molecular diffusion will be seen again with respect to boundary layer phenomena and convective diffusion in Chapter 5.

2.7 ELECTRIC FIELDS AND CURRENTS IN CONDUCTING BIOLOGICAL MEDIA

2.7.1 Steady Conduction

In conducting biological media in which steady ("dc"; time-independent) currents flow, the EQS system of Section 2.5.1 is further simplified since the time derivative in the conservation of current law is identically zero, giving the following laws and associated boundary conditions:

$$\nabla \cdot \epsilon \boldsymbol{E} = \rho_e, \qquad \boldsymbol{n} \cdot (\epsilon_1 \boldsymbol{E}_1 - \epsilon_2 \boldsymbol{E}_2) = \sigma_s \qquad (2.99)$$

$$\nabla \times \boldsymbol{E} = 0, \qquad \boldsymbol{n} \times (\boldsymbol{E}_1 - \boldsymbol{E}_2) = 0 \qquad (2.100)$$

$$\nabla \cdot \boldsymbol{J} = 0, \qquad \boldsymbol{n} \cdot (\boldsymbol{J}_1 - \boldsymbol{J}_2) = 0 \qquad (2.101)$$

For a linear, isotropic homogeneous conducting medium in which there is negligible convection and diffusion (i.e., $\boldsymbol{J} = \sigma \boldsymbol{E}$), Equations (2.100) and (2.101) immediately combine to give Laplace's equation. This is true even if there is net charge in the system (e.g., caused by a nonuniform permittivity ϵ). Thus, the potential and field for steady state conduction problems satisfy Laplace's equation. Figure 2.11 shows a very important and common example of a current in a conducting solution near an insulating material ($\sigma = 0$), in this case an insulating cylindrical post within a microchannel containing physiological saline in which an applied electric field (and current) act to simultaneously electrophorese DNA molecules down the channel and stretch the molecules as they migrate [17]. In the example below, we solve for the form of the electric field and current density in the channel (the problem of molecular electrophoresis is taken up in Chapter 6).

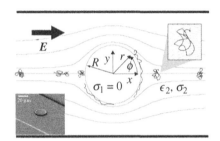

Figure 2.11 Schematic of a negatively charged DNA molecule electrically driven into an insulating obstacle (moving right to left) by an electric field, the distribution of which is given by (2.103) [17].

Example 2.7.1 Current in a Microfabricated Channel with an Insulating Post One of the most powerful approaches to finding solutions of Laplace's (or Poisson's) equation is the method of separation of variables [18, 19]. For any given problem, we first choose the most appropriate coordinate system based on the geometric configuration of the system and the surfaces on which boundary conditions are imposed. Next, we identify solutions of Laplace's equation in the desired coordinate system, e.g., by using separation of variables (see Appendix B, Table B.7 for a list of variable-separable solutions in cartesian, polar, and spherical coordinates). Finally, we use superposition to find the total potential such that $\nabla^2 \Phi^{\text{TOT}} = 0$ is satisfied in all regions of space, and the potential Φ^{TOT} (or $\boldsymbol{n} \cdot \nabla \Phi^{\text{TOT}}$) takes on the required values specified at the boundaries (or interfaces) of the system. Thus, Φ^{TOT} is that combination of individual Laplacian potential functions chosen from Table B.7 needed to satisfy the boundary conditions.

The problem of steady conduction in a uniform conducting medium reduces to a solution of Laplace's equation (2.84). We assume that the electric field and current density within the channel vary with x and y (r and ϕ in cylindrical coordinates), but are independent of z. We also assume that the insulating post is small enough compared with the channel width that the electric field is essentially uniform (x-directed) far from the post. Therefore, for the potential in the

EXAMPLE

conducting fluid (region 2), we try a potential having the form of the superposition of a uniform field and a dipole field, since the uniform field will satisfy the boundary condition far from the post, and the dipole solution will hopefully satisfy the boundary condition at the surface of the post, $r = R$:

$$\Phi_2 = -E_0 r \cos \phi + B \frac{R}{r} \cos \phi \qquad (2.102)$$

This potential is chosen from the cyclindrical coordinate solutions of Table B.7 with $n = 1$ and corresponds to the electric field (in polar coordinates),

$$\boldsymbol{E}_2 = \boldsymbol{i}_r \left(E_0 \cos \phi + B \frac{R}{r^2} \cos \phi \right) + \boldsymbol{i}_\phi \left(-E_0 \sin \phi + B \frac{R}{r^2} \sin \phi \right) \qquad (2.103)$$

The first (E_0) terms in (2.103) automatically satisfy the boundary condition at $r = \infty$. To find the constant B, we note that for steady state conduction, the condition (2.87) applies, and the conductivity of the post, σ_1 is zero; hence, there is no normal electric field (no normal current density) at $r = R$. Therefore, $B = -RE_0$. Since the boundary conditions surrounding the fluid medium are fully specified, we can solve for the potential (and the electric field and current) in the fluid region 2 without considering the fields within the post (region 1). Of course, there is no current density in the insulating post. However, there is an electric field inside, which could be calculated since we now know the potential at $r = R$, which fully defines the field in region 1 by the Uniqueness Theorem. Since that field has no effect on the DNA molecules, we can ignore it.

Example 2.7.2 Spherical Conductor (σ^i) in a Medium (σ^o) with a Uniform "dc" Applied Electric Field For the situation shown schematically in Figure 2.12, with $\epsilon^i = \epsilon^o \equiv \epsilon_0$, find Φ, $\boldsymbol{E}$ and $\boldsymbol{J}$ for $r < R$ and $r > R$. $\boldsymbol{E}$ is uniform for $r \to \infty$. Sketch $\boldsymbol{E}$ and $\boldsymbol{J}$ for $\sigma^i > \sigma^o$, $\sigma^o > \sigma^i$, $\sigma^i = 0$, and $\sigma^i = \infty$.

For this *steady state conduction problem*, the continuity law (2.63) requires that $\nabla \cdot \boldsymbol{J} = \nabla \cdot (\sigma \boldsymbol{E}) = 0$ in each region. That is, $\sigma \nabla \cdot \boldsymbol{E} + \boldsymbol{E} \cdot \nabla \sigma = 0$ in each region. Since σ^o and σ^i are both uniform and $\boldsymbol{E} = -\nabla \Phi$, Φ is a solution of Laplace's equation inside and outside the sphere. In each region, therefore,

$$\left(\frac{1}{r^2} \right) \frac{\partial}{\partial r} \left(r^2 \frac{\partial \Phi}{\partial r} \right) + \frac{1}{r^2 \sin \theta} \frac{\partial}{\partial \theta} \left(\sin \theta \frac{\partial \Phi}{\partial \theta} \right) = 0 \qquad (2.104)$$

Assume a variable-separable solution $\Phi(r, \theta) = \Theta(\theta) R(r)$ and insert into (2.104) to find $R(r) = A r^n + B r^{-(n+1)}$ and $\Theta(\theta) = P_n(\cos \theta)$ for Legendre polynomials independent of ϕ. Knowing that the uniform field at $r \to \infty$ has the form $-E_0 r \cos \theta$, we expect that only the $P_{n=1}$ polynomial will be able to satisfy the boundary condition. Physically, the applied field induces a dipolar surface charge on the sphere and we therefore assume solutions of the form

$$\Phi^o = -E_0 r \cos \theta + \frac{B}{r^2} \cos \theta \quad (r \geq R) \qquad (2.105)$$

$$\Phi^i = -A r \cos \theta = -A z \quad (r \leq R) \qquad (2.106)$$

EXAMPLE

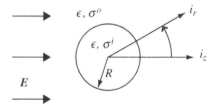

Figure 2.12 A spherical conductor embedded in a medium, with a uniform "dc" applied electric field.

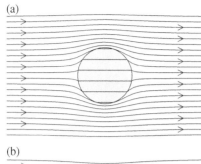

(a)

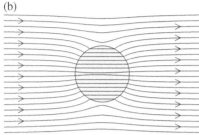

(b)

Figure 2.13 Current distribution around a conducting sphere: (a) conductivity lower than surrounding media ($\sigma^i < \sigma^o$); (b) conductivity higher than surrounding media ($\sigma^i > \sigma^o$)

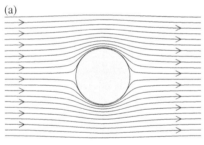

(a)

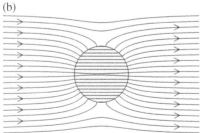

(b)

Figure 2.14 Current distribution limits around (a) a perfectly insulating sphere ($\sigma^i = 0$); (b) a perfectly conducting sphere ($\sigma^i \to \infty$).

We note that these potentials correspond to the spherical coordinate Laplacian solutions in Appendix B, Table B.7. The corresponding fields $\boldsymbol{E} = -\nabla\Phi$ are (see Appendix B)

$$\boldsymbol{E}^o = \frac{\boldsymbol{J}^o}{\sigma^o} = E_0(\boldsymbol{i}_r \cos\theta - \boldsymbol{i}_\theta \sin\theta) + \frac{B}{r^3}(\boldsymbol{i}_r 2\cos\theta + \boldsymbol{i}_\theta \sin\theta)$$

(2.107)

$$\boldsymbol{E}^i = \frac{\boldsymbol{J}^i}{\sigma^i} = A(\boldsymbol{i}_r \cos\theta - \boldsymbol{i}_\theta \sin\theta)$$

(2.108)

The field solutions (2.105)–(2.108) are subject to the boundary conditions

$$\lim_{r\to\infty} \boldsymbol{E}^o \to \boldsymbol{i}_z E_0 = (\boldsymbol{i}_r \cos\theta - \boldsymbol{i}_\theta \sin\theta)E_0$$

(2.109)

$$E^i \text{ must remain finite as } r \to 0$$

(2.110)

$$\boldsymbol{n} \times (\boldsymbol{E}^o - \boldsymbol{E}^i) = 0 \quad \text{at} \quad r = R$$

(2.111)

$$\boldsymbol{n} \cdot (\boldsymbol{J}^o - \boldsymbol{J}^i) = 0 \quad \text{at} \quad r = R$$

(2.112)

The boundary conditions (2.109) and (2.110) were automatically satisfied by the assumed form of solution (2.105) and (2.106). Application of (2.111) and (2.112) to (2.105)–(2.108) yields

$$A = E_0 \frac{3\sigma^o}{\sigma^i + 2\sigma^o}, \quad B = E_0 R^3 \frac{\sigma^i - \sigma^o}{\sigma^i + 2\sigma^o}$$

(2.113)

$$\Phi^i = -E_0 \frac{3\sigma^o}{\sigma^i + 2\sigma^o} r \cos\theta \quad (r \le R)$$

(2.114)

$$\Phi^o = -E_0 \left[r - \frac{R^3}{r^2}\left(\frac{\sigma^i - \sigma^o}{\sigma^i + 2\sigma^o}\right) \right] \cos\theta \quad (r \ge R)$$

(2.115)

$\boldsymbol{E}$ and $\boldsymbol{J}$ follow from (2.107) and (2.108).

Figures 2.13(a,b) and 2.14(a,b) show sketches of the current density for the cases $\sigma^i < \sigma^o$, $\sigma^i > \sigma^o$, and $\sigma^i = 0$, $\sigma^i = \infty$, respectively.

2.7.2 Nonsteady Conduction: EQS and MQS

The use of EQS and MQS fields to stimulate bone growth provides a very interesting and sometimes controversial set of examples. Since the work of Fukada and co-workers [20] on the electromechanical properties of bone, many researchers have speculated that mechanical stress-generated electrical potentials may play a vital role in the internal feedback mechanism that is the biophysical control of bone growth and remodeling (for a good summary, see [21]). It is therefore not surprising that many research groups have been actively engaged in the study of the electrical stimulation of fracture healing. Clinical investigations have included the use of implanted electrodes and the concomitant application of ac or dc currents [22], as well as noninvasive methods using capacitively coupled electric fields and inductively coupled magnetic fields (Figure 2.15). Several devices are still used in clinical practice.

Electromagnetic stimulation of osteogenesis must be considered at both the macroscopic and the microscopic (molecular) levels. Much recent interest has been devoted to the interaction of electromagnetic fields with biological macromolecules and cells, and the resulting effects on molecular structure and function and cellular response. *However, we should not lose sight of the necessity to first evaluate, in a true engineering sense, the macroscopic distribution of fields within physiological tissues, the latter viewed in the sense of conducting, dielectric media.* This is extremely important with respect to designing experiments and evaluating and interpreting experimental results. With this goal in mind, we now discuss two examples concerning noninvasive techniques for the stimulation of osteogenesis. These examples have been chosen in order to illustrate the use of EQS field systems and of MQS field systems. Many questions remain unanswered regarding the actual biological effects of applied fields.

Example 2.7.3 Use of EQS Fields   The rabbit tibia–electric field experiment of Figure 2.16 can be modeled by the idealized geometry of Figure 2.17, in which an infinitely long cylindrical conducting rod is centrally located between two infinite parallel plate electrodes. (With plate spacing d much less than plate width W, fringing at the electrode edges can be neglected as far as the field in the vicinity of the rod is concerned; therefore, the discussion is valid for finite-sized electrodes.) The conductivity σ and dielectric constant ϵ will be assumed to be that of the average value for the tissue space involved. Thus, the complexity of the true nonhomogeneous tissue has been suppressed in order to characterize the experiment in terms of simple analytical solutions that can be examined in the relevant physical limits.

The electric field between the plates and inside the conductor can be found by recognizing that the electroquasistatic formulation of (2.58)–(2.63) is valid for frequencies ω such that $\ell, R \ll \lambda = c/(\omega/2\pi)$. In practice, this might correspond to frequencies as high as 100 MHz. The electric fields in regions (1) and (2) of Figure 2.17 must satisfy Laplace's equation (2.84) in each region, subject to the boundary conditions (2.75)–(2.77) at $r = R$, and to the condition at $r \to \infty$, where $\boldsymbol{E}_1 \to E_0(\boldsymbol{i}_r \cos \phi - \boldsymbol{i}_\phi \sin \phi)$.

EXAMPLE

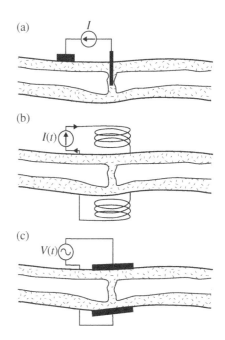

Figure 2.15 Methods for the application of electric and magnetic fields: (a) implanted electrode; (b) inductive coupling; (c) capacitive coupling.

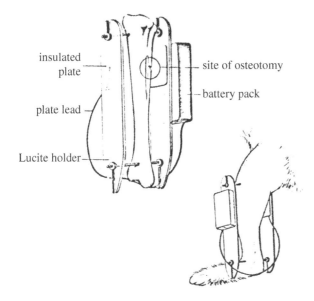

insulated plate

site of osteotomy

battery pack

plate lead

Lucite holder

Figure 2.16 Diagram of battery pack, lead connections, and insulated plates with respect to osteotomy site [23]. Polarities of the applied plate potentials were determined by appropriate plate lead connections to the battery pack. The insert demonstrates completed unit secured to rabbit tibia.

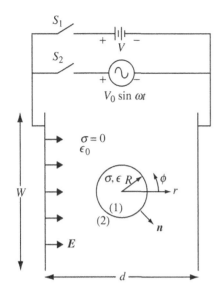

Figure 2.17 Capacitively coupled electric field experiment.

We first consider the transient problem in which switch S_2 is open and S_1 is closed at $t = 0$; this corresponds to the conditions of the rabbit tibia experiment of Figure 2.16. We assume solutions to Laplace's equation in regions (1) and (2) of the form (Table B.7)

$$\Phi_1 = -E_0 r \cos \phi + B \frac{R}{r} \cos \phi \tag{2.116}$$

$$\Phi_2 = -A \frac{r}{R} \cos \phi \tag{2.117}$$

which correspond to the electric fields (in polar coordinates)

$$\boldsymbol{E}_1 = \boldsymbol{i}_r \left(E_0 \cos \phi + B \frac{R}{r^2} \cos \phi \right) + \boldsymbol{i}_\phi \left(-E_0 \sin \phi + B \frac{R}{r^2} \sin \phi \right) \tag{2.118}$$

$$\boldsymbol{E}_2 = A \frac{1}{R} \left(\boldsymbol{i}_r \cos \phi - \boldsymbol{i}_\phi \sin \phi \right) \tag{2.119}$$

Application of the boundary conditions (2.75)–(2.77) at $r = R$ leads to a differential equation for the electric field inside the conducting cylinder for $t > 0$. The solution to this differential equation (which you should derive) is

$$E_2 = \frac{2\epsilon_0}{\epsilon_0 + \epsilon} E_0 e^{-t/\tau_r} (\boldsymbol{i}_r \cos \phi - \boldsymbol{i}_\phi \sin \phi) \tag{2.120}$$

where

$$\tau_r = \frac{\epsilon + \epsilon_0}{\sigma} \tag{2.121}$$

To arrive at (2.120), it has been assumed that there is no initial surface charge on the cylinder, i.e., $\sigma_s(r = R) = 0$ at $t = 0^+$. We see that the effective charge relaxation time τ_r is not simply ϵ/σ, but has been modified owing to the coupling of fields inside and outside the cylindrical rod. In the limit $\epsilon \to \epsilon_0$, the $\boldsymbol{E}$ field inside at $t = 0^+$ is just $\boldsymbol{E}_0$, as one would expect, since the relaxation process has not had time to occur. After $t = 0^+$, charge is induced at the surface of the cylinder in a time τ_r. This charge terminates the $\boldsymbol{E}$ field lines emanating from the electrodes, and therefore σ_s effectively shields out the electric field from the conducting rod in a few relaxation time constants τ_r. This can be seen by solving for σ_s using (2.76) after $\boldsymbol{E}_1$ and $\boldsymbol{E}_2$ have been found.

To examine the transient rabbit tibia experiment in terms of (2.120) and (2.121), we assume that $\epsilon \simeq 80\epsilon_0$, i.e., the dielectric constant of water. The average conductivity σ should lie between that of blood ($\sigma \simeq 0.5\,\Omega^{-1}\,\text{m}^{-1}$) and that of bone, about one or two orders of magnitude less; to be on the conservative side, we choose $\sigma \simeq 0.1\,\Omega^{-1}\,\text{m}^{-1}$. Then

$$\tau = \frac{\epsilon + \epsilon_0}{\sigma} \simeq 7 \times 10^{-9}\,\text{s} \tag{2.122}$$

As the electric field in the actual experiment of Figure 2.16 was switched on for 21 days, it is apparent that the $\boldsymbol{E}_2$ is shielded out from the leg for a significant portion of time!

Time-varying electric fields can exist in the conducting rod to an extent depending on the rate of change of the applied $\boldsymbol{E}$. In the sinusoidal steady state (S_1 open + S_2 closed in Figure 2.17), the boundary conditions (2.75) and (2.85)—the latter being the combination of (2.76) and (2.77)—are used with the potential solutions (2.116) and (2.117). The sinusoidally varying $\boldsymbol{E}$ inside the conductor becomes (show that this is true)

$$\boldsymbol{E}_2 = \frac{\frac{2j\omega\epsilon_0}{\sigma}E_0}{\frac{j\omega(\epsilon+\epsilon_0)}{\sigma}+1}\left(\boldsymbol{i}_r \cos\phi - \boldsymbol{i}_\phi \sin\phi\right) \qquad (2.123)$$

In the limit $\omega \to 0$, $\boldsymbol{E}_2 \to 0$ in agreement with the physical notion that electric fields are shielded out of homogeneous conductors by induced surface charges. In the opposite limit of high enough frequency such that $\omega(\epsilon+\epsilon_0)/\sigma \gg 1$, $\boldsymbol{E}_1$ and $\boldsymbol{E}_2$ are self-consistent with the "capacitive dividing" nature of the dielectric rod and free space. While the conductivity σ is unimportant in the calculation of $\boldsymbol{E}_2$ in this frequency range, Ohmic dissipation due to the displacement current may lead to significant heating problems. For intermediate frequencies, the extent to which the electric field can penetrate into the rod depends on the relaxation process governed by the ratio $\omega(\epsilon+\epsilon_0)/\sigma$. The breakpoint frequency in this case is $\omega \sim \sigma/(\epsilon+\epsilon_0) = 1.43 \times 10^8\,\mathrm{s}^{-1}$ or 23 MHz.

Example 2.7.4 Use of MQS Fields The experiment pictured in Figure 2.15(b) and 2.18 involves the use of time-varying magnetic fields to induce currents in the area of long bone fractures or nonunions. Such experiments might be modeled by the corresponding magnetic field system of Figure 2.19, where the conducting rod has a permeability approximately equal to that of free space, μ_0. The distribution of magnetic field and current in the rod is governed by (2.95) and (2.69). Analytical solution in the geometry of Figure 2.19 is complex since (2.95) is a vector rather than a scalar diffusion equation. With regard to certain aspects of the present example, a one-dimensional geometry and analysis might be appropriate, as is discussed in Problem 2.6 (and several illustrated examples concerning magnetic diffusion in homogeneous conductors can be found in [24]). However, we can still make some relevant calculations in the context of Figure 2.19. To find whether or not the magnetic field actually penetrates the conducting rod (tissue), (2.98) is used to calculate the skin depth δ, which can then be compared with a characteristic length of the system, in this case the rod radius R. Approximate experimental parameters of interest in human clinical experiments are

$$R = 0.05\,\mathrm{m}$$

$$f = 10^3\text{--}10^5\,\mathrm{Hz}$$

$$\mu = \mu_0 = 4\pi \times 10^{-7}\,\mathrm{H\,m^{-1}}$$

$$\sigma = 0.1\text{--}1.0\,\Omega^{-1}\,\mathrm{m^{-1}}$$

EXAMPLE

Figure 2.18 Diagram of a dog with electromagnetic coils in position flanking medially and laterally the tibular osteotomies [14]. Note the unpowered coils on the opposite leg. The vest carries batteries (24 V) and pulse-shaping circuits.

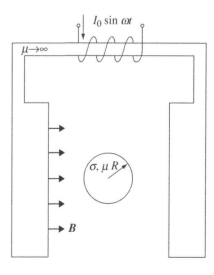

Figure 2.19 Use of MQS fields.

$$\delta = \sqrt{\frac{2}{\omega\mu\sigma}} = \begin{cases} 1.6\text{--}4.9 \, \text{m} & \text{at } 10^5 \, \text{Hz} \\ 16\text{--}49 \, \text{m} & \text{at } 10^3 \, \text{Hz} \end{cases}$$

Since $R \ll \delta$, it is reasonable to assume that the imposed magnetic field is almost negligibly perturbed by the currents that are induced in the conducting tissue; an exception includes high-frequency components of the pulse trains that are now used in certain clinical instruments. Said another way, the magnetic diffusion time (2.97) based upon the characteristic length R is very short compared with the period of the excitation:

$$\frac{R^2}{2(1/\mu\sigma)} \ll \frac{1}{\omega} \tag{2.124}$$

The induced $\boldsymbol{J}$ is relatively small. To estimate the induced $\boldsymbol{J}$ in the "tissue," one can model the rod of Figure 2.19 as a single-turn conducting loop and calculate an averaged current density inside the loop using Faraday's law (2.71), with $\boldsymbol{B}(t) = \boldsymbol{B}_0(t)$, the imposed magnetic field. To prove that $\boldsymbol{B}(t)$ in the tissue is not significantly perturbed by the induced current, i.e., that $\boldsymbol{B}(t) \simeq \boldsymbol{B}_0(t)$, an estimate of the induced magnetic field (which we will call $\boldsymbol{B}(t)$) can be calculated from Ampère's law (2.69) and the ratio $\boldsymbol{B}_1(t)/\boldsymbol{B}_0(t)$ evaluated. To get a better idea of the distribution of $\boldsymbol{J}$ and the possible significance of Ohmic heating, we continue this discussion in terms of a simple planar model in Problem 2.6.

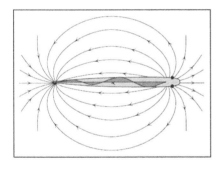

Figure 2.20 Electric field of the electric fish *Gymnarchus niloticus*.

Example 2.7.5 Electric Fish   The African freshwater fish *Gymnarchus niloticus* literally becomes the current dipole source of Figure 2.20, and is thought to live via $\nabla^2\Phi = 0$ [25, 26]. Figures 2.20–2.22 illustrate the electric field properties of this fish and the experimental arrangement for studying it [25, 26].

2.8 SUMMARY

We summarize Maxwell's equations in Tables 2.7 and 2.8. Table 2.7 shows Maxwell's equations for linear media. Table 2.8 shows the quasistatic forms of Maxwell's equations, again for linear media. Note that certain primary equations in electrostatic systems become "secondary" in magnetostatic systems, and vice versa.

Figure 2.21 Objects in the electric field of *Gymnarchus* distort the path of electric current: (a) insulating; (b) conducting.

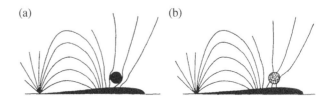

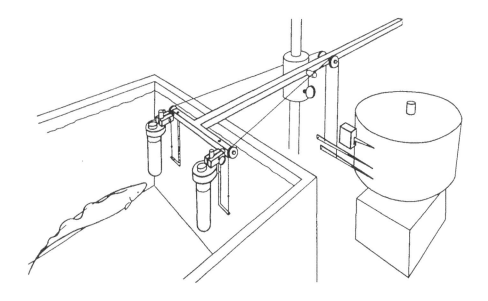

Figure 2.22 Experimental arrangement for studying *Gymnarchus.*

Table 2.7 Maxwell's equations for linear media.

Name	Integral form	Differential form
Gauss' law	$\oint_S \epsilon \boldsymbol{E} \cdot d\boldsymbol{a} = \int_V \rho_e$	$\nabla \cdot \epsilon \boldsymbol{E} = \rho_e$
Ampère's law	$\oint_C \boldsymbol{H} \cdot d\boldsymbol{s} = \int_S \boldsymbol{J} \cdot d\boldsymbol{a} + \dfrac{d}{dt} \int_S \epsilon \boldsymbol{E} \cdot d\boldsymbol{a}$	$\nabla \times \boldsymbol{H} = \boldsymbol{J} + \dfrac{\partial \epsilon \boldsymbol{E}}{\partial t}$
Faraday's law	$\oint_C \boldsymbol{E} \cdot d\boldsymbol{s} = -\dfrac{d}{dt} \int_S \mu \boldsymbol{H} \cdot d\boldsymbol{a}$	$\nabla \times \boldsymbol{E} = -\dfrac{\partial \mu \boldsymbol{H}}{\partial t}$
Magnetic flux	$\oint_S \mu \boldsymbol{H} \cdot d\boldsymbol{a} = 0$	$\nabla \cdot \mu \boldsymbol{H} = 0$
Charge conservation	$\oint_S \boldsymbol{J} \cdot d\boldsymbol{A} = -\dfrac{d}{dt} \int_V \rho_e \, dV$	$\nabla \cdot \boldsymbol{J} = -\dfrac{\partial \rho_e}{\partial t}$

Table 2.8 Quasistatic laws for linear media.

Electroquasistatic (EQS)	Magnetoquasistatic (MQS)
$\nabla \cdot \epsilon \boldsymbol{E} = \rho_e$	$\nabla \times \boldsymbol{H} = \boldsymbol{J}, \nabla \cdot \boldsymbol{J} = 0$
$\nabla \times \boldsymbol{E} = 0$	$\nabla \cdot \mu \boldsymbol{H} = 0$
$\nabla \cdot \boldsymbol{J} = -\dfrac{\partial \rho_e}{\partial t}$	$\nabla \times \boldsymbol{E} = -\dfrac{\partial \mu \boldsymbol{H}}{\partial t}$
Secondary	
$\nabla \times \boldsymbol{H} = \boldsymbol{J} + \dfrac{\partial \epsilon \boldsymbol{E}}{\partial t}$	$\nabla \cdot \epsilon \boldsymbol{E} = \rho_e$
$\nabla \cdot \mu \boldsymbol{H} = 0$	

2.9 PROBLEMS

Problem 2.1 Conductivity of Heterogeneous Media [27]
The case of a spherical particle of conductivity σ^i and radius R embedded in a medium of conductivity σ^o was treated in Example 2.7.2. When a uniform electric field is applied by electrodes whose spacing and diameter are each much greater

than R, the potential $\Phi(r \geq R)$ was given by (2.115). Assume $\epsilon^i = \epsilon^o \equiv \epsilon_0$.

(a) Let there be n spheres of conductivity σ^i, radius R embedded in a (larger) spherical region (radius a) of a medium of conductivity σ^o. A uniform $\mathbf{E}$ is applied by electrodes whose spacing and diameter are much greater than a. The n little spheres are spaced far enough apart from each other so that their induced dipole field components do not significantly interact with each other. Argue that the potential far enough away from the n spheres has the form

$$\Phi(r > a) = -E_0 \left[r - \frac{va^3}{r^2} \left(\frac{\sigma^i - \sigma^o}{\sigma^i + 2\sigma^o} \right) \right] \cos\theta \qquad (2.125)$$

where v is the volume ratio $n(\frac{4}{3}\pi R^3)/(\frac{4}{3}\pi a^3)$. Note that v must be $\ll 1$ for the assumption of noninteracting particle fields to hold.

(b) Let σ' be defined as the effective conductivity of the heterogeneous spherical region within $r = a$. Then (2.115) gives the form of the potential for $r \geq a$, with $\sigma^i \to \sigma'$. Compare this result with (2.125) (far enough away from $r = 0$) and show that

$$\sigma' = \sigma^o \left[\frac{\sigma^i + 2\sigma^o + 2v(\sigma^i - \sigma^o)}{\sigma^i + 2\sigma^o - v(\sigma^i - \sigma^o)} \right] \qquad (2.126)$$

This is the effective conductivity for a compound medium composed of spherical particles of σ^i material in σ^o material. σ' is independent of the exact spacing of the particles, and is only a function of the volume ratio v.

Problem 2.2 Coulter RBC Counter Figure 2.23 is a schematic of the Coulter method for counting red blood cells (RBCs) and measuring mean cell volume (MCV) [28]. Coulter devices are widely used in clinics and laboratories today, along with flow cytometers (see Problem 5.5). RBCs pass one at a time through a circular glass orifice having a typical diameter of $2b = 50\,\mu\text{m}$ and a length-to-diameter ratio $L/2b \simeq 1.0$. A constant current is applied between electrodes on both sides of the orifice. A cell passing through the orifice increases the effective resistance between the electrodes and produces a concomitant voltage pulse height $\Delta V = i\,\Delta R$ measured across the electrodes.

Figure 2.23 Coulter counter.

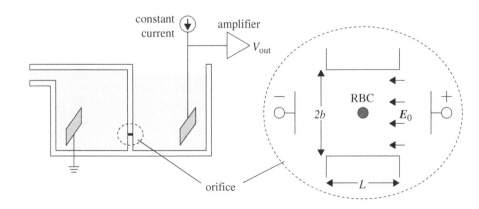

Use the results of Example 2.7.2 and Problem 2.1(b) to find an approximate expression for ΔV in terms of MCV $= \frac{4}{3}\pi R^3$ for the case of a spherical cell. Assume that:

(1) The cell is spherical and its membrane is insulating.
(2) The cell volume $\frac{4}{3}\pi R^3$ is much less than $\pi b^2 L$.
(3) The electric field in the orifice is approximately uniform in the absence of the cell; with the cell centered in the orifice, E is still approximately uniform in the neighborhood of the orifice inner walls and at the inlet and outlet of the orifice.

In practice, errors arising from these assumptions have been computed, and calibration factors for cell shape and field nonuniformity can be measured.

Problem 2.3 Electrical Double Layer It has long been known that charge separation naturally occurs at phase boundaries in electrochemical systems and at biological interfaces. This separation leads to the formation of an electrical "double layer" of charge, most often with one sign of charge in each adjacent phase. The resulting charge density and electric field strength can be of such magnitude as to play a crucial role at electrochemical (e.g., electrode/electrolyte) interfaces, as well as macromolecular, fibrillar and membrane/electrolyte interfaces.

In Figure 2.24, we picture an idealized metal electrode/electrolyte interface where the metal is known to have a net surface charge σ_d at $x = 0$. This leads to a net space charge of mobile ions in the adjacent electrolyte phase. We wish to find the *equilibrium* potential and space charge distribution in the electrolyte. (With this information we will be in a position to calculate interaction forces between charged particles—metal or biological; see e.g., Chapters 4 and 6 for related problems.)

(a) For the one-dimensional model of Figure 2.24, write Poisson's equation for the electrolyte region $x \geq 0$. Assume that the distribution of free charge (electrolyte ions) varies only in the x direction.

PROBLEM

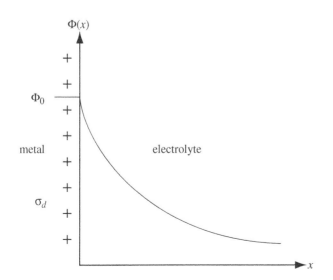

Figure 2.24 Idealized electrode/electrolyte interface.

Further, assume that the distribution of all mobile ions can be adequately represented by Boltzmann statistics, so that the probability of finding a given ion of species i and valence z_i at position x can be written as $\exp[-z_i F\Phi(x)/RT]$, and therefore

$$c_i = c_{i0} e^{-z_i F\Phi(x)/RT} \tag{2.127}$$

where c_{i0} is the bulk concentration of the ith species.

(**b**) Show that your answer to part (a) reduces to the limiting form

$$\frac{d^2\Phi(x)}{dx^2} = \kappa^2 \Phi(x) \tag{2.128}$$

for the case of a mono–monovalent, symmetrical electrolyte ($z_+ = -z_-$ and $c_{+0} = c_{-0} = c_0$) in the limit $zF\Phi/RT \ll 1$. (This form is sometimes referred to as the linearized Poisson–Boltzmann equation or linear Debye–Huckel approximation.) Express the Debye length $\equiv 1/\kappa$ in terms of the relevant parameters, and evaluate $1/\kappa$ for the case of 0.1 M and 0.001 M solutions of NaCl.

(**c**) Find the electric potential $\Phi(x)$ and space charge $\rho_e(x)$ for $x \geq 0$. What is the physical significance of the Debye length?

(**d**) Using the boundary condition at $x = 0$ relating surface charge σ_d and electric field, find σ_d in terms of $\Phi(x = 0)$ and other material and geometric parameters. Show that in the limit of small surface potential $\Phi(x = 0)$, the diffuse electrical double layer may be modeled as a planar capacitor. What is the "thickness" of the capacitor?

(**e**) Now returning to Poisson's equation of part (a), write the charge density in terms of hyperbolic function(s), i.e., do not linearize. Integrate the second-order differential equation once and repeat part (d) for arbitrary surface potential $\Phi(0)$. Show that in the limit of small $\Phi(0)$, the expression for σ_d becomes the same as in part (d). Integrate a second time to find a transcendental relation for $\Phi(x)$.

Problem 2.4 Current Dipole One of the simplest models used in introductory discussions of electrocardiography is the centric-dipole model. The torso is pictured as a uniform conducting sphere with the heart represented as a dipolar current source in the center, as in Figure 2.25(a). We wish to derive the potential on the surface of the sphere.

(**a**) First, consider the situation of a spherically symmetric point source of current I at the origin, as shown in Figure 2.25(b). In spherical coordinates, find the distribution of current density $\boldsymbol{J}$ and electric potential in the neighborhood of the source.

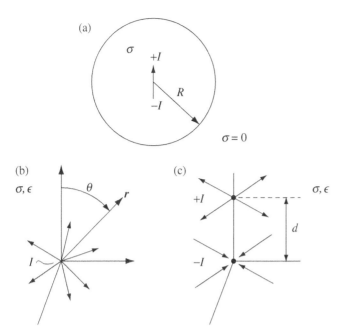

Figure 2.25 (a) Centric dipole model of the heart at the center of a spherical "torso." (b) Point current source at the origin of a conducting medium of infinite extent. (c) Point current dipole at the origin of a conducting medium of infinite extent.

(**b**) Suppose there exists a point current source I and a sink of equal magnitude spaced by the distance d as shown in Figure 2.25(c). Under the assumption that d is small compared with the dimensions of interest, find the potential distribution around this current dipole source–sink combination. This can be derived by analogy with the potential of a charge dipole of Example 2.5.1.

(**c**) Finally, the dipole current source of part (b) is placed at the origin of a conducting sphere of radius $R \gg d$, as shown in Figure 2.25(a). The sphere is surrounded by a perfect insulator. Find the distribution of current density and the potential at $r = R$. This is the potential measured in electrocardiography.

Problem 2.5 Uniqueness of Electric Field Solutions   In solving boundary value problems, we must always ask whether or not there is a unique solution corresponding to the specified boundary conditions. For example, a scalar potential $\Phi(\boldsymbol{r})$ that is known to satisfy Laplace's equation $\nabla^2\Phi = 0$ in a region R of space is *uniquely* specified in R if the potential is known everywhere on the surface bounding R. Another set of boundary conditions that leads to a unique solution is the specification of Φ on part of S and $\boldsymbol{n} \cdot \nabla\Phi$ on the remainder of S.

For the class of problems concerning regions of nonuniform conductivity $\sigma(\boldsymbol{r})$, the steady state boundary value problem does not reduce to a solution of Laplace's equation, but to a solution of

$$\nabla \cdot [\sigma(\boldsymbol{r})\boldsymbol{E}] = 0 = \nabla \cdot [\sigma(\boldsymbol{r})\nabla\Phi] \qquad (2.129)$$

The general case of time-varying conduction problems reduces to (2.129) when the charge relaxation process occurs much more rapidly than other times of interest—the limit of "instantaneous charge relaxation."

PROBLEM

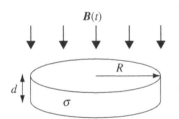

Figure 2.26 Region R spanned by surface $S = S' + S''$, and having nonuniform conductivity.

Prove that the conduction problem defined by

$$\nabla \cdot (\sigma \nabla \Phi) = 0 \qquad (2.130)$$

has a unique solution in a region R of space for the following specifications of boundary conditions (Figure 2.26):

$$\left.\begin{array}{ll} \Phi & \text{is known on } S' \\ \boldsymbol{n} \cdot \nabla \Phi & \text{is known on } S'' \end{array}\right\} \begin{array}{l} S' + S'' \text{ span the} \\ \text{entire surface } S \\ \text{surrounding } R \end{array}$$

$$(2.131)$$

Problem 2.6 Osteogenesis with Time-Varying Magnetic Fields To make a simple, order-of-magnitude estimate of the induced current density and Ohmic heating using the technique shown in Example 2.7.5, consider the planar model of the "tissue space" shown in Figure 2.27. A pair of Helmholtz coils above and below a homogeneous material (σ, μ_0) produces a magnetic flux density $\boldsymbol{B}_0(t)$. For simplicity, assume that the applied $\boldsymbol{B}_0(t)$ is constant for $r \leq R$ and zero for $r > R$. (In fact, $\boldsymbol{B}$ is smoothly continuous at $r = R$, but the precise distribution need not be considered for the purposes of this problem).

(a) Make a simple calculation that suggests that the imposed field $\boldsymbol{B}_0(t)$ is not significantly perturbed by the conducting tissue. This essentially decouples Ampère's law (2.69) from Faraday's law (2.71) and obviates the need to solve the complete magnetic diffusion equation. Therefore, estimate the induced current $\boldsymbol{J} \sim J_\phi(r)\boldsymbol{i}_\phi$ by using Faraday's law. (Assume that the material extends far enough in the r direction that the distribution of current is not significantly affected by the exact shape and extent of the boundaries.) What is the magnitude and position of the maximum J_ϕ?

(b) Find the Ohmic power dissipation in the region $r < R$.

(c) How much time is needed to raise the temperature of the volume $r < R$, $0 \leq z \leq d$ by 1°C, assuming there is no heat flow out of the region due to thermal conduction (obviously a worst case)? (Assume that the specific heat of the tissue is about $1 \, \text{cal g}^{-1} \, \text{deg}^{-1}$, i.e., that of water.)

(d) Describe how you would alter the calculation of part (c) to include thermal conduction away from the heated zone into the surrounding tissue space. (What is the form of the term to be added, where is it added, and what are the relevant parameters?)

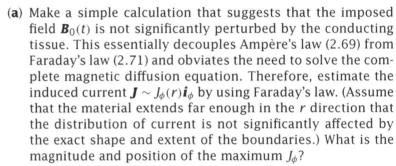

Figure 2.27 Homogeneous material: conductivity $\sigma \approx 1 \, \Omega^{-1} \, \text{m}^{-1}$ and permeability μ_0; applied $\boldsymbol{B}(t) = -\boldsymbol{i}_z B_0 \cos \omega t$ (uniform in the region $r < R$) $f = 10^3 – 10^5$ Hz, $R = 5$ cm, $B_0 = 20$ G, and $d = 10$ cm.

Problem 2.7 Electrosurgical Cutting and Coagulation In electrosurgery, radiofrequency (rf) power is fed to a "cutting" electrode. The patient is placed on top of a return (ground) electrode. The model of Figure 2.28(a) shows a half-cylindrical "patient" infinitely long in the z direction and having uniform σ and ϵ. Assume that the system is electroquasistatic at $\omega = 10^6 \, \text{rad s}^{-1}$. The concentric cylindrical electrodes are perfect conductors and the amplitude of the supply voltage is constant. Neglect fringing.

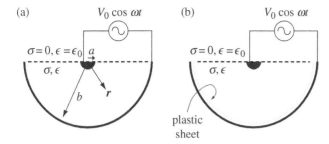

Figure 2.28 Hemicylindrical "patient" model ($\sigma = 1\,\Omega^{-1}\,\mathrm{m}^{-1}$, $\epsilon = 80\epsilon_0$, $b = 100a$, and $f = 10^6\,\mathrm{rad\,s^{-1}}$): (a) with direct connection to ground; (b) insulated from ground by plastic sheet ($\sigma = 0$, $\epsilon = 2\epsilon_0$, and thickness $d = 10^{-4}b$).

(**a**) Find analytical expressions for, and then calculate, the capacitance per unit length and conductance per unit length presented to the electrodes.

(**b**) Find an analytical expression for the time-average power dissipated per unit length in the "patient" of Figure 2.28(a).

(**c**) If the patient and the ground electrode are separated by a sheet of plastic of thickness $d = 10^{-4}b$ and having $\sigma = 0$, $\epsilon = 2\epsilon_0$ (see Figure 2.28(b)), will the current delivered by the electrodes change significantly? Justify your answer.

(**d**) One effect of power dissipation is to make the conductivity and dielectric constant of the body nonuniform. Assume that $\sigma = \sigma_0(2r/b)$ and $\epsilon = 80\epsilon_0(2r/b)$ in Figure 2.28(a). Find an analytical expression for the magnitude of the electric field and volume free charge density as a function of position inside the body.

2.10 REFERENCES

[1] Haus HA and Melcher JR (1989) *Electromagnetic Fields and Energy.* Prentice Hall, Upper Saddle River, NJ. This book provides an excellent and extensive treatment of electromagnetic fields and media, with many examples. It is also available online at http://web.mit.edu/6.013_books/www/book.html.

[2] Jackson JD (1999) *Classical Electrodynamics*, 3rd ed. Wiley, New York, Section 1.2.

[3] Maxwell JC (1954) *Treatise on Electricity and Magnetism*, 2 Volumes. Dover Publications, New York (originally published 1873).

[4] Thomas JM (1991) *Michael Faraday and the Royal Institution.* Adam Hilger, Bristol, UK. This book gives an excellent account of Faraday's life and contributions.

[5] Tanford C (1961) *Physical Chemistry of Macromolecules.* Wiley, New York.

[6] Von Hippel A (ed.) (1995) *Dielectric Materials and Applications.* Artech House, Boston. This book contains excellent discussions of the atomic and molecular mechanisms of polarization and the measurement of macroscopic dielectric properties of materials.

[7] Nye JF (1985) *Physical Properties of Crystals: Their Representation by Tensors and Matrices.* Oxford University Press, Oxford.

[8] Cole KS (1965) Electrodiffusion models for the membrane of squid giant axon. *Physiol. Rev.* **45**, 340–379.

[9] Presman A (1970) *Electromagnetic Fields and Life.* Plenum Press, New York.

[10] Illinger KH (ed.) (1981) *Biological Effects of Nonionizing Radiation.* ACS Symposium Series, Volume 157. American Chemical Society, Washington, DC.

[11] Osepchuk JM (ed.) (1983) *Biological Effects of Electromagnetic Radiation.* IEEE Press, New York.

[12] Woodson HH and Melcher JR (1968) *Electromechanical Dynamics*, Part I: *Discrete Systems.* Wiley, New York, Appendix B.

[13] Rossini PM, Rossini L, and Ferreri F (2010) Brain–behavior relations: transcranial magnetic stimulation (a review). *IEEE Eng. Med. Biol. Mag.* **29**, 84–96.

[14] Bassett CAL, Pilla AA, and Pawluk RJ (1977) A non-operative salvage of surgically-resistant pseudarthroses and non-unions by pulsing electromagnetic fields: a preliminary report. *Clin. Orthop. Rel. Res.* **124**, 128–143.

[15] Woodson and Melcher [12], Chapter 6.

[16] Penfield P and Haus HA (1967) *Electrodynamics of Moving Media.* MIT Press, Cambridge, MA.

[17] Randall GC and Doyle PS (2005) DNA deformation in electric fields: DNA driven past a cylindrical obstruction. *Macromolecules* **38**, 2410–2418.

[18] Haus and Melcher [1], Chapter 5.

[19] Lin CC and Segel LA (1988) *Mathematics Applied to Deterministic Problems in the Natural Sciences.* SIAM, Philadelphia. Chapter 4. This gives an excellent description of the theoretical background for these techniques, as well as illustrative examples.

[20] Fukuda E and Yasuda I (1957) On the piezoelectric effect of bone. *J. Phys. Soc. Jpn* **12**, 1158–1162.

[21] Bassett CAL (1971) Biophysical principles affecting bone structure. In *Biochemistry and Physiology of Bone*, 2nd ed., Volume III (Bourne GH, ed.). Academic Press, New York, pp. 1–76.

[22] Brighton CT, Friedenberg ZB, Zemsky LH, and Pollis PR (1975) Direct-current stimulation of non-union and congenital pseudarthrosis. Exploration of its clinical application. *J. Bone Joint Surg. Am.* **57**, 368–377.

[23] Bassett CAL and Pawluk RJ (1975) Non-invasive methods for stimulating osteogenesis. *J. Biomed. Mater. Res.* **9**, 371–374.

[24] Woodson HH and Melcher JR (1968) *Electromechanical Dynamics*, Part II: *Fields, Forces, and Motion*. Wiley, New York, Chapter 7.

[25] Arnegard ME, McIntyre PB, Harmon LJ, et al. (2010) Sexual signal evolution outpaces ecological divergence during electric fish species radiation. *Am. Nat.* **176**, 335–356.

[26] Lissman HW (1963) Electric location by fishes. *Sci. Am.* **208**(3), 50–59.

[27] Maxwell [3], p. 435.

[28] Coulter WH (1953) Means for counting particles suspended in a fluid. *US Patent* 2656508.

Electrochemical Coupling and Transport

3.1 ION TRANSPORT IN A BINARY ELECTROLYTE

The motions of positive and negative ions in a convecting electrolyte do not occur independently of one another. Migration (due to an applied E) and diffusion (due to gradients in total electrolyte concentration) are both processes that tend to separate positive and negative species—the former because of the opposite sign of charge and the latter because of differing diffusion coefficients. However, any slight separation results in very large electric fields that give rise to conduction currents tending to restore charge neutrality. Thus, the tendency towards carrier separation (leading to net space charge) is balanced by the charge relaxation process. Electrolyte solutions that are of interest to us have sufficient conductivity that relaxation time constants of interest are short, approximately 10^{-6}–10^{-9} s (physiological ionic strength being about 0.15 molar). Therefore, we will find that electrolyte solutions can be considered electrically neutral over characteristic distances much greater than an electrical Debye length. This last statement can be motivated physically by the results of Problem 2.3, where $\Phi(x)$ and $\rho_e(x)$ were calculated for an imposed steady state charge separation (that of an interfacial double layer). The net space charge in the electrolyte phase and the associated electric field were found to decay towards zero over a distance equal to a few Debye lengths.

One of the goals of this section is to formulate a system of governing equations that will be sufficient to characterize the concentration profiles of chemical species, $c_i(\mathbf{r}, t)$, as well as the charge and potential profiles, $\rho_e(\mathbf{r}, t)$ and $\Phi(\mathbf{r}, t)$, for a wide variety of problems involving electrolyte transport phenomena. For example, there are many problems involving biological and hydrated scaffold membranes that support concentration gradients. It is often crucial to know the intramembrane concentrations and electrical potential profiles, where the membrane may be subjected to an electric field and may be moving or deforming, as depicted in the one-dimensional schematic of Figure 3.1. As another example, Figure 3.2 shows a pioneering method for measuring the flow velocity of blood in a vessel by injecting a "pulse" of concentrated salt solution into the blood and measuring the conductivity versus time at a fixed downstream position [1]. Owing to diffusional spreading of the salt in the convecting fluid stream, the exact space–time profile of mobile salt species is necessary to interpret the measurement. This early example has state-of-the-art counterparts involving microfluidics for diagnostic protein and cell separations.

The intent of this section is not to pursue the detailed solution to examples such as these, but rather to formulate the general system of governing equations. From these equations, two very important relations will be derived describing *coupled diffusion* and *charge relaxation* processes, *valid when the spatial dimensions of interest are much larger than a Debye length*. Once again, the question of competition between migration, diffusion, and convection of ionic species comes into play in the description of the macroscopic fluxes and spatial distributions of

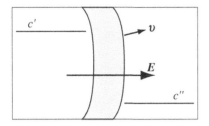

Figure 3.1 Membrane permeable to ionic and fluid species in general, subjected to a field E and concentration gradient.

these species. This competition can lead to even more complex interactions when the characteristic dimensions of interest are on the order of or less than a Debye length.

For the purpose of simplifying the mathematical description, we restrict the following discussion to the case of a binary electrolyte, such as NaCl in water, with no other mobile species present. (The concentrations of H^+ and OH^- can be considered unimportant for now, since $[H^+] = [OH^-] = 10^{-7}$ M at neutral pH.) Further, we will consider only salts that are strong electrolytes and therefore completely dissociated in aqueous solution. Therefore, we do not yet consider chemical reactions ($R_i \to 0$) and their associated chemical kinetic equations.

In order to specify the carrier concentrations, we first need to write the flux and continuity relations for both the positive and negative ionic species. From (1.5) and (2.37), with $\boldsymbol{N}_i = (z_i/|z_i|)u_i c_i \boldsymbol{E} - D_i \nabla c_i + c_i \boldsymbol{v}$, we have

$$\boldsymbol{N}_+ = +u_+ c_+ \boldsymbol{E} - D_+ \nabla c_+ + c_+ \boldsymbol{v} \tag{3.1}$$

$$\boldsymbol{N}_- = -u_- c_- \boldsymbol{E} - D_- \nabla c_- + c_- \boldsymbol{v} \tag{3.2}$$

$$\frac{\partial c_+}{\partial t} = -\nabla \cdot \boldsymbol{N}_+ \tag{3.3}$$

$$\frac{\partial c_-}{\partial t} = -\nabla \cdot \boldsymbol{N}_- \tag{3.4}$$

where $\boldsymbol{v}$ is the molar-average velocity (1.10), which essentially reduces to that of the solvent, water, and z_i is the valence of the i ionic species, taken to be $|z_i| = 1$ here. We now restrict the discussion by considering examples in which the velocity $\boldsymbol{v}$ is either zero or known at the outset; i.e., we decouple the fluid mechanical dynamics from the diffusion–migration problem. This is an extremely useful restriction for a wide range of problems in membrane transport (in which intramembrane fluid motion is often so small that it can be neglected) and a class of convective–diffusion problems in which the fluid motion can be solved first without regard to and uninfluenced by the distribution of chemical species (see Chapter 5). Since the three components of the electric field E in (3.1)–(3.4) are generally unknown, we must add Gauss' law and Faraday's law:

$$\nabla \cdot \epsilon \boldsymbol{E} = \rho_e(\boldsymbol{r}, t) = F\left[z_+ c_+(\boldsymbol{r}, t) + z_- c_-(\boldsymbol{r}, t)\right] \tag{3.5}$$

$$\nabla \times \boldsymbol{E} = 0 \tag{3.6}$$

where we recall that (3.5) and (3.6) are really three independent equations; only two of the three scalar equations in the vector equation (3.6) are independent. Therefore, (3.1)–(3.6) are 11 scalar equations in 11 unknowns: c_+, c_-, the three components each of $\boldsymbol{N}_+$ and $\boldsymbol{N}_-$, and E_x, E_y, E_z. This system of equations constitutes a complete description of $c_\pm(\boldsymbol{r}, t)$, $\rho_e(\boldsymbol{r}, t)$, and $\Phi(\boldsymbol{r}, t)$, provided the fluid motion is known at the outset. If the fluid motion is not known and is in fact influenced by the electric field and the chemical species, then the relevant fluid mechanical equations of motion would have to be added to (3.3)–(3.6) and the velocity and pressure distributions would have to be found in a self-consistent manner along with all of the other unknowns. Finally, we

Figure 3.2 Method for measuring flow velocity of blood in a vessel. A region of concentrated salt, A, is injected at $t = 0$ into an otherwise fairly uniform, dilute electrolyte and the concentration profile is measured downstream for $t > 0$ by the detector, B.

note that conservation of charge (2.21) and the constitutive law for current density (2.41) are not independent equations, since they can be derived from (3.1)–(3.4) above.

We continue the discussion of transport in a binary electrolyte by using (3.1)–(3.6) to derive general relations describing coupled salt diffusion and charge relaxation in the medium. We recall that the processes of migration and diffusion both tend to separate positive and negative species. We also recognize that for most electrolyte concentrations of physiological interest, there is a large background of positive and negative carriers, so that we may define the total concentrations $c_+(\mathbf{r}, t)$ and $c_-(\mathbf{r}, t)$ as sums of background and perturbation components:

$$c_+(\mathbf{r}, t) = c_{+0}(\mathbf{r}, t) + \tilde{c}_+(\mathbf{r}, t) \tag{3.7}$$

$$c_-(\mathbf{r}, t) = c_{-0}(\mathbf{r}, t) + \tilde{c}_-(\mathbf{r}, t) \tag{3.8}$$

Owing to the relatively large magnitude of the background concentrations, the perturbation quantities $\tilde{c}_\pm$ (which are associated with the slight tendency for charge carrier separation) will normally be much less in magnitude than the background concentrations in the bulk of a *single binary electrolyte*:

$$\tilde{c}_\pm \ll c_{\pm 0} \tag{3.9}$$

The Electroneutrality Condition

The requirement of *bulk electroneutrality* for the electrolyte as a whole manifests itself in equal background charge densities:

$$F\left(z_+ c_{+0} + z_- c_{-0}\right) = 0 \tag{3.10}$$

The statement of bulk charge neutrality (3.10) does *not* preclude the existence of net charge

$$\rho_e = F\left(z_+ c_{+0} + z_- c_{-0}\right) + F\left(z_+ \tilde{c}_+ + z_- \tilde{c}_-\right) \tag{3.11}$$

but we expect (from the results of Problem 2.3) that ρ_e will be much less in magnitude than that associated with the background density of either species for dimensions much *larger* than a Debye length:

$$\rho_e = \left|F\left(z_+ \tilde{c}_+ + z_- \tilde{c}_-\right)\right| \ll \rho_0 \equiv \left|F z_\pm c_{\pm 0}\right| \tag{3.12}$$

We will call the inequality (3.12) a *quasineutrality* condition. For a $z : z$ binary electrolyte, where

$$z_+ = -z_- = z \tag{3.13}$$

and therefore

$$c_{+0} = c_{-0} = c_0 \tag{3.14}$$

the quasineutrality condition (3.12) can be rewritten in terms of perturbation concentrations as

$$\frac{\left|\tilde{c}_+ - \tilde{c}_-\right|}{c_0} \ll 1 \tag{3.15}$$

In the early literature, it is common to see the use of the so-called electroneutrality assumption, $\tilde{c}_+ \simeq \tilde{c}_-$, in order to simplify the solution of certain classes of nerve–muscle membrane and electrolyte transport problems [2]. However, the stricter statement of quasineutrality is (3.15), which should be evaluated to check its validity in any given application.

3.1.1 Coupled Diffusion in a Binary Electrolyte

With the purpose of examining the effect of the electrical interaction between positive and negative species on the diffusion of an electrolyte, we begin by combining the continuity and flux relations (3.1)–(3.4) for a 1:1 electrolyte:

$$\frac{Dc_+}{Dt} = \nabla \cdot (D_+ \nabla c_+) - \nabla \cdot (u_+ c_+ \boldsymbol{E}) \tag{3.16}$$

$$\frac{Dc_-}{Dt} = \nabla \cdot (D_- \nabla c_-) + \nabla \cdot (u_- c_- \boldsymbol{E}) \tag{3.17}$$

where the rate of change is written in terms of the convective (material) derivative (D/Dt) for an observer moving with the fluid, and becomes $(\partial/\partial t)$ in the absence of fluid motion. We can see the effect of the coupling of positive and negative species by multiplying (3.16) by D_- and (3.17) by D_+, and adding the two equations:

$$\frac{D}{Dt}(D_- c_+ + D_+ c_-) = \nabla \cdot [D_- D_+ (\nabla c_+ + \nabla c_-)] - \nabla \cdot \left[\frac{F}{RT} D_- D_+ (c_+ - c_-) \boldsymbol{E} \right] \tag{3.18}$$

where use has been made of the Einstein relation (2.38) and the definition (3.13) in the last term of (3.18),

$$\frac{D_i}{u_i} = \frac{RT}{zF} \tag{3.19}$$

The last term in (3.18) may be neglected when the following inequality is valid:

$$\frac{|D_- D_+ \boldsymbol{E} (c_+ - c_-)|}{|(RT/F)D_- D_+ (\nabla c_+ + \nabla c_-)|} \ll 1 \tag{3.20}$$

When the inequality (3.20) is valid, (3.18) can be simplified by recalling that $\tilde{c}_\pm \ll c_0$; thus, for small perturbations, (3.18) becomes

$$\frac{D}{Dt} c_0 = D^* \nabla^2 c_0 + \nabla \cdot \underbrace{\left[\frac{F}{RT} \frac{D_+ D_-}{D_+ + D_-} (c_- - c_+) \boldsymbol{E} \right]}_{\text{negligible}} \tag{3.21}$$

where

$$\boxed{D^* = \frac{2D_+ D_-}{D_+ + D_-}} \tag{3.22}$$

According to (3.21), the electrolyte diffuses as a whole relative to the convecting fluid with an *effective diffusion coefficient* D^*. Although the right-most term of (3.21) with its dependence on $\boldsymbol{E}$ may be negligible, it is the internal or self $\boldsymbol{E}$ induced by the tendency toward unequal diffusion of carriers that leads to the overall *coupled* diffusion process. For the case $D_+ = D_- = D$ (almost true for KCl, KNO$_3$, and certain other salts), $D^* \to D$ in (3.22), and no self-field would be generated. If, in addition, there is no applied field present, then the inequality (3.20) is certainly satisfied with $\boldsymbol{E} = \boldsymbol{E}_{\text{self}} + \boldsymbol{E}_{\text{applied}} \equiv 0$, and (3.21) with $D^* \to D$ describes electrolyte diffusion with no other approximations.

In order to estimate the relative magnitudes of the terms in the inequality (3.20), use is made of the assumptions that $\tilde{c}_\pm \ll c_0$ and that

∇c_0 can be approximated by c_0 divided by a characteristic length δ. Then (3.20) becomes

$$\left| \left(\frac{E\delta}{2RT/F} \right) \left(\frac{\rho_e}{\rho_0} \right) \right| \ll 1 \qquad (3.23)$$

For characteristic lengths much larger than a Debye length, the quasineutrality approximation $|\rho_e/\rho_0| \ll 1$ is valid because of charge relaxation effects, and (3.23) then gives an upper bound on E for the electrolyte diffusion equation (3.21) to remain valid:

$$|E| \ll \left| \left(\frac{2RT/F}{\delta} \right) \left(\frac{\rho_0}{\rho_e} \right) \right| \qquad (3.24)$$

We must remember that E in general is the sum of any applied field plus any induced, "internal" field resulting from unequal rates of carrier diffusion (e.g., diffusion potentials arising from concentration gradients across membranes—see Section 3.2). For the latter case, $|E\delta| \approx RT/F$ (i.e., the migration potential balances the diffusion or thermal potential RT/F). Thus, we can see that one way in which the inequality breaks down occurs when $|E\delta| \approx RT/F$ *and* $\delta \simeq 1/\kappa$ (Debye length)—the latter condition implying that ρ_e can be on the order of ρ_0. We conclude that the diffusion model (3.21) is valid as long as we are *not* interested in dimensions on the order of $1/\kappa$. (Note that the above discussion will be clarified further in the context of a concrete example in Section 3.2.)

3.1.2 Charge Relaxation and Quasineutrality in a Binary Electrolyte

To evaluate the effect of concentration gradients on the charge relaxation process examined initially in Section 2.6, we now calculate the net charge density ρ_e. The following discussion will also lead to further clarification of the validity of the quasineutrality assumption (3.12). Another motivation for finding ρ_e is the possibility of an applied electric field coupling to the bulk electrolyte via the electrical force density $\rho_e E$. We should not be surprised to find that in reasonably homogeneous electrolyte solutions of physiological interest, relaxation times are so fast that bulk ρ_e and therefore $\rho_e E$ forces are negligible. However, inhomogeneous regions of electrolyte leading to net ρ_e often occur; prime examples are space charge regions or interfacial double layers associated with membranes and charged macromolecules.

With this in mind, we find ρ_e by multiplying (3.16) by $z_+ F$ and (3.17) by $z_- F$, and adding the two equations (let $|z_+| = |z_-| = z$ as before):

$$\frac{\mathrm{D}\rho_e}{\mathrm{D}t} = -\nabla \cdot [\sigma(\boldsymbol{r}, t)E] + \nabla \cdot [zF(D_+ \nabla c_+ - D_- \nabla c_-)] \qquad (3.25)$$

where

$$\sigma(\boldsymbol{r}, t) = zF[u_+ c_+(\boldsymbol{r}, t) + u_- c_-(\boldsymbol{r}, t)] \qquad (3.26)$$

With $c_0 \gg \tilde{c}$, (3.25) reduces to

$$\frac{\mathrm{D}\rho_e}{\mathrm{D}t} = -\nabla \cdot \sigma_0 E + \nabla \cdot [(D_+ - D_-) zF \nabla c_0] \qquad (3.27)$$

The last term in (3.27) is related to the electric field and the associated space charge due to diffusion of ions with unequal diffusivities. When $D_+ = D_-$, a concentration gradient ∇c_0 will result in diffusion with no attendant space charge.

In general, the last term of (3.27) can be neglected in the limit when the migration term dominates the diffusion term:

$$\left| \frac{zF(D_+ - D_-)\nabla c_0}{zF(u_+ + u_-)c_0|\boldsymbol{E}|} \right| \ll 1 \tag{3.28}$$

Use of the Einstein relation (2.38) in (3.28) will show that the latter is dimensionally the same inequality (2.46) seen earlier:

$$|\boldsymbol{E}\delta| \gg \left| \left(\frac{RT}{zF} \right) \left(\frac{D_+ - D_-}{D_+ + D_-} \right) \right| \tag{3.29}$$

When (3.29) is satisfied, (3.27) becomes

$$\frac{\mathrm{D}\rho_e}{\mathrm{D}t} + \frac{\sigma_0}{\epsilon} \underbrace{\nabla \cdot \epsilon \boldsymbol{E}}_{\equiv \rho_e} = -\boldsymbol{E} \cdot \nabla \sigma_0 \tag{3.30}$$

Equation (3.30) tells us that when $\boldsymbol{E} \cdot \nabla \sigma_0 \simeq 0$, the net charge, ρ_e relaxes in the same manner seen in the discussion of Section 2.6 (according to (2.79)), but now with respect to a *convecting electrolyte*. (In many applications, $\boldsymbol{E} \cdot \nabla \sigma_0$ is indeed negligible in (3.30); see Problem 3.1.) Once again, the picture may break down when we focus on dimensions on the order of $1/\kappa$, where $|\boldsymbol{E}|\delta \approx RT/F$ and the inequality (3.29) may not be satisfied. Under such conditions, the full equation (3.25) must be used to find $\rho_e(\boldsymbol{r}, t)$. Said another way, diffusion across a characteristic length δ can induce a self-field on the order of $(1/\delta)(RT/zF)[(D_+ - D_-)/(D_+ + D_-)]$ when the ion species have unequal diffusion coefficients. For migration to dominate diffusion in the determination of the space charge $\rho_e(\boldsymbol{r}, t)$, the applied fields must be greater than the self-field. When the characteristic distance is as small as the Debye length, the applied field magnitude must be very large, i.e., $\gg 10^7 \, \mathrm{V\,m^{-1}}$. As this is rarely achievable in highly conducting electrolytes when $\boldsymbol{E}_{\mathrm{applied}}$ leads to net current, diffusion cannot be neglected in the calculation of ρ_e inside a double layer.

One might question the relevance of (3.30) in homogeneous electrolyte systems where $\tau_r \sim \epsilon/\sigma_0 \sim 10^{-7}$–$10^{-9}$ s and therefore $\rho_e \simeq 0$. In such cases, we will find that it is the boundary condition associated with (3.30) that assumes the greatest importance in physiological systems. This boundary condition (see Section 2.6) is associated with surface or interfacial zones where surface or double layer charge densities can play a dominant role.

In summary, the transport of ions in a binary electrolyte may be characterized by the *coupled diffusion* and *charge relaxation* equations (3.21) and (3.30), respectively, when distances of interest are much *larger* than a Debye length *and* when the total electric field satisfies the inequalities

$$\left| \left(\frac{D_+ - D_-}{D_+ + D_-} \right) \left(\frac{RT/F}{\delta} \right) \right| \ll |\boldsymbol{E}| \ll \left| \left(\frac{2RT/F}{\delta} \right) \left(\frac{\rho_0}{\rho_e} \right) \right| \tag{3.31}$$

If we are interested in distances on the order of $1/\kappa$ or if (3.31) is violated, then we must resort to the complete diffusion and charge equations (3.18) and (3.25), respectively. In most physiological systems of interest, $1/\kappa$ is small enough that the internal $\boldsymbol{E}$ (where $\boldsymbol{E}_{\mathrm{in}}\delta \approx \boldsymbol{E}_{\mathrm{in}} \cdot 1/\kappa \approx RT/F$) is usually much larger than that achievable by external application of an electric field. Later in this chapter, we will calculate $\rho_e(\boldsymbol{r}, t)$ in several examples dealing with charged and neutral membranes; it will be seen that quasineutrality as expressed by (3.12) is an excellent assumption as long as the membrane thickness δ is much greater than a Debye length.

3.2 COUPLED STEADY STATE DIFFUSION ACROSS A NEUTRAL MEMBRANE: THE DIFFUSION POTENTIAL

A diffusion potential (or liquid junction potential) will naturally arise in the nonequilibrium configuration of Figure 3.3—that of irreversible diffusion of ions across a membrane. The diffusion of oppositely charged (hydrated) ions having unequal diffusivities tends to produce a charge separation with its concomitant electric field. This self-induced field, in turn, gives rise to conduction or ohmic (migration) currents that counterbalance the unequal diffusive flows. Here specifically, the induced field acts in such a direction as to retard the faster-diffusing ion and speed the slower one. In the steady state, this electric field manifests itself as a potential difference across the membrane, the so-called junction or membrane potential. As the physical concepts described above relate directly to the electrolyte transport phenomena discussed in Section 3.1, we should not be surprised that an expression for the diffusion potential can be derived from these previous results.

To examine the case of coupled, steady state diffusion across a neutral membrane, we refer to Figure 3.3. This system is categorized as a steady state ($\partial/\partial t \to 0$), nonequilibrium system, the latter nomenclature signifying the presence of nonequilibrium transport processes. The bulk electrolyte concentrations to the left and right of the membrane are c' and c'', respectively. These steady state concentrations are assumed to be constant right up to the membrane interfaces (accomplished by continual recirculation and vigorous stirring on both sides of the membrane). We are interested in finding the steady state diffusion potential $\Delta\Phi_{\text{diff}} \equiv \Phi'' - \Phi'$, as well as $\bar{c}(x)$, $\Phi(x)$, and $\rho_e(x)$ inside the membrane. For simplicity, we treat the case where the electrolyte is the same on both sides, and is symmetric ($|z_+| = |z_-| = z$).

At the outset, we assume that a steady state has been reached, following the initial assembly of the system pictured in Figure 3.1. (The transient diffusion process that leads to such a steady state is examined in Problem 3.3; it should be evident that the transient diffusion times involved are rate-limiting—they are longer than any other physical time constant of interest, such as the charge relaxation time associated with (3.30). Thus, as the concentration profile $\bar{c}(x)$ inside the membrane evolves into its steady state value, the small but finite charge distribution can be considered to be in a state of "instantaneous relaxation"). We will also assume that quasineutrality as expressed by (3.12) is valid throughout all space. Once the potential $\Phi(x)$ has been determined, $\rho_e(x)$ can be explicitly calculated to see if it is consistent with the quasineutrality approximation, $|\rho_e/\rho_0| \ll 1$. Finally, convective transport of ions will be neglected *inside* the membrane. The validity of neglecting convection must be checked for each membrane or physiological tissue of interest. It is a reasonable assumption inside many types of membranes.

The approach that will now be used to derive $\bar{c}(x)$ and $\Phi(x)$ is based on the coupled diffusion and relaxation equations (3.18) and (3.27), respectively. Alternative derivations can be found (starting directly with the flux equations (3.1) and (3.2), or using, instead, a quasithermodynamic approach), which will be addressed in the examples at the end of this section. However, we wish to show the general applicability of the concepts discussed in the previous section.

First, we must carefully consider which terms, if any, can be neglected in the coupled diffusion and charge relaxation equations. With regard to the latter, we recall that the second term on the right-hand side of

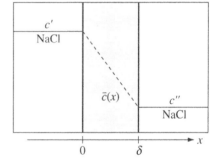

Figure 3.3 Coupled diffusion across a neutral membrane. $\bar{c}(x)$ is the concentration inside the membrane.

(3.27) is the divergence of a current associated with net space charge due to the differing diffusivities. But in the steady state ($D\rho_e/Dt \to 0$), it was reasoned that this current is responsible for the diffusion potential; i.e., this current is balanced by the ohmic current in the other right-hand term in (3.27). Thus, with no applied electric field, we expect that the **E** in the other right-hand term is probably not much greater or much less than, but is on the order of, the "thermal field" $(RT/F)/\delta$. Therefore, the inequalities (3.28) and (3.29) are not satisfied in general, and both the migration and diffusion terms in (3.27) must be kept. On the other hand, we suspect that if membrane thickness $\delta \gg 1/\kappa$, then quasineutrality is a reasonable assumption. Therefore, the inequality (3.23) is satisfied as long as $|\rho_e/\rho_0| \ll 1$. Coupled diffusion is apparently well described by (3.21) with the effective diffusion coefficient $\bar{D}^*$ inside the membrane (the overbar notation implies that any interaction between the membrane material and the permeating ions can be lumped into the effective intramembrane diffusion coefficient). In summary, the diffusion of electrolyte across the membrane of Figure 3.3 can be described by the simplified coupled diffusion equation (3.21), but the charge and field distributions can only be inferred by using the complete "charge conservation" or relaxation equation (3.27).

Starting with (3.27) in the steady state, the right-hand side becomes

$$\nabla \cdot \left[\sigma_0(\mathbf{r})\mathbf{E} - (z\bar{D}_+ - z\bar{D}_-)F\nabla \bar{c}_0(\mathbf{r})\right] = \nabla \cdot \mathbf{J}_{\text{total}} = 0 \qquad (3.32)$$

Integration of (3.32) yields

$$\mathbf{J}_{\text{total}} = \mathbf{J}_+ + \mathbf{J}_- = \mathbf{J}_{\text{migration}} + \mathbf{J}_{\text{diffusion}} = \text{const} \equiv 0 \qquad (3.33)$$

where the final equality results from the physical constraint that there is no net current flow in the system. (Any measurement of $\Delta\Phi_{\text{diff}}$ had best be made with a very high-input-impedance electrometer or differential amplifier.) Equation (3.33) gives directly the electric field $E_x(x) = -d\Phi/dx$, where a one-dimensional model is used henceforth:

$$\frac{d\Phi(x)}{dx} = -\frac{zF(\bar{D}_+ - \bar{D}_-)d\bar{c}_0(x)/dx}{zF(\bar{u}_+ + \bar{u}_-)\bar{c}_0(x)} \qquad (3.34)$$

The denominator of (3.34) comes from the conductivity σ_0 for a symmetric $z : z$ electrolyte,

$$\sigma_0(x) = zF(\bar{u}_+ + \bar{u}_-)\bar{c}_0(x) \qquad (3.35)$$

Equation (3.34) can be integrated directly, incorporating the Einstein relation to substitute for the mobilities:

$$\Phi(x) = -\frac{RT}{zF}\left(\frac{\bar{D}_+ - \bar{D}_-}{\bar{D}_+ + \bar{D}_-}\right)\ln\left[\frac{\bar{c}_0(x)}{\bar{c}_0(x=0)}\right] \qquad (3.36)$$

In (3.36), it is assumed that the potential at $x = 0$ is zero; i.e., the reference is defined such that $\Phi(x=0) = 0$. Thus, the $\bar{c}_0(x=0)$ in the logarithmic term of (3.36) corresponds to the integration constant evaluated to satisfy this reference potential definition.

$\Phi(x)$ would be known if $\bar{c}_0(x)$ in (3.36) were known, and the coupled diffusion equation (3.21) could be used to predict $\bar{c}_0(x)$. By neglecting the migration (**E**) term in (3.21), we have essentially decoupled the solution of the concentration profile $\bar{c}_0(x)$ from that of the potential profile $\Phi(x)$.

In the steady state,

$$\frac{d^2 \bar{c}_0(x)}{dx^2} = 0 \qquad (3.37)$$

Integrating twice subject to the boundary conditions on concentration

$$\bar{c}(0) = c', \qquad \bar{c}(\delta) = c'' \qquad (3.38)$$

gives the intramembrane salt concentration

$$\bar{c}_0(x) = (c'' - c')\frac{x}{\delta} + c' \qquad (3.39)$$

Using (3.39) in (3.36) results in the potential profile

$$\Phi(x) = -\frac{RT}{zF}\left(\frac{\bar{D}_+ - \bar{D}_-}{\bar{D}_+ + \bar{D}_-}\right)\ln\left[\left(\frac{c'' - c'}{c'}\right)\frac{x}{\delta} + 1\right] \qquad (3.40)$$

The diffusion potential $\Delta\Phi_{\text{diff}} \equiv \Phi(\delta) - \Phi(0) \equiv \Phi'' - \Phi'$ follows immediately:

$$\boxed{\Delta\Phi_{\text{diff}} = -\frac{RT}{zF}\left(\frac{\bar{D}_+ - \bar{D}_-}{\bar{D}_+ + \bar{D}_-}\right)\ln\left(\frac{c''}{c'}\right)} \qquad (3.41)$$

Note the dependence of $\Delta\Phi_{\text{diff}}$ on the valence z (explain). In addition, we see that (3.41) contains no information about the exact spatial dependence of $\bar{c}_0(x)$, or about the thickness or nature of the membrane (the exception is the inclusion of any effect of the membrane on diffusivities of ions, expressed by $\bar{D}_i$). In general, the thickness δ plays no role in *steady state* diffusion potentials; but we should expect δ to be important in any *transient* phenomena (see Problem 3.3).

Equation (3.41) may be rewritten in terms of the individual ion transport numbers t_i, which are defined inside the membrane as

$$t_\pm \equiv \frac{|z_\pm|\,\bar{D}_\pm\,\bar{c}_\pm}{\displaystyle\sum_i |z_i|\,\bar{D}_i\,\bar{c}_i} \qquad (3.42)$$

(In bulk solution $\bar{c}_i \to c_i$ and $\bar{D}_i \to D_i$.) The transport numbers essentially express the fraction of total current or flux carried by each individual mobile species. The assumption of bulk electroneutrality, $\bar{c}_+(x) \simeq \bar{c}_-(x) = \bar{c}_0(x)$ can be used in (3.42), allowing the diffusion potential to be written as

$$\Phi'' - \Phi' = \frac{RT}{zF}(t_- - t_+)\ln\left(\frac{c''}{c'}\right) \qquad (3.43)$$

As an example, consider the case where $c'' = 0.001$ M NaCl and $c' = 0.002$ M NaCl:

$z = 1$
$R = 8.314\,\mathrm{W\,s\,K^{-1}\,mol^{-1}}$
$T = 300\,\mathrm{K}$
$F = 96{,}500\,\mathrm{C\,Eq^{-1}}$
$t_- = 0.61$ (dilute solution)
$t_+ = 0.39$
$RT/F = 25.69\,\mathrm{mV}$

$$\Phi'' - \Phi' = 59.2(t_- - t_+)\log \tfrac{1}{2} = -3.92\,\mathrm{mV} \qquad (3.44)$$

Note the sign of the potential: the right side is more negative than the left. This agrees with physical reasoning, since $u_- > u_+$ (or $D_- > D_+$) and $c' > c''$. The negative ion diffuses more rapidly, and a "diffusion potential" results that acts to retard its movement while accelerating that of the positive ion. The *electric field* associated with $\Delta\Phi_{\mathrm{diff}}$ therefore serves as a *coupling agent* in the diffusion of electrolyte from left to right. But this is exactly the description of coupled diffusion derived a bit more abstractly in Section 3.1. If the coupled diffusion process is really identical to that of Section 3.1, we should expect to find that the individual ion fluxes can be written in terms of the effective diffusion coefficient $\bar{D}^*$ (see Example 3.2.1). Note, finally, that $\Delta\Phi_{\mathrm{diff}} \simeq 0$ when $D_+ \simeq D_-$, as is the case for special salts such as KCl and KNO_3.

It is important to realize that the diffusion potential discussed in this section is essentially a property of an electrolyte solution. The membrane simply serves as a barrier to geometrically define the concentration gradient. The diffusion potential (3.41) is seen to depend only on the bulk boundary concentrations c' and c'' and the carrier diffusivities $\bar{D}_+$ and $\bar{D}_-$. If there are no specific interactions between membrane molecules and the diffusing species, the diffusivity of carrier inside the membrane, $\bar{D}_i$, is the same as that in bulk solution, D_i. Therefore, if it were possible to take away the membrane and maintain the gradient, $\Delta\Phi_{\mathrm{diff}}$ would remain unchanged—it would then be equivalent to the idealized "junction potential" of physical chemistry. Several varieties of *charge-neutral* polymeric membranes exhibit diffusion potentials that agree quite well with the prediction of (3.41). Membranes made from packed beds of various molecules or cast from aggregates of long-chain polyelectrolyte molecules in their charge-neutral state may also give steady state diffusion potentials approximately satisfying (3.41). Of course, the ionizable charge groups of biological membranes can be found in their charged state. In this case, the model posed in this section is inadequate, and a reformulation of the fundamental laws describing transport across a charged membrane must be made. It may be possible to recast the model in such a way as to represent the effect of the electrostatic interaction in terms of modified intramembrane diffusion coefficients $\bar{D}_i$ or equivalent transport numbers $\bar{t}_i$ (Example 3.2.2). Alternatively, more comprehensive charged membrane models may be necessary (see, e.g., Section 3.5).

Example 3.2.1 The Coupled Diffusivity $\bar{\mathbf{D}}^*$ Show that the theoretical expressions for carrier fluxes inside the membrane, $\bar{N}_\pm(x)$, can be written in terms of a single, lumped diffusion term where the diffusion coefficient is $\bar{D}^*$—the effective diffusion coefficient inside the membrane having the form of (3.22). Therefore, show that the theoretical expressions for the fluxes $\bar{N}_\pm(x)$ corresponding to the case study of Section 3.2 are consistent with the coupled diffusion equation (3.18) of Section 3.1.

EXAMPLE

Starting with $\bar{N}_+(x)$, we use (3.1) to write

$$\bar{N}_+ = -\bar{u}_+ \bar{c}_+(x) \frac{d\Phi(x)}{dx} - \bar{D}_+ \frac{d\bar{c}_+}{dx} \tag{3.45}$$

Equation (3.34) can be incorporated directly into (3.45), assuming $\bar{c}_+(x) \sim \bar{c}_-(x) = \bar{c}_0(x)$:

$$\bar{N}_+ = \frac{+\bar{u}_+ \bar{c}_0(x) \left(\dfrac{\bar{D}_+ - \bar{D}_-}{\bar{u}_+ + \bar{u}_-} \right) \dfrac{d\bar{c}_0(x)}{dx}}{\bar{c}_0(x)} - D_+ \frac{d\bar{c}_0}{dx} \tag{3.46}$$

Use of the Einstein relation (2.38) in (3.46) gives

$$\bar{N}_+ = \left(+\bar{D}_+ \frac{\bar{D}_+ - \bar{D}_-}{\bar{D}_+ + \bar{D}_-} - \bar{D}_+ \right) \frac{d\bar{c}_0}{dx} \tag{3.47}$$

$$\bar{N}_+ = -\left(\frac{2\bar{D}_+ \bar{D}_-}{\bar{D}_+ + \bar{D}_-} \right) \frac{d\bar{c}_0}{dx} = -\bar{D}^* \frac{d\bar{c}_0}{dx} \tag{3.48}$$

An analogous relation can be derived for $\bar{N}_-$.

Example 3.2.2 The Limit of Ideal Semipermeability Many biological membranes, as well as synthetic polymeric membranes, are selectively permeable to certain species. That is, only certain species are found to be able to cross the membrane, while the transport of all other species is blocked. Several different mechanisms giving rise to "semipermeability" are known, the most obvious being steric hindrance or filtration of particles that are too large to fit through the pores of the membrane. Another important mechanism results from the existence of fixed-charge groups attached to the membrane. For example, nerve, muscle, and other cell membranes contain proteins, glycoproteins, lipids, and lipoproteins with charged species attached to their macromolecular backbones. These charge groups can modulate the flow of electrolyte ions across the membrane. Various membrane models account for fixed charge either explicitly or in terms of interfacial partition coefficients, which are essentially empirical factors (see Problem 3.8).

A heavily charged membrane may totally block the transport of ions having like charge, called "co-ions." On the other hand, the oppositely charged "counter-ions" find little opposition in crossing the membrane. The membrane is *selectively permeable* to counter-ions, a characteristic that is a general, nonspecific electrostatic property of the membrane.

We now consider a permselective membrane interposed between two solutions of the same binary electrolyte having different concentrations—the configuration of Figure 3.3. In this case, ionic transport and the transmembrane potential are quite different than that described in Section 3.2. The ideally permselective membrane will presumably block the flow of the co-ion. But as soon as the flow of one ion is stopped, electroneutrality requires that its oppositely charged partner in the solution also be stopped. In this ideal semipermeable limit, irreversible diffusive flows have been completely halted; bulk transport is stopped. Only the ion to which the membrane *is* permeable reaches a true equilibrium across the membrane.

For a chemical species in true thermodynamic equilibrium across the membrane, the flux $N_i \equiv 0$ for that species. This allows direct calculation of the transmembrane potential. Assuming the concentration cell configuration of Figure 3.3, with $z_+ = -z_- = z$,

$$N_\pm = \mp \left(u_\pm c_\pm \frac{d\Phi}{dx} \right) - D_\pm \left(\frac{dc_\pm(x)}{dx} \right) \equiv 0 \qquad (3.49)$$

$$d\Phi = \mp \left(\frac{D_\pm}{u_\pm} \right) \frac{dc_\pm(x)}{c_\pm(x)} \qquad (3.50)$$

Integrating between $x = 0$ and $x = \delta$ in Figure 3.3 and using the Einstein relation, the transmembrane potential for a membrane permeable only to the positive or negative ion, respectively, is

$$\boxed{\Phi'' - \Phi' = \mp \frac{RT}{|z|\,F} \ln\left(\frac{c_0''}{c_0'} \right)} \qquad (3.51)$$

when $c_+' = c_-' = c_0'$ and $c_+'' = c_-'' = c_0''$.

For example, with $c_0' = 0.1\,\text{M NaCl}$, $c_0'' = 0.01\,\text{M NaCl}$, we find that $\Phi'' - \Phi' \simeq -59.2\,\text{mV}$ for an "anion exchange" membrane (permeable to the negative ion). Note that the result (3.51) is identical to that of (3.43) in the limit ($t_+ \to 0$, $t_- \to 1$) for an anion exchanger and ($t_+ \to 1$, $t_- \to 0$) for a cation exchanger. Physically, this is the permselective limit for that example.

Example 3.2.3 Solute Permeability: A Phenomenological Representation for a Sometimes-Confusing Parameter From Example 3.2.1, the flux at any point outside ($\boldsymbol{N}_i$) or inside ($\bar{\boldsymbol{N}}_i$) a membrane, respectively, can be written as

$$\boldsymbol{N}_i = -D^* \nabla c_0, \quad \bar{\boldsymbol{N}}_i = -\bar{D}^* \nabla \bar{c}_0 \qquad (3.52)$$

where (3.52) represent coupled diffusion in a binary electrolyte as before. These flux equations are point-by-point relations; flux is proportional to concentration gradient at every point. As $\nabla \bar{c}_0$ is *not* an experimentally accessible parameter, other "phenomenological" representations of membrane diffusion are often used. For example, c_0' and c_0'' *are* measurable, and (3.52) can be represented alternatively by

$$N = -D^* \frac{c_0'' - c_0'}{\delta_m} = \frac{D^*}{\delta_m}(c_0' - c_0'') = P^* \Delta c_0 \qquad (3.53)$$

$$P^* \equiv D^*/\delta_m \qquad (3.54)$$

$$\bar{N} = \frac{\bar{D}^*}{\delta_m}(\bar{c}_0' - \bar{c}_0'') = \bar{P}^* \Delta \bar{c}_0 \qquad (3.55)$$

$$\bar{P}^* \equiv \bar{D}^*/\delta_m \qquad (3.56)$$

The permeabilities P^* and $\bar{P}^*$ relate measured flux to measured differences in concentration.

EXAMPLE

For ideal membranes of the type pictured in Figure 3.4(a), (3.53) and (3.54) are sufficient. If there are no specific or electrostatic interactions between the membrane and the solutes, then $\bar{D}_i \simeq D_i$ and $\bar{D}^* \simeq D^*$. Further, $c_i'(0^-) = \bar{c}_i'(0^+)$ and $\bar{c}_i''(\delta^-) = c_i''(\delta^+)$. Hence, for a very highly swollen, neutral membrane, P^* and $\bar{P}^*$ are essentially equivalent, as are (3.53) and (3.55).

In reality, most membranes have solid constituents through which diffusion cannot occur (as pictured in Figure 3.4(b)), and the permeability must be redefined accordingly. We define the ratio of membrane water volume to total, swollen membrane volume (solid plus water) as $\bar{V}_w \equiv V_p/V_m = $ volume fractional water content or porosity (sometimes called ϕ), and the fractional solid content as $\bar{V}_s = 1 - \bar{V}_w$, where V_p is the total membrane pore volume and V_m is the total swollen membrane volume.

The membrane is modeled conceptually as being made up of solid polymeric regions and twisted aqueous pores that meander through the solid membrane matrix. The average diffusion path through the membrane may sometimes be represented as a series of uniform cylindrical pores of length $\delta_p = \nu\delta_m$ and radius r_0 that traverse the membrane of thickness δ_m at an angle, as shown in Figure 3.5. The *tortuosity factor* ν is defined as δ_p/δ_m. An advantage of pore theory is that it allows definite relations to be developed between measured transport properties and structural parameters of the membrane model. In this way, we will be able to account quantitatively for the effects of steric and frictional interactions between the diffusing solute and the pore walls of the membrane matrix. Pore theory will also allow an analysis of the hydrodynamic permeability of the membrane in terms of an equivalent cylindrical pore radius (see Chapters 5 and 6).

For diffusion through the pores, the flux per unit pore cross-sectional area A_p in the direction of the pore axis will be given by

$$\bar{N}_{A_p} = \frac{\bar{D}}{\nu\delta_m}\Delta\bar{c} \tag{3.57}$$

where $\bar{D}$ is the diffusivity of the solute in the membrane pore. For a neutral, noninteracting solute having radius a much less than the effective pore radius r_0, the solute diffusivity will be about the same as its diffusivity in free solution.

We now relate the flux per unit pore cross-sectional area A_p to the flux per unit area of the pore at the membrane surface A', as shown in Figure 3.5. From simple geometry, it is clear that A' is simply the projection of the pore cross-sectional area onto the membrane surface:

$$A' = \frac{A_p}{\cos\theta} = A_p\nu \tag{3.58}$$

Hence, the flux N through the membrane per unit area A' is

$$\bar{N}_{A'} = \bar{N}_{A_p}\frac{A_p}{A'} = \frac{\bar{D}}{\nu^2\delta_m}\Delta\bar{c} \tag{3.59}$$

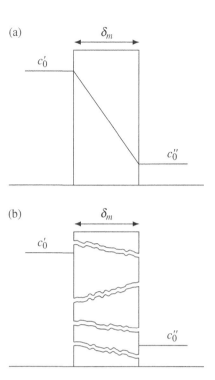

Figure 3.4 Uncharged membranes.

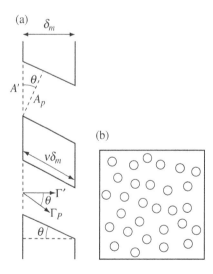

Figure 3.5 Pore model of membrane: (a) side view; (b) front view. A_p is the cross-sectional pore area, A' the projection of A_p on the membrane surface, and ν the tortuosity.

Therefore, the effective diffusivity in the membrane depends on the tortuosity:

$$D_{\text{eff}} = \frac{\bar{D}}{v^2}$$ (3.60)

The projected pore area of a single pore, A', can be related to the measurable volume fraction of water in the membrane, $\bar{v}_w$. In the assumed isoporous model, the total volume of water in the membrane will be the product of the number of pores, N, and the volume of each pore. The volume of each pore is simply

$$V_{\text{pore}} = A_p v \delta_m = A' \delta_m$$ (3.61)

and the total membrane volume

$$V_m = A_m \delta_m$$ (3.62)

where A_m is the total membrane area. Hence, by definition, the membrane volume fraction of water is

$$\bar{V}_w = \frac{NA' \delta_m}{A_m \delta_m} = \frac{NA'}{A_m}$$ (3.63)

where NA'/A_m is simply the fraction of total membrane surface area that is available for diffusion.

3.3 ELECTRODES AND THE MEASUREMENT OF MEMBRANE POTENTIALS

Having discussed the general nature of the steady state diffusion potential, we now consider the additional complications involved in measuring it (Figure 3.6). Two essential issues are the choice of electrodes and the characterization and modeling of the electrode/electrolyte interface. At our disposal are common electrodes such as platinum, glass (pH), silver chloride (Ag/AgCl), and calomel (Hg/Hg$_2$Cl$_2$) (Figures 3.7 and 3.8). A salt bridge is sometimes used to isolate the electrode from direct contact with the electrolyte bath (Figure 3.7). (A salt bridge is simply a "length" of gel or fluid containing a concentrated solution of a salt such as KCl. The electrode is placed in contact with one side of the bridge, and the other side is placed in the bath whose potential is to be measured.) An electrometer connected to the electrodes will record the sum of *three* potential drops: two drops at the electrode/solution interfaces as well as the desired membrane potential difference. The problem is to distinguish the membrane potential from the other two, and this requires the use of electrodes whose interfacial potential drops can be well-characterized.

The most desirable features of an electrode used to *measure* equilibrium and steady state (nonequilibrium) potentials across membranes, are as follows:

(1) Use of the electrode (and the act of measuring) should not cause a significant departure from existing equilibrium or steady state conditions. (It is good practice to use a very high-input-impedance electrometer (10^{13}–10^{15} Ω) to ensure that negligible current is drawn by the measurement circuit through the electrode/electrolyte and membrane/electrolyte interfaces.)

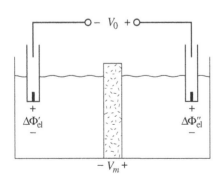

Figure 3.6 Schematic showing the convention for defining the electrode interfacial potentials ($\Delta\Phi'_{\text{el}}$ and $\Delta\Phi''_{\text{el}}$), the transmembrane potential ($V_m = \Phi'' - \Phi'$), and the total measured potential V_0.

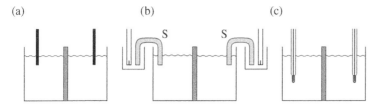

Figure 3.7 System of two electrolyte solutions separated by a membrane. Measurement electrodes are (a) platinum; (b) calomel electrodes with salt bridges (S); (c) silver–silver chloride.

(2) The electrode should ideally be "nonpolarizable"; i.e., the electrode/solution potential drop should be unaffected by current flow through the interface. (This implies zero electrode impedance and is never fully achieved in practice.)

(3) The electrode should be completely reversible (given finite electrode impedance, the *V–I* characteristic should be ohmic in the measurement range of interest, with no hysteretic effects, as in Figure 3.9). This *can* be achieved within certain limits.

(4) The electrode should be immune to the presence of stirring (convection) at the electrode/bulk electrolyte interface.

Bare metal electrodes such as platinum do not have properties (2) and (3), and are generally unsatisfactory as measuring electrodes. This is due to the potential drop across the electrical "double layer," or charge dipole layer, that naturally occurs at a metal/solution interface (see Chapter 6 and Problem 2.3). This drop is relatively uncontrollable and often much larger in magnitude than the membrane potential that is in series with it. (Metal electrodes are often used as "working" electrodes to apply current, where relatively large current densities are required.) Figure 3.9 compares reversible and irreversible electrode properties.

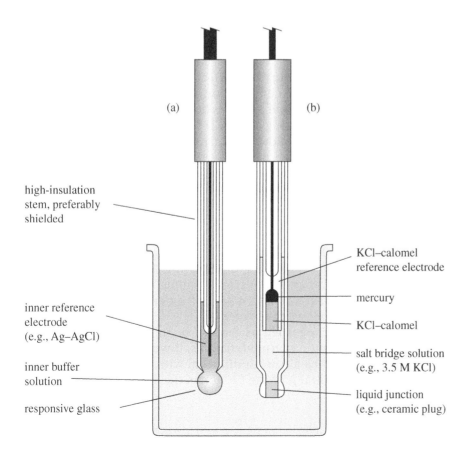

Figure 3.8 Schematic of (a) glass electrode for pH measurement; (b) calomel (mercury/calomel) electrode.

high-insulation stem, preferably shielded

inner reference electrode (e.g., Ag–AgCl)

inner buffer solution

responsive glass

KCl–calomel reference electrode

mercury

KCl–calomel

salt bridge solution (e.g., 3.5 M KCl)

liquid junction (e.g., ceramic plug)

Figure 3.9 Micropolarization tests for electrode reversibility: (a) hysteresis; (b) reversible [3].

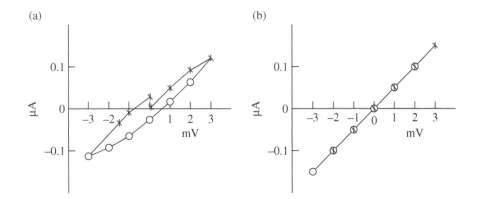

In order to overcome the problems of irreversibility and polarizability, it is desirable to use an electrode whose electrode/solution potential is well defined and controlled by a single chemical reaction. Both Ag/AgCl and calomel electrodes have this property. For example, the potential drop $\Delta\Phi$ at an Ag/AgCl–solution interface has been found empirically to be well characterized over a wide range by the relation [3]

$$\Delta\Phi(\text{Ag/AgCl}) = \Delta\Phi°(\text{Ag/AgCl}) - \frac{RT}{F}\ln a_{\text{Cl}} \qquad (3.64)$$

where $\Delta\Phi°$ is the so-called "standard electrode potential" and is a function of the composition of the solid phases and other terms independent of Cl^- concentration. A similar expression applies to a calomel electrode.

The equilibrium expression (3.64) reflects the fact that the AgCl reacts *reversibly* with the Cl^- ions in the solution; $\Delta\Phi$ is a direct function of Cl^- ion concentration. Obviously, the electrolyte solutions must contain Cl^- ions if we wish to use such electrodes. Calculation of the total potential difference between two Ag/AgCl electrodes (e.g., Figure 3.7(c)) requires knowledge of the Cl^- concentration at each electrode. The $\Delta\Phi°$ terms will theoretically subtract out for electrodes of identical composition; practically, any difference in composition will result in a dc offset potential, which must be accounted for in order to calculate the membrane potential. Finally, a functioning Ag/AgCl electrode also satisfies requirement (1) above—the small amount of current that may be passed during a potential measurement does not cause significant departures from the situation described by (3.64). Chlorided platinum electrodes, known as "platinum black," also have this property. They are characterized by an extremely low interfacial impedance for ac applications.

Membrane potentials can be measured with Ag/AgCl electrodes alone, or with concentrated KCl salt bridges coupled either to calomel or Ag/AgCl electrodes. The use of *bridges* has the effect of reducing the effective electrode/solution junction potential to a negligibly small value for a certain range of electrolyte ionic strength [4]. *For the latter case, the measured total potential difference can be equated approximately to the membrane potential.*

While (3.64) can be regarded as empirical, this relation can also be motivated by equilibrium thermodynamics. If we assume that the measuring electrode reaches a true equilibrium with the solution, then the electrochemical potential $\tilde{\mu}$ of the Cl^- ion is the same in the solution as in the AgCl salt:

$$\tilde{\mu}_{\text{Cl}}^{\text{AgCl}} = \tilde{\mu}_{\text{Cl}^-}^{\text{soln}}, \quad \tilde{\mu} \equiv \mu + z_i F \Phi \qquad (3.65)$$

The electrochemical potential is the sum of the chemical potential μ and the electrical potential $z_i F \Phi$. (The chemical potential is the partial molal Gibbs free energy, i.e., the rate of change of Gibbs free energy with the number of moles of substance i, holding temperature, pressure, and the number of moles of all other species constant. Note that $\tilde{\mu}$, μ, and $F\Phi$ have the units of energy.) Equation (3.65) becomes

$$\mu_{Cl^-}^{AgCl} - F\Phi^{AgCl} = \mu_{Cl^-}^{soln} - F\Phi^{soln} \tag{3.66}$$

$$F(\Phi^{AgCl} - \Phi^{soln}) = (\mu_{Cl^-}^{\circ})^{soln} + RT \ln a_{Cl^-}^{AgCl} - RT \ln a_{Cl^-}^{soln} \tag{3.67}$$

$$\Delta\Phi = \Delta\Phi^{\circ} - \frac{RT}{F} \ln a_{Cl^-}^{soln} \tag{3.68}$$

where, by convention, the activity in the solid phase, $a_{Cl^-}^{AgCl} \equiv 1$.

The assumption of a true thermodynamic equilibrium for an "open circuit" electrode measurement can be closely achieved in practice, using electrometers that have very high-input-impedance differential amplifiers. Even in the presence of finite current flow at the electrode, (3.68) can be valid (see Example 3.3.1), as long as the total Cl^- flux is much less than the individual migration and diffusion terms at the interface. With the total flux approximated as zero, the familiar Nernstian relation (3.68) between the electrode potential and the Cl^- ion activity in solution follows.

The total measured potential V_0 in Figure 3.6 is seen to be the sum of two electrode potentials and the diffusion potential:

$$V_0 = (\Delta\Phi_{el}'' - \Delta\Phi_{el}') + \Phi'' - \Phi' \tag{3.69}$$

$$V_0 = -\frac{RT}{F} \ln\left(\frac{c''}{c'}\right) + \frac{RT}{zF}(t_- - t_+) \ln\left(\frac{c''}{c'}\right) \tag{3.70}$$

where concentrations have replaced the activities in V_0, assuming the limit of dilute solutions. Equation (3.70) is thus the total measured potential difference for the case of steady state diffusion of a $z:z$ electrolyte across a *neutral* membrane, using Ag/AgCl electrodes.

The first term of (3.70) (the electrode potential difference) can be reduced to a negligible quantity in certain cases by the use of saturated KCl salt bridges interposed between the electrode and electrolyte. For example, the concentration cell in Figure 3.6 would become

$$\text{Ag/AgCl} \left| \begin{array}{c} \text{sat. KCl} \\ \text{bridge} \\ {\scriptstyle (1)'} \end{array} \right| \begin{array}{c} c' \\ \text{NaCl} \\ {\scriptstyle (2)'} \end{array} \left\| \text{membrane} \right\| \begin{array}{c} c'' \\ \text{NaCl} \\ {\scriptstyle (2)''} \end{array} \left| \begin{array}{c} \text{sat. KCl} \\ \text{bridge} \\ {\scriptstyle (1)''} \end{array} \right| \text{Ag/AgCl} \tag{3.71}$$

The potential drops at junctions $(1)'$ and $(1)''$ are equal, since the KCl concentrations in both salt bridges are presumably equal. Thus, they will cancel in the expression for the overall potential difference V_0. The drops at $(2)'$ and $(2)''$ will each be negligibly small, provided that c' and c'' are much less than the concentration of saturated KCl (about 4.2 M). This can be seen by deriving the "multi-ionic" diffusion potential at each KCl | NaCl junction in a manner analogous to Section 3.2, based on the general development of Section 3.1. The resulting potential drop is of the form (for the left-hand junction)

$$\Delta\Phi = \frac{RT}{F}\left[\frac{c'(u_{Na^+} - u_{Cl^-}) - 4.2(u_{K^+} - u_{Cl^-})}{c'(u_{Na^+} + u_{Cl^-}) - 4.2(u_{K^+} + u_{Cl^-})}\right] \ln\left[\frac{c'(u_{Na^+} + u_{Cl^-})}{4.2(u_{K^+} + u_{Cl^-})}\right] \tag{3.72}$$

where c' in (3.72) is the molar concentration of NaCl. For $c' \ll 4.2\,\mathrm{M}$, the coefficient of the ln term will be very small. The introduction of salt bridges can simultaneously bring in undesirable effects, such as KCl contamination of the electrolyte baths and unstable readings when there is electrolyte streaming past the bridges owing to stirring.

Some pioneering references that describe fundamental and practical aspects of electrodes, and are still important in modern practice, include [3–7].

Example 3.3.1 Ag/AgCl Electrode Potential—An Alternative Approach An expression for the interfacial potential drop at a reversible chloride electrode (e.g., Ag/AgCl) was found by using a thermodynamic equilibrium approach (see (3.65) and (3.68)). The electrode potential has been found *empirically* to obey the expression (3.68) over a wide range of Cl^- activity (concentration). Equation (3.68) can also be derived from the Cl^- flux equation in a manner analogous to that of Example 3.2.2.

Chloride ions can pass reversibly from the electrolyte phase to the solid AgCl phase and vice versa. In the interfacial region of Figure 3.10,

$$N_{Cl^-} = +u_- c_- \frac{d\Phi}{dx} - D_- \frac{dc_-}{dx} \tag{3.73}$$

But, for the case of a true equilibrium between the chloride ions in solution and those associated in the solid phase,

$$N_{Cl^-} \equiv 0 \tag{3.74}$$

Therefore, (3.73) can be rewritten using the Einstein relation and integrated between the solution and solid phases:

$$\int_{soln}^{AgCl} d\Phi = \frac{RT}{F} \int_{soln}^{AgCl} \frac{dc_-(x)}{c_-(x)} \tag{3.75}$$

$$\Phi^{AgCl} - \Phi^{soln} = \frac{RT}{F} \ln c_{Cl^-}^{AgCl} - \frac{RT}{F} \ln c_{Cl^-}^{soln} + \Delta\Phi^\circ \tag{3.76}$$

In (3.76), the first term on the right-hand side is zero by convention. (The activity of an ion in a solid phase is taken to be 1 by thermodynamic convention. Here, we have been using concentrations rather than activities, the latter being equal to the concentration for dilute solutions.) $\Delta\Phi^\circ$ is a constant defined as the standard electrode potential at unit Cl^- ion activity (concentration). Therefore, (3.76) assumes a form identical to (3.68):

$$\Delta\Phi = \Delta\Phi^\circ - \frac{RT}{F} \ln c_{Cl^-}^{soln} \tag{3.77}$$

As stated in Section 3.3, it is important to realize that (3.73) with $N \simeq 0$ may adequately describe the electrode interface even when current is drawn—a nonequilibrium situation. As long as the net flux associated with finite current is small compared with the individual migration and diffusion terms in (3.73), this small flux will be equal to the difference of two very large terms. Then (3.73) may still be adequately described by $N \simeq 0$; i.e., the interfacial potential and the ion distribution may

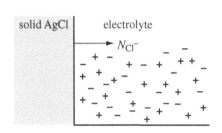

Figure 3.10 AgCl/electrolyte interface.

not be significantly perturbed as long as the current is small enough! To obtain an order-of-magnitude estimate of how large a current can be drawn, we calculate the migration term in (3.73) assuming $c_{Cl^-} \simeq 0.1$ M, and $d\Phi/dx \rightarrow \Delta\Phi/(1/\kappa) \simeq 25$ mV/1 nm:

$N_{migration}$

$$\simeq \frac{(0.1\,\mathrm{mol\,L^{-1}})(10^{-3}\,\mathrm{L\,cm^{-3}})(8\times10^{-4}\mathrm{cm^2\,V^{-1}\,s^{-1}})(2.5\times10^{-2}\,\mathrm{V})}{10^{-7}\,\mathrm{cm}}$$

$$= 20\times10^{-3}\,\mathrm{mol\,cm^{-2}\,s^{-1}}$$

$$J = NF = 2\times10^3 \mathrm{A\,cm^{-2}}$$

For an electrode area of $0.1\,\mathrm{mm}^2 = 10^{-3}\,\mathrm{cm}^2$,

$$J(\mathrm{area}) = I = 2\,\mathrm{A}$$

Therefore, if the current drawn is $\ll 2$ A, the use of (3.73) and (3.74) is not a bad approximation! This is easily achieved using high-input-impedance electrometers.

3.4 DONNAN EQUILIBRIUM AND THE DONNAN POTENTIAL: CHARGED MEMBRANES AND TISSUES IN EQUILIBRIUM

3.4.1 Boltzmann Distribution for Ions in Equilibrium

In bulk solutions and other neutral media, concentration gradients of electrolyte ions will give rise to diffusion fluxes tending to wipe out the gradients. However, media containing fixed charges (e.g., membranes and other biological tissues having fixed-charge groups) can lead to situations in which concentration gradients of *mobile* ionic species are maintained in a true equilibrium. Such situations arise as a result of the need to preserve bulk (macroscopic) electroneutrality, which counterbalances the tendency of the mobile ions to diffuse down their concentration gradients. This is yet another example of a balance between ionic conduction and diffusion currents. An analogous inter-action occurs in the case of a nonuniformly doped single crystal of semiconductor material in thermal equilibrium with its surroundings: the resulting spatially nonuniform distribution of holes and electrons leads to an equilibrium "built-in" $\boldsymbol{E}$ field (and associated contact potential), which we now see in the context of charged membranes and tissues.

Let us examine the distribution of all mobile species, $c_i(\boldsymbol{r})$, and that of the potential, $\Phi(\boldsymbol{r})$, that would exist when in *thermal equilibrium* throughout the electrolyte-swollen tissue. (In Section 3.6, we will discuss nonequilibrium phenomena due to bulk diffusion, chemical reactions, and applied electric fields.) We first imagine assembling the system by placing a tissue or membrane specimen in an aqueous electrolyte and waiting for the establishment of equilibrium, including dissociation reactions that give rise to fixed-charge groups. We should not be surprised to find that $c_i(\boldsymbol{r})$ and $\Phi(\boldsymbol{r})$ are related by the Boltzmann distribution, whether or not the fixed-charge groups are uniformly or nonuniformly distributed. This can be seen by recalling that, in equilibrium, the flux of each individual mobile species is identically zero both inside and outside the tissue

$$\bar{N}_i = -\bar{D}_i \nabla \bar{c}_i - \bar{c}_i \bar{u}_i \frac{z_i}{|z_i|} \nabla \Phi \equiv 0 \qquad (3.78)$$

where $\bar{D}_i$ and $\bar{u}_i$ are the diffusivity and mobility inside the tissue (no overbar for the analogous equation outside). This statement follows from the *Law of Detailed Balance*, which states that *in equilibrium*, every physical process proceeds on the average at exactly the same rate as its own inverse. In the present case, the physical process under consideration is that of net solute flux. Equation (3.78) states that the migration and diffusion fluxes identically balance for *each individual species* in equilibrium. (Note that, thermodynamically, (3.78) is equivalent to the statement that the electrochemical potential $\tilde{\mu}_i$ of the ith species does not vary in space: $\nabla \tilde{\mu}_i = \nabla[\mu_i^\circ + RT \ln \bar{c}_i(\bar{r}) + z_i F \Phi(\bar{r})] = 0$.)

The desired relation between $\bar{c}_i(x)$ and $\Phi(x)$ follows directly from (3.78). (Note that in the *steady state*, $Dc_i/Dt = 0$ in the continuity equations (3.3) and (3.4). Therefore, $\nabla \cdot \mathbf{N}_i = 0$, or $\mathbf{N}_i = $ constant. The additional constraint imposed by the equilibrium condition is that the constant is identically zero for each species.) Integration of (3.78) gives

$$\Phi(\mathbf{r}) = -\frac{\bar{D}_i}{\bar{u}_i \, (z_i/|z_i|)} \ln \bar{c}_i(\mathbf{r}) + A \qquad (3.79)$$

If we define the reference potential to be located in the bulk electrolyte phase outside the hydrated tissue, i.e., where the bulk electrolyte concentration is c_{i0}, then

$$\Phi(\mathbf{r}) = -\frac{\bar{D}_i}{\bar{u}_i \, (z_i/|z_i|)} \ln\left[\frac{\bar{c}_i(\mathbf{r})}{c_{i0}}\right] \qquad (3.80)$$

Therefore, if we know the concentration $\bar{c}_i(\mathbf{r})$ at a point $\mathbf{r}$ inside the tissue, (3.80) can be used to relate the potential $\Phi(\mathbf{r})$ at point $\mathbf{r}$ to the reference potential in the outside bulk electrolyte phase. Equivalently, the concentration of the ith species at any point inside the charged tissue is related to the potential at that point:

$$\bar{c}_i(\mathbf{r}) = c_{i0} \exp\left[-\frac{\bar{u}_i(z_i/|z_i|)}{\bar{D}_i} \Phi(\mathbf{r})\right] \qquad (3.81)$$

If we incorporate the Einstein relation (2.38), which applies in the limit of ideal solutions, then (3.81) shows that ions inside the tissue will be distributed according to Boltzmann statistics:

$$\begin{aligned} \bar{c}_+(\mathbf{r}) &= c_{+0} e^{-|z_+|F\Phi(\mathbf{r})/RT} \\ \bar{c}_-(\mathbf{r}) &= c_{-0} e^{+|z_-|F\Phi(\mathbf{r})/RT} \end{aligned} \qquad (3.82)$$

Of course, we could have postulated (3.82) from the outset, knowing that the *Boltzmann model describes the distribution for any noninteracting particles that are in thermal equilibrium* at temperature T. Had we used this logic, (3.81) and (3.82) would provide a "derivation" of the Einstein relation (the approach used by Einstein).

For the case in which the charged tissue is in equilibrium with a single binary electrolyte, we know that $c_{+0} = \bar{c}_{-0} \equiv c_0$ in the solution outside the tissue. It is not surprising to observe, therefore, that

$$c_+ c_- = c_0^2 \qquad (3.83)$$

outside the tissue in the bulk fluid phase—a condition we could have argued simply from electroneutrality. However, the equilibrium distribution represented by (3.82) shows that

$$\bar{c}_+(\mathbf{r})\bar{c}_-(\mathbf{r}) = c_0^2 \qquad (3.84)$$

inside the tissue, at each point in space, whether the internal fixed charge density is uniform or nonuniform.

Note that the form of (3.83) and (3.84) is strikingly similar to that found in expressions for the dissociation equilibria of certain molecules. For example, water is a "weak electrolyte" whose dissociation reaction can be written as

$$H_2O \overset{k}{\underset{k'}{\leftrightarrow}} H^+ + OH^- \tag{3.85}$$

$$\frac{d[H_2O]}{dt} = k'[H^+][OH^-] - k[H_2O] \tag{3.86}$$

where (3.86) is the first-order, reversible kinetic equation for the reaction (3.85). Because the dissociation of water occurs so rapidly compared with most other physical time constants associated with diffusion and reaction processes of physiological interest, we usually take $d/dt \to 0$ in (3.86) and approximate (3.86) by the "instantaneous" equilibrium expression

$$[H^+][OH^-] = \frac{k}{k'}[H_2O] \simeq \text{const} \equiv k_w \tag{3.87}$$

Thus, the "ion product" in (3.87) comes from a *reaction-kinetics argument* or a *reaction equilibrium* statement, while that of (3.83) and (3.84) arises from an equilibrium due to a *balance of electrical and thermal forces* (i.e., the Boltzmann equilibrium). Equations (3.83) and (3.84) apply to the distribution of all mobile *ions*, whether these ions derive from "weakly" or "strongly" dissociated salts. Since H^+ and OH^- ions must also obey (3.83) and (3.84) in equilibrium, we note that (3.87) would be satisfied in a self-consistent manner. In general, (3.83) and (3.84) must be distinguished from relations such as (3.87) (see Example 3.4.1).

Equations (3.83) and (3.84) are especially important and applicable in the interfacial regions between tissue or membrane and bulk solution. We expect that tissue charge will give rise to a contiguous region of net space charge on the solution side of the interface, as in Problem 2.3; i.e., $c_+ \neq c_-$ in the space-charge region. However, the existence of thermal equilibrium requires that $c_+(\boldsymbol{r})$ and $c_-(\boldsymbol{r})$ in this electrolyte space-charge region be related by (3.83). Thus, the Boltzmann distribution (3.82) will be of central importance in the study of membrane, electrode, double layer and other interfacial phenomena.

Although (3.81) and (3.82) are strictly valid only in thermal equilibrium, they may be approximately valid *even in the presence of nonequilibrium phenomena* such as net flux due to finite current or bulk diffusion. This will be true for the many important situations in physiology and electrochemistry in which $N_i \neq 0$, but nevertheless the *net* flux N_i is much less than the individual migration and diffusion components of the total flux associated with an ensemble of species (*i*):

$$|\boldsymbol{N}_i| \ll |D_i \nabla c_i| \quad \text{and} \quad |N_i| \ll |c_i u_i \nabla \Phi| \tag{3.88}$$

This is often found to be the case at membrane/electrolyte and electrode/electrolyte interfaces even in the presence of current, i.e., net transport across the interface (see Example 3.3.1). For situations in which the inequalities (3.88) are valid, the true equilibrium condition (3.78) may be replaced by a *quasiequilibrium* condition

$$-D_i \nabla c_i - c_i u_i \frac{z_i}{|z_i|} \nabla \Phi \simeq 0 \tag{3.89}$$

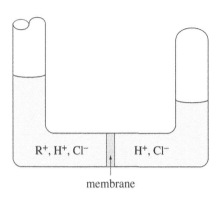

Figure 3.11 Membrane permeable to H^+ and Cl^- but not R^+.

The use of (3.89) will often enable the fairly accurate (though not exact) calculation of $\Phi(\mathbf{r})$ and $c_i(\mathbf{r})$ at interfacial regions and inside the bulk tissue phase, even in the presence of small but finite current or diffusion. Some important case studies involving this quasiequilibrium assumption are taken up in Section 3.5.

3.4.2 The Donnan Equilibrium

Figure 3.11 depicts a membrane known to be permeable to H^+ and Cl^- ions, but impermeable to the positively charged macromolecular species R^+. Such a situation could arise when a "chloride salt" of a protein, RCl, exists on one side of the membrane. Dissociation of RCl produces two species; while the small, hydrated chloride ion may easily pass through the membrane, the R^+ molecule may be too large to permeate the membrane pores. Figure 3.12 shows the interfacial region of a *uniformly* charged tissue or membrane placed in an electrolyte bath. The membrane fixed charge groups might arise from dissociation of macromolecules comprising the membrane, from the specific adsorption of impurities or charged species in the electrolyte bath, or from other processes.

The systems pictured in Figures 3.11 and 3.12 will exhibit a *true* equilibrium with respect to the distribution of permeable (mobile) species across the membrane (phase boundary) of Figure 3.11 and in the membrane of Figure 3.12. This equilibrium results from the opposing tendencies of diffusion (due to concentration gradients) and migration (due to a local violation of electroneutrality). In equilibrium, electroneutrality must be satisfied in the bulk of each phase. However, this will lead to concentration gradients of mobile species. Diffusion is balanced by migration, as manifest in the equilibrium *Donnan potential differences* $\Delta\Phi_D$ that exists across the phase boundary. In 1911, FG Donnan [8] first proposed such a balance between migration and diffusion with respect to a system of the kind shown in Figure 3.11. The configuration of Figure 3.12 is essentially equivalent to that of Figure 3.11 with membrane fixed charge taking the part of the mobile but impermeable species, R^+. We first derive the relation between $\Delta\Phi_D$ and the ratio of bulk concentrations of each mobile species across the phase boundary.

For the case of a one-dimensional model with a mono–monovalent electrolyte providing the only mobile species (e.g., H^+, Cl^-), (3.82) becomes

$$c_+(x) = c_0 e^{-F\Phi(x)/RT}$$
$$c_-(x) = c_0 e^{+F\Phi(x)/RT}$$

(3.90)

Figure 3.12 Fixed charge membrane/electrolyte interface. The variation of potential $\Phi(x)$ across the interface is superimposed on the diagram. $\bar{c}$ refers to the concentration of electrolyte inside the fixed charge membrane and c_0 to the concentration of the electrolyte external to the membrane.

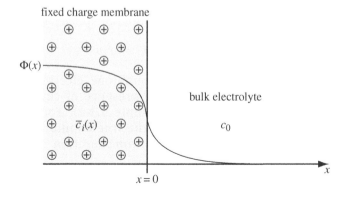

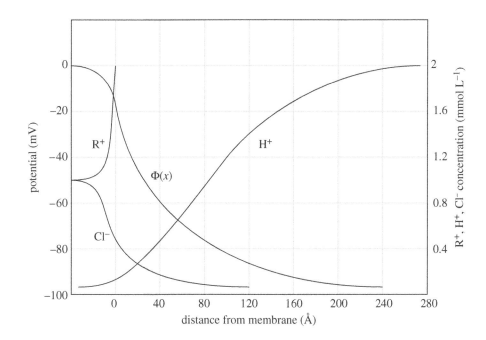

Figure 3.13 Potential and concentrations versus distance.

where $c_{+0} = c_{-0} = c_0$ is the bulk electrolyte concentration in the electrolyte phase of Figure 3.12 or the right-hand phase of Figure 3.11. Note that the reference potential is that in the bulk right-hand phase for both cases. Once again, (3.90) represents a true equilibrium for each mobile (permeable) species: $N_i \equiv 0$. The bulk mobile ion concentration in the left-hand phase, $\bar{c}_i$, where $\Phi = \Delta\Phi_D$, is therefore

$$\bar{c}_+ = c_0 e^{-F\Delta\Phi_D/RT}$$
$$\bar{c}_- = c_0 e^{F\Delta\Phi_D/RT} \qquad (3.91)$$

From (3.91),

$$\Delta\Phi_D \equiv -\frac{RT}{F}\ln\left(\frac{\bar{c}_+}{c_0}\right) = -\frac{RT}{F}\ln\left(\frac{c_0}{\bar{c}_-}\right) \qquad (3.92)$$

The detailed spatial distributions of $\Phi(x)$, $c_i(x)$, and $\bar{c}_i(x)$ can be found by incorporating (3.90) into Poisson's equation, written in both regions $x > 0$ and $x < 0$, and subject to the boundary conditions

(a) $\Phi \to 0$ for $x \to +\infty$

(b) $E = -\dfrac{\partial\Phi}{\partial x} \to 0$ for $x \to -\infty$

(c) Φ continuous at $x = 0$

(d) $\dfrac{\partial\Phi}{\partial x}$ continuous at $x = 0$

Conditions (a) and (b) are equivalent to the assumption of electroneutrality in the bulk of each phase, with the reference or zero of potential being ascribed to the right-hand phase.

The results of a numerical calculation of $\Phi(x)$, $c_i(x)$, and $\bar{c}_i(x)$ for the conditions $|\Delta\Phi_D| = 100\,\text{mV}$, $\bar{c}_- = 0.1\,\text{M}$, and $c_0 = 2.136 \times 10^{-3}\,\text{M}$ are shown in Figure 3.13 [9, 10]. The potential $\Phi(x)$, which is a solution of the Poisson–Boltzmann equation in both phases, takes the form of a "double-double layer," which extends into each phase by a distance that can be estimated from the Debye length $1/\kappa$ corresponding to the bulk concentration in each phase.

The examples of Figures 3.11 and 3.12 are concerned with a single mono–monovalent electrolyte in equilibrium within or across a membrane. Of course, there are many important examples in biology involving equilibria with multi-ionic solutions containing multivalent ions. It should be apparent that in thermal equilibrium, the same Boltzmann equations (3.82) describe the distributions of all such ions. Therefore, the Donnan potential $\Delta\Phi_D$ across a membrane interface or phase boundary is related to the bulk concentration of the ith species on both sides of the phase boundary, $\bar{c}_i$ and c_i, respectively, by using (3.80):

$$\Delta\Phi_D = -\frac{1}{z_i}\frac{RT}{F}\ln\left(\frac{\bar{c}_i}{c_i}\right) \qquad (3.93)$$

Thus, for the case of multi-ionic solutions, each mobile ion in equilibrium across the phase boundary satisfies the relation (neglecting activity coefficients for a dilute solution)

$$\boxed{\left(\frac{\bar{c}_+}{c_+}\right)^{1/z_+} = \left(\frac{\bar{c}_-}{c_-}\right)^{1/z_-} = \text{const} = e^{-F\Delta\Phi_D/RT}} \qquad (3.94)$$

The ratios of internal to external concentration for *all* permeable species in equilibrium are related by (3.94) to $\Delta\Phi_D$.

In Problem 3.9, the analysis of the Donnan equilibrium will be extended to show the relation between the membrane fixed-charge density $\bar{\rho}_m$ and the concentrations $\bar{c}_i$ and c_{i0}, as well as the relation describing the osmotic pressure induced across the membrane.

Example 3.4.1 Internal pH Most biological tissues have acidic and basic side groups attached to their constituent macromolecular backbones (Figure 3.14), as described in Section 1.4. These groups will dissociate in a manner consistent with the pH and ionic strength of the electrolyte bath with which they are in equilibrium. Such dissociation will leave the macromolecule with a *net fixed charge*, with the dissociated "counter-ions" nearby in the aqueous phase; the counter-ions form an "ionic atmosphere" about the fixed-charge groups. This occurs, for example, with the dissociable side groups of amino acids—the building blocks of globular and fibrous proteins.

One important physicochemical method used in an attempt to measure the number of dissociable groups present in a given tissue or membrane specimen is that of a pH titration. The titration curves for amino acids (e.g., Figure 1.13) are examples, as previously mentioned, where the number of dissociable groups is related to the pH of the soaking electrolyte, i.e., the *external bath pH*. However, it is often important to know the *internal pH*, $\bar{c}_{H^+}(x)$ in the specimen, since $\bar{c}_{H^+}(x)$ is directly involved in reaction equilibria with the dissociable groups. The Donnan equilibrium affords one method of finding $\bar{c}_{H^+}$ by combining an experimental measurement with the theory of Section 3.4. Figure 3.14 depicts a membrane having many fixed carboxyl (–COOH) groups that dissociate to an extent determined by internal H^+ concentration. However, it is usually only the external pH that is available for measurement. If the internal concentration of Cl^- or Na^+ were known, then Donnan theory

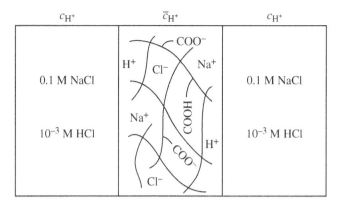

Figure 3.14 Charged membrane in equilibrium with electrolyte bath.

could be used to calculate the internal pH. From (3.94),

$$\frac{\bar{c}_{H^+}}{c_{H^+}} = \frac{\bar{c}_{Na^+}}{c_{Na^+}} = \frac{c_{Cl^-}}{\bar{c}_{Cl^-}}$$ (3.95)

and therefore

$$c_{H^+} c_{Cl^-} = \bar{c}_{H^+} \bar{c}_{Cl^-}$$ (3.96)

Thus, if the internal chloride ion concentration were measurable, then

$$\overline{pH} = pH + \log\left(\frac{\bar{c}_{Cl^-}}{c_{Cl}}\right)$$ (3.97)

Freeman and Maroudas [11] used radioactively labeled $^{22}Na^{36}Cl$ to calculate the amount of dissociated carboxyl groups in specimens of human articular cartilage by using (3.94). They compared their results with an ideal titration curve, as shown in Figure 3.15.

It is worthwhile to point out that the theory of pH titration of polyelectrolyte molecules ultimately attempts to relate the number of dissociated groups to the *external bath pH*, since it is only the external pH that is directly measurable. Thus, for a polyelectrolyte solution

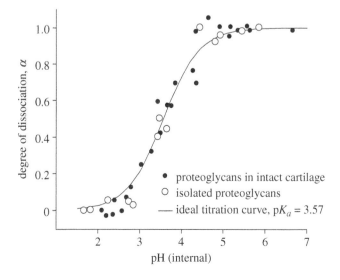

Figure 3.15 Degree of dissociation of proteoglycan carboxyl groups versus pH.

with a fraction of dissociated groups α_n and a fraction of undissociated groups $1 - \alpha_n$ of the nth family, titration theory results in the expression [12, 13]

$$\overline{\text{pH}} = \text{p}K_n^\circ + \log\left(\frac{\alpha_n}{1 - \alpha_n}\right) + f(\Phi) \qquad (3.98)$$

where K_n° is the intrinsic dissociation equilibrium constant for groups of the nth family when there is no electrostatic interaction between polyelectrolyte molecules. $\overline{\text{pH}}$ is the *internal*, "smoothed-out" or macroscopic pH, which can be related to the external pH by means of one additional relation like that of (3.97).

The last term of (3.98), $f(\Phi)$, accounts for the possible importance of nano-scale "nearest neighbor" electrostatic interactions. When there are many charged groups situated closely along an individual polyelectrolyte molecule, the ionization of one charge group will be affected by the ionization state (electrostatic free energy) of its nearest neighbors. In other words, the addition or removal of a proton requires more or less energy owing to the electrostatic potential energy associated with neighboring dissociated groups. The form of the term $f(\Phi)$ can be derived in a manner analogous to that of the Debye–Hückel correction for ion activities [12].

In summary, there are distinct consequences of fixed charge $\bar{\rho}_m$ on the titration behavior of polyelectrolyte materials. First, the macroscopic Donnan effect requires that $\overline{\text{pH}} \neq$ external pH. Hence, the dissociation of charge groups in a membrane matrix will be found to occur at a measured *external pH* that is different than would be measured for a *dilute solution* of the same polyelectrolyte macromolecules not organized as a closely packed matrix. An additional molecular-scale potential barrier ($f(\Phi)$) to the dissociation process is produced by the close aggregation of charged groups; i.e., there is non-negligible interaction between neighboring polyelectrolyte charge groups.

Example 3.4.2 Donnan Equilibria and Nonuniformly Charged Materials; Indirect Measurement of $\bar{\rho}_m(r)$ using Donnan Theory and Experiment

(a) Donnan Equilibrium in the Bulk In general, the distribution of ions within a nonuniformly charged tissue can be found using the combination of electroneutrality and the Boltzmann distribution. For multi-ionic solutions, Donnan equilibrium is described by the relations

$$\boxed{\bar{\rho}_m(x) + \sum_i z_{i+} F \bar{c}_{i+}(x) + \sum_j z_{j-} F \bar{c}_{j-}(x) = 0} \qquad (3.99)$$

$$\boxed{\left(\frac{\bar{c}_+(x)}{c_+}\right)^{1/z_{i+}} = \left(\frac{\bar{c}_-(x)}{c_-}\right)^{1/z_{i-}} = e^{-F\Delta\Phi_D/RT} = \text{const for all } i}$$

$$(3.100)$$

If $\bar{\rho}_m(x)$ is known *a priori*, this information can be used directly in the electroneutrality equation. If, instead, $\bar{\rho}_m(x)$ depends on the internal pH and ionic strength at the site of the fixed-charge groups (as is most often the case with biological tissues), then $\bar{\rho}_m(x)$ and $\bar{c}_i(x)$ must be solved simultaneously

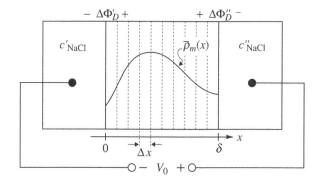

Figure 3.16 Nonuniformly charged tissue.

and self-consistently, for example via a titration iostherm in terms of the pK's of the charge groups.

(b) Interfacial Equilibria For a tissue with fixed charge density that varies with position, the interfacial Donnan potential drops also vary with position. For the one-dimensional model of Figure 3.16, with $\bar{\rho}_m = \bar{\rho}_m(x)$, (3.93) can be used at $x = 0$ and $x = \delta$ to obtain

$$\Delta\Phi'_D = +\frac{RT}{F}\ln\left[\frac{\bar{c}'_{Cl}(0^+)}{c'_{Cl}(0^-)}\right] = -\frac{RT}{F}\ln\left[\frac{\bar{c}'_{Na}(0^+)}{c'_{Na}(0^-)}\right] \qquad (3.101)$$

where (0^+) and (0^-) are several Debye lengths away from $x = 0$; i.e., $\bar{c}(0^+)$ and $c(0^-)$ are macroscopic, averaged concentrations. Similarly,

$$\Delta\Phi''_D = +\frac{RT}{F}\ln\left[\frac{\bar{c}''_{Cl}(\delta^-)}{c''_{Cl}(\delta^+)}\right] = -\frac{RT}{F}\ln\left[\frac{\bar{c}''_{Na}(\delta^-)}{c''_{Na}(\delta^+)}\right] \qquad (3.102)$$

For the case when $|\bar{\rho}_m(x)| \gg c'_0,\ c''_0$ for all x, the counter-ion concentration inside the membrane is approximately $\bar{\rho}_m(x)$ for all x. If we consider the case of a tissue in equilibrium with a single binary electrolyte, with $\bar{\rho}_m(x) > 0$ (positive fixed charge), and further that $c' = c'' = c_0$, which must be true for a true equilibrium to exist (no bulk diffusion across the tissue), then

$$\bar{c}_{Cl}(x) \simeq \bar{\rho}_m(x)/F \qquad (3.103)$$

$$\bar{c}_{Na}(x) \simeq \frac{c_0^2}{|\rho_m(x)/F|} \qquad (3.104)$$

Note that this is true as long as $\bar{\rho}_m(x)$ is relatively smooth compared with a Debye length, so that $\rho_e(total)$ is still very small.

However, $\bar{c}'_i(0^+) \neq \bar{c}''_i(\delta^-)$, since $\bar{\rho}_m(0^+) \neq \bar{\rho}_m(\delta^-)$ in general. Therefore,

$$\bar{c}'_{Cl}(0^+) \simeq |\bar{\rho}_m(0^+)/F|$$

$$\bar{c}'_{Na}(0^+) \simeq \frac{c_0^2}{|\bar{\rho}_m(0^+)/F|} \qquad (3.105)$$

$$\bar{c}''_{Cl}(\delta^-) \simeq \left| \bar{\rho}_m(\delta^-)/F \right|$$

$$\bar{c}''_{Na}(\delta^-) \simeq \frac{c_0^2}{\left| \bar{\rho}_m(\delta^-)/F \right|} \tag{3.106}$$

(c) Measured Potential V_0 The total measured potential in Figure 3.16 is the sum of both electrode potentials, the equilibrium potential drop across the tissue $\Delta\Phi_m \equiv \Phi''(\delta^-) - \Phi'(0^+)$, and the two Donnan interfacial potentials:

$$V_0 = \Delta\Phi_{el} + \Delta\Phi'_D - \Delta\Phi''_D + \Delta\Phi_m \tag{3.107}$$

We again consider the case $c''_{NaCl} = c'_{NaCl}$. For reversible chloride electrodes such as Ag/AgCl or "platinum black" electrodes, $\Delta\Phi_{el}$ as well as all the other terms in (3.107) can be written in terms of Cl^- ion concentrations (activities):

$$V_0 = -\frac{RT}{F} \ln\left(\frac{c''_{Cl}}{c'_{Cl}}\right) + \frac{RT}{F} \ln\left(\frac{\bar{c}'_{Cl}(0^+)}{c'_{Cl}}\right) - \frac{RT}{F} \ln\left(\frac{\bar{c}''_{Cl}(\delta^-)}{c''_{Cl}}\right) + \Delta\Phi_m \tag{3.108}$$

From (3.80),

$$\Delta\Phi_m = +\frac{RT}{F} \ln\left[\frac{\bar{c}''_{Cl}(\delta^-)}{\bar{c}'_{Cl}(0^+)}\right] \tag{3.109}$$

Incorporation of (3.109) into (3.108) gives

$$V_0 \equiv 0 \tag{3.110}$$

The result (3.110) is physically reasonable, because we know that the Cl^- ion is in thermal equilibrium in all phases throughout the system. An argument based on the laws of thermodynamics would lead to the conclusion that we cannot measure the potential drop across a system in complete equilibrium. (By analogy, one cannot make an equilibrium measurement of the contact potential of a p–n junction; any attempt to measure such a contact potential must disturb the equilibrium in order to obtain a nonzero reading.)

The reader should be convinced that the use of other electrodes (e.g., calomel) would also result in a $V_0 = 0$ measurement as long as $c' = c''$. Of course, if $c' \neq c''$, then bulk diffusion occurs across the tissue. This is the subject of Section 3.5. We should expect the presence of a finite diffusion potential. In general, this potential will not be the same as that described in Section 3.2 because of finite tissue fixed charge, $\bar{\rho}_m \neq 0$.

(d) Measurement of $\bar{\rho}_m(x)$ Using Donnan Theory Values of $\bar{\rho}_m(x)$ versus distance from the surface of articular cartilage can be obtained by use of the Donnan theory [14]. Slices of the tissue having thickness Δx (see Figure 3.16) were equilibrated with radioactive 0.015 M ^{22}NaCl. For the case $|\bar{\rho}_m|/F \gg 0.015$ M, the Donnan theory predicts that co-ions are essentially excluded from the tissue matrix. The counter-ion (Na^+) concentration is approximately equal to that of the fixed

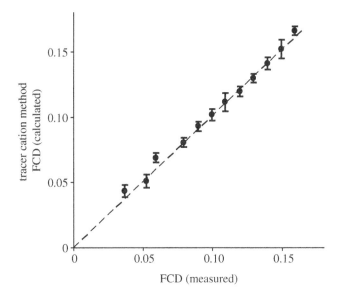

Figure 3.17 Correlation between fixed charge density (FCD) measured by the tracer cation method and that from uronic acid and hexosamine analysis [14].

charge density (averaged over Δx):

$$F\langle \bar{c}_{Na}\rangle_{\Delta x} \simeq \langle \bar{\rho}_m\rangle_{\Delta x} \qquad (3.111)$$

Since this is really a measurement of $^{22}Na^+$ concentration and therefore only an indirect measurement of $\bar{\rho}_m$, a check on the results was provided by an independent chemical analysis of each slice. This analysis essentially provided a counting of each of the sections of the macromolecule that was thought to contain dissociated groups (in cartilage, the glycosaminoglycans contain COO^- and SO_3^-). The results of the chemical analysis were shown to predict a fixed charge density very close to that predicted by the Donnan theory, as seen in Figure 3.17.

3.5   FIXED CHARGE MEMBRANE MODELS: STEADY STATE DIFFUSION POTENTIALS ACROSS CHARGED MEMBRANES

When a membrane has sufficient charge to block some but not all of the flow of an electrolyte across it, the expressions derived in Sections 3.2 and 3.3 for the *steady state* diffusion potentials no longer apply (i.e., neither the solution for a neutral membrane nor that for an ideal semipermeable membrane).

The expression for V_0 is now more complicated than that of (3.70) of Section 3.3. In addition to the terms already there, other terms must be added to account for a redistribution of electrolyte concentration *within* the membrane in order to preserve electroneutrality, as required by the membrane's fixed charge. One approach to this problem will be treated below, in a manner that combines the results of several previous examples. It embodies the basic concepts of the *Teorell–Meyer–Sievers* theory of fixed charge membranes [15, 16].

Although the TMS theory is applicable to many charged membranes systems, there are several simplifying assumptions:

Figure 3.18 Potentials measured across a fixed charge membrane. $\bar{c}'$ and $\bar{c}''$ refer to concentration of electrolyte internal to the membrane at left- and right-hand interfaces. c' and c'' refer to external electrolyte concentrations.

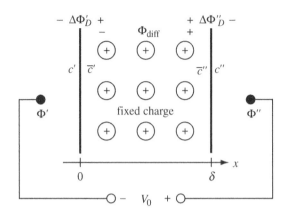

(1) The fixed charge is distributed *uniformly* within the membrane: $\bar{\rho}_m$ is independent of x.

(2) Activity coefficients are assumed constant across the membrane.

(3) Changes in potential due to pressure (e.g., differential swelling of the membrane) are assumed negligible.

(4) No allowance is made for water transport.

(5) Ratios of positive ion mobilities to negative ion mobilities are assumed constant.

(6) Donnan equilibrium still applies at each membrane/electrolyte interface (see below).

With reference to the schematic of Figure 3.18, the approach used in the TMS model parallels that of Section 3.2 in that the fundamental system of equations is the same (with modifications due to fixed charge). The flux and continuity relations are analogous (neglecting convection inside the membrane and assuming $\partial/\partial t \to 0$):

$$\bar{\boldsymbol{N}}_i = -\bar{D}_i \nabla \bar{c}_i - \frac{z_i}{|z_i|} \bar{u}_i \bar{c}_i \nabla \Phi \tag{3.112}$$

$$\bar{\boldsymbol{J}}_i = z_i F \bar{\boldsymbol{N}}_i, \quad \bar{\boldsymbol{J}}_{\text{total}} = \sum \bar{\boldsymbol{J}}_i \tag{3.113}$$

$$\nabla \cdot \bar{\boldsymbol{J}}_i = 0 = \nabla \cdot \bar{\boldsymbol{J}}_{\text{total}} \tag{3.114}$$

where (3.114) is analogous to (3.32). Bulk electroneutrality inside the membrane and outside in the electrolyte phase can be written respectively as

$$\sum z_i F \bar{c}_i + \bar{\rho}_m = 0 \tag{3.115}$$

$$\sum z_i F c_i = 0 \tag{3.116}$$

The electroneutrality conditions (3.115) and (3.116) can be used to decouple Gauss' law from the self-consistent, simultaneous solution of (3.112)–(3.114), as in Section 3.1. (It is important to realize that the "elimination" of Gauss' law in this manner makes it easier to solve mathematically for $\bar{c}_i(x)$ and V_0; the exact solution with Gauss' law is nonlinear in general.) Incorporation of (3.115) and (3.116) is often found to yield accurate theoretical predictions of V_0 compared with experimental measurements. However, (3.115) is certainly not very useful in predicting the exact field inside the membrane (see Problem 3.5). The open circuit measurement of V_0 is defined such that no net current flows across the membrane:

$$\boldsymbol{J}_{\text{total}} = 0 \tag{3.117}$$

As previously mentioned, the TMS theory pictures the membrane's fixed charge as leading to a redistribution of electrolyte ions in order to preserve electroneutrality. In general, there will be an increase in counter-ion and a decrease in co-ion concentration as compared with the bulk electrolyte. As a result, there will be concentration gradients at both membrane/electrolyte interfaces. These gradients, in turn, give rise to potential gradients that counterbalance the interfacial diffusion fluxes.

If the electrolyte concentrations on both sides of the membrane are equal ($c_i' = c_i''$), then there are no net ion fluxes across the membrane, and a true *Donnan equilibrium* will exist within the membrane and at each interface (see Section 3.4.2 and Example 3.4.6). If $c_i'' \neq c_i'$, then net diffusional fluxes will occur; however, if the magnitude of the net flux $\mathbf{N}_i$ is small enough to satisfy (3.88) (which is most often the case), then *quasiequilibrium* may still be assumed for the two interfaces. *This is the assumption of the TMS theory* as depicted in Figure 3.18. The total membrane potential $\Phi'' - \Phi'$ can now be represented as the sum of a diffusion potential $\Delta\Phi_{\text{diff}}$ and two Donnan potentials:

$$\Phi'' - \Phi' = \Delta\Phi_{D1} - \Delta\Phi_{D2} + \Delta\Phi_{\text{diff}} \tag{3.118}$$

The Diffusion Potential Term $\Delta\Phi_{\text{diff}}$

The diffusion potential can be calculated by using an approach similar to that of Section 3.2, but using internal concentrations $\bar{c}_i$ rather than external ones. The interfacial $\bar{c}_i$ are then calculated by means of the Donnan theory. (The effect of membrane fixed charge is assumed only to redistribute internal concentrations; diffusive processes are not interfered with in any way (e.g., by ion binding). In general, any mobile-ion–fixed-charge interaction could be accounted for by an empirical value for $\bar{D}_i$.) There are several ways of accounting for $\Delta\Phi_{\text{diff}}$. The simplest approach incorporates the Henderson equation for a liquid junction potential, written for the general case of multi-ion solutions as

$$\Delta\Phi_{\text{diff}} = \frac{-(RT/F)\sum \bar{u}_i (z_i/|z_i|)\left(\bar{c}_i'' - \bar{c}_i'\right)}{\sum \bar{u}_i |z_i| (\bar{c}_i'' - \bar{c}_i')} \ln\left(\frac{\sum \bar{u}_i |z_i| \bar{c}_i''}{\sum \bar{u}_i |z_i| \bar{c}_i'}\right) \tag{3.119}$$

which is written using mobilities and concentrations *inside* the membrane.

The Henderson equation (3.119) is formulated under the assumption of a linear concentration profile for *each* mobile ion in the membrane (analogous to (3.39)), which is then inserted into an equation for $\Delta\Phi$ of the form (3.36), summed over all species and integrated from $x = 0$ to $x = \delta$ [17]. The Henderson equation also incorporates the assumption of constant mobility values across the membrane, and the substitution of concentrations for activities.

The Donnan Potential Terms

The Donnan potentials $\Delta\Phi_D$ can be found from (3.89) if $\bar{c}_i$ is known. Alternatively it is often convenient to express $\Delta\Phi_D$ in terms of c_i and $\bar{\rho}_m$ if the latter is known rather than $\bar{c}_i$. According to (3.91), the ratio $\bar{c}_i/c_i$ for each permeable species is fixed by $\Delta\Phi_D$. Therefore, for the left-hand interface, we may write, for positive ions,

$$r' \equiv \frac{\bar{c}_{1+}'}{c_{1+}'} = \frac{\bar{c}_{2+}'}{c_{2+}'} = \frac{\bar{c}_{3+}'}{c_{3+}'} = \cdots = \frac{\bar{c}_{i+}'}{c_{i+}'} = \frac{\sum \bar{c}_{i+}'}{\sum c_{i+}'} \equiv \frac{\bar{c}_+'}{c_+'} \tag{3.120}$$

where the final equality represents a sum of concentrations. Similarly, for the negative ions at the left-hand interface,

$$r' \equiv \frac{c'_{1-}}{\bar{c}'_{1-}} = \frac{c'_{2-}}{\bar{c}'_{2-}} = \frac{c'_{3-}}{\bar{c}'_{3-}} = \cdots = \frac{c'_{i-}}{\bar{c}'_{i-}} = \frac{\sum c'_{i-}}{\sum \bar{c}'_{i-}} \equiv \frac{c'_{-}}{\bar{c}'_{-}} \qquad (3.121)$$

The summed internal concentrations, $\bar{c}'_{-}$ and $\bar{c}'_{+}$ in (3.120) and (3.121) can be eliminated in favor of $\bar{\rho}_m$ by using the electroneutrality conditions (3.115) and (3.116). Thus, (3.120) can be rewritten as

$$r' = \sqrt{1 + \left(\frac{\bar{\rho}_m/F}{2c'}\right)^2} - \frac{\bar{\rho}_m/F}{2c'} \qquad (3.122)$$

Similarly, at the right-hand interface,

$$r'' = \sqrt{1 + \left(\frac{\bar{\rho}_m/F}{2c''}\right)^2} - \frac{\bar{\rho}_m/F}{2c''} \qquad (3.123)$$

Therefore, the sum of the two Donnan potentials as shown in Figure 3.18 is

$$\Delta\Phi_{D1} - \Delta\Phi_{D2} = \left[\frac{-RT}{F}\ln\left(\frac{\bar{c}'_{+}}{c'_{+}}\right) + \frac{RT}{F}\ln\left(\frac{\bar{c}''_{+}}{c''_{+}}\right)\right] = \frac{RT}{F}\ln\left(\frac{r''}{r'}\right) \qquad (3.124)$$

The total measured potential difference V_0 for the case of Ag/AgCl electrodes is now

$$V_0 = \Delta\Phi_{\text{diff}} + \underbrace{\frac{RT}{F}\ln\left(\frac{r''}{r'}\right)}_{\text{Donnan}} - \underbrace{\frac{RT}{F}\ln\left(\frac{c''_{\text{Cl}}}{c'_{\text{Cl}}}\right)}_{\text{electrode}} \qquad (3.125)$$

Equation (3.125) for the measured steady state potential across a *uniformly charged* membrane should be compared with (3.70), which corresponds to the measured V_0 for a *neutral* membrane.

There have been several other theories developed that do not rely on all of the assumptions that are incorporated into the TMS theory. Scatchard [18] believed that the internal structure of the membrane was not significant, and left the expression for the junction potential in terms of an integral of the transport numbers $\bar{t}_i$ expressed in terms of activities; i.e., in terms of changes in the total free energy as each phase boundary is crossed. This approach allowed him to include the effect of water transport on the membrane potential; an effect neglected in the TMS approach. The theory of irreversible thermodynamics has also been applied to the problem [19]. This theory is more rigorous than the others, since equilibrium thermodynamics is not being incorporated into the description of an inherently nonequilibrium process. However, the TMS theory has the advantage of lending itself to calculations, such as determining the concentration of charge $\bar{\rho}_m(x)$ in the membrane (see Example 3.5.7). In addition, the TMS model is the only one that provides a relatively concise *physical* explanation for charged membrane behavior.

Example 3.5.1 Use of the TMS Theory to Measure the Average Fixed Charge Density of Polyelectrolyte Materials The use of biomaterials in surgical, tissue engineering, drug delivery, and other applications often requires knowledge of the effective fixed charge density of the material. By effective charge, we mean that associated with the intact specimen bathed in its natural chemical environment. It is this effective charge that is reflected in measurable macroscopic material properties such as resistance, transport, transmembrane potential, and mechanical response to applied electric fields. Such information is necessary for applications involving electrochemical or electromechanical transduction mediated by charged groups, and is useful as a reflection of those chemical attributes that involve charged groups. Examples of the latter include crosslink density and the stability in electrolyte baths of polyelectrolyte composite materials.

With respect to other charge-measuring techniques, it is known that predictions based on dilute molecular solution analysis generally do not agree with those involving intact specimens, owing to the consequences of structural organization (e.g., blockage of some functional groups and electrostatic interactions between groups). In addition, a knowledge of only the total number of available dissociable groups does not necessarily reflect the effective charge density corresponding to a given electrolyte environment, i.e., the dissociation equilibrium that exists for a given environment. Static electrokinetic measurements made with biomaterials relate solely to hydrodynamically effective charge densities (see Chapter 6), since they reflect only that portion of electrolyte counter-ion charge in the mobile region of the electrical double layer. Further, they are often plagued by the inability to distinguish rate processes associated with electrodes and electrolyte bath from those of the specimen.

The present example concerns the measurement of transmembrane diffusion potentials as a relatively simple nondestructive electrochemical technique for charge and transport characterization of many biomaterials in their natural state. It is shown that measured transmembrane potentials V_0 can be used to calculate effective fixed charge densities of charged membranes in an imposed concentration gradient, over a range of pH and neutral salt conditions. Therefore, the calculated $\bar{\rho}_m$ is accurate only to within the confines of the theoretical model and its associated assumptions. In practice, it is the *comparison* of V_0 or $\bar{\rho}_m$ from sample to sample, or for the same sample undergoing successive stages of chemical processing, that can be very useful in materials measurement design problems.

In addition, data presented here suggest the usefulness of real time or continuous monitoring of $V_0(t)$ as a technique for following the time course of chemical reactions involving membrane charged groups. Such information can help to shed light on the various rate processes involved.

The material used in the experiments presented here [20] is a membrane of collagen, representative of a broad class of cell-membrane-associated, pericellular, and extracellular matrix constituents containing dissociable groups. The dissociation of collagen fixed charge groups is primarily pH-dependent, but is also affected by the ionic strength of the electrolyte bath

Figure 3.19 Measurement configuration: (a) membrane of thickness δ; (b) poly(methyl methacrylate) chamber; (c) Ag/AgCl electrodes, (d) ports for addition of reagents and monitoring of pH and temperature, (e) magnetic stirring bars. c' and c'' are bulk solution concentrations; $\Delta\Phi'_D$, $\Delta\Phi''_D$, and $\Delta\Phi_{\text{diff}}$ are the two interfacial Donnan potentials and the diffusion potential, respectively, having the polarities indicated; V_0 is the total measured potential difference.

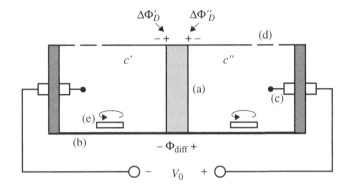

(e.g., Figure 1.12). The films were mounted in a Plexiglas chamber as shown in Figure 3.19. The membrane chamber was initially filled with electrolytes such that a gradient in neutral salt concentration was imposed across the membrane. The pH was equal on both sides throughout the course of an experiment, but was changed from an initial value of about 11.5 (3.0) to a final value of about 3.0 (11.5) by pipetting successive increments of HCl (NaOH) into the stirred baths, each within less than 5 s. The curves of Figure 3.20 typify the change in measured potential $V_0(t)$ (using AgCl electrodes) accompanying such step changes in pH in both the low- and high-pH regions away from the central isoelectric regime.

In order to calculate $\bar{\rho}_m$ from the measured V_0, the latter was modeled (using the TMS approach) as the sum of a diffusion potential $\Delta\Phi_{\text{diff}}$ internal to the membrane, two Donnan potentials $\Delta\Phi_D$ at the membrane/electrolyte interfaces, and two electrode/electrolyte potentials $\Delta\Phi_{\text{el}}$, respectively, exactly as in (3.125). $\Delta\Phi_{\text{diff}}$ was modeled by the Henderson equation (3.119). The relation between c_i of (3.125) and the interfacial $\bar{c}_i$ of (3.119) is determined by the requirement of bulk electroneutrality in the membrane, as embodied in the Donnan equilibrium, and therefore by (3.120)–(3.124). *The assumption of electroneutrality is justified since the membrane thickness δ is much greater than a Debye length $1/\kappa$ for the experiments discussed.*

Representative results of measured steady state V_0 and calculated $\bar{\rho}_m$ appear in Figures 3.21 and 3.22, respectively. The calculation of $\bar{\rho}_m$ was accomplished in terms of the unknowns $\bar{\rho}_m$ and $\bar{c}_i$. With $\Delta\Phi_{\text{diff}}$ initially set to zero in (3.125), an initial value for $\Delta\Phi'_D - \Delta\Phi''_D$ was found in terms of $\Delta\Phi_{\text{el}}$ and the measured V_0 at each pH (the $\Delta\Phi_{\text{el}}$ was calculated from the imposed bulk concentration c_{Cl^-}). Equations (3.122) and (3.124) and their analogs at the right interface were then rearranged so as to solve for an initial value of $\bar{\rho}_m$ in terms of $\Delta\Phi'_D - \Delta\Phi''_D$, c', and c''. The $\bar{\rho}_m$ in turn was used to initialize $\bar{c}_i$ and then $\Delta\Phi_{\text{diff}}$ in (3.112), (3.121), and (3.119), respectively. A new value of $\Delta\Phi'_D - \Delta\Phi''_D$ was then found using (3.125) with the calculated value of $\Delta\Phi_{\text{diff}}$. Five iterations were generally sufficient for convergence.

In keeping with the TMS approach, the $\bar{u}_i$ in (3.112) were assumed to be the same as that in bulk solution. However, the $\bar{c}_i$ were treated as unknowns along with $\bar{\rho}_m$, to be determined by the Donnan theory via the iterative technique outlined above. Changes in the measured V_0 due to changes in membrane swelling would thus be reflected in the calculated

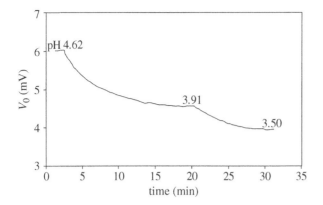

Figure 3.20 Transmembrane potential V_0 versus time resulting from step decreases in pH (equal on both sides) for a collagen membrane in an imposed concentration gradient; $c'_{NaCl} = 0.06$ M and $c''_{NaCl} = 0.03$ M. The pH values indicated in the figure are the steady state values prior to the next addition of acid. Final (steady state) potentials are plotted in Figure 3.21 (open circles).

values of $\bar{\rho}_m$ and $\bar{c}_i$. The effect of osmotic and electroosmotic water transport across the membrane on V_0, which is neglected in the TMS theory, has been assumed negligible for the case at hand, based on independent measurements of steady state hydrodynamic and electrokinetic transport properties of the membrane. Finally, the calculation of $\bar{\rho}_m$ was based on the assumption that solution volumes were sufficiently large that changes in external concentration could be neglected over the course of an experiment.

Several conclusions concerning the collagen membrane used here may be drawn by interpreting Figures 3.21 and 3.22 in terms of fixed charge membrane theory. We note that measured potentials V_0 directly reflect collagen charge density, bath ionic strength, and the degree of collagen swelling or porosity consistent with both. In general, the higher the ratio of $\bar{\rho}_m$ to bath ionic strength, the closer V_0 should approach the value attained by (3.125) for the limit of an ideal semipermeable membrane where co-ion transport becomes negligibly small (~ 0 mV and $\sim (RT/F)[-\ln(a''_{Cl^-}/a'_{Cl^-}) - \ln(a''_{Na^+}/a'_{Na^+})]$ for a positive and negative membrane, respectively). Often interpreted

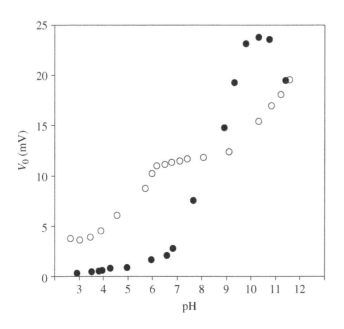

Figure 3.21 Steady state transmembrane potential V_0 versus bath pH for a single membrane of hide corium collagen. Open circles: initial NaCl concentrations were $c'_{NaCl} = 0.06$ M and $c''_{NaCl} = 0.03$ M, and the pH was decreased in steps from about 11.5 to about 3. Filled circles: initial concentrations were $c'_{NaCl} = 0.003$ M, $c''_{NaCl} = 0.001$ M, and the pH was increased from about 3 to about 11.5.

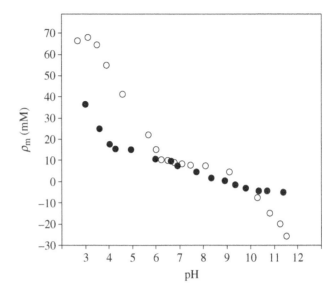

Figure 3.22 Calculated collagen net fixed charge density $\bar{\rho}_m$ versus pH, based on the measured V_0 of Figure 3.21 and (3.119)–(3.125).

in terms of Donnan exclusion, this effect is seen in Figure 3.21 as expected, where the V_0 (which includes the electrode potentials) for the low-ionic-strength case is smaller at low pH and larger at high pH than that of the higher-ionic-strength case. This does not necessitate a higher $\bar{\rho}_m$ at low ionic strength; in fact, just the opposite is seen here in Figure 3.22. The reversal of V_0 at high pH (filled circles) is due primarily to the significant perturbation in ionic strength and change in total concentration gradient due to the addition of base, another indication that V_0 is sensitive to an effective rather than stoichiometric charge density.

Of course, the exact value of $\bar{\rho}_m$ in Figure 3.22 will reflect the assumptions and limitations of the model used in the calculation. While the assumptions of the TMS theory may be violated to a greater or lesser extent in a given experiment, the major point to be stressed is that the trends of the V_0 and the $\bar{\rho}_m$ thus calculated, as well as comparative data for different membranes and experimental conditions, can be quite instructive as a guide in material design. With these limitations in mind, we note that the $\bar{\rho}_m$ of Figure 3.22 show charging trends that are characteristic of the few titration curves of intact collagen specimens available for comparison in the literature (see Figure 1.12). One notable trend is the increase in $|\rho_m|$ with bath ionic strength at a fixed pH in the low- and high-pH regions. This may be interpreted in terms of Donnan exclusion, and the Debye screening effect of the neutral salt. Such an interpretation is equivalent to and consistent with that based on the effect of ionic strength on osmotic swelling.

The utility of a diffusion potential measurement rests not only on its ease, but also on the added possibility of following the time course of chemical reactions and other processes involving the dissociable groups. For example, the data of Figure 3.20 show the time course of $V_0(t)$ in response to step changes in pH at approximately constant ionic strength, which reflects the combined effects of diffusion (of added reagent into the membrane) and chemical reaction (see Section 1.6 and Problem 1.8). In an attempt to distinguish between reaction and diffusion, $V_0(t)$ was measured in response to step changes in bath ionic strength while keeping the pH and the overall concentration gradient c'/c'' constant. Bath conditions were chosen such that those changes in ionic strength were known to leave essentially unperturbed the total number of ionized groups (though changes in $V_0(t)$ and ρ_m would follow from variations in membrane water content). The resulting rise times of these $V_0(t)$ were typically an order of magnitude less than those of Figure 3.20. In magnitude, they were on the order of the time for diffusion of ions across a distance equal to that of the swollen membrane. Thus, these two groups of experiments were rate-limited by diffusion-controlled reaction (Figure 3.20) and diffusion processes, respectively.

Finally, these experiments showed that transmembrane potentials varied with the degree of crosslinking (induced thermally or by means of glutaraldehyde) in a manner consistent with the belief that crosslinks involve charged groups. Second, the measured V_0 was found to be sensitive to the binding of the glycosaminoglycan chondroitin-6-sulfate to these same collagen films. (These latter films were developed by Yannas, Burke and co-workers [21] for use as a skin substitute.) In this case, V_0 reflected an average $\bar{\rho}_m$ for the composite material. The additional negative groups of the glycosaminoglycan were detected by the potential measurement over a wide range of pH.

3.6 NONSTEADY ELECTROLYTE TRANSPORT PHENOMENA IN MEDIA CONTAINING FIXED CHARGE GROUPS: ELECTRODIFFUSION

The fundamentals of electrodiffusion are important in far-ranging applications, from processes at the nanoscale (transport across cell membranes) to the macro-tissue scale (e.g., transport of charged contrast agents into charged tissues for clinical imaging studies). In Section 3.1, coupled diffusion and charge relaxation relations were developed for the case of a binary electrolyte solution. These relations exemplify the interplay between ionic migration and diffusion currents, the importance of which was first proposed by Nernst [22] and Planck [23] in the late nineteenth century. In Section 3.2, the transport of electrolyte ions across neutral media was described in terms of the coupled diffusion formulation of Section 3.1. Neutral membranes and tissues were viewed as biomaterials that could support space-varying concentrations by acting essentially as diffusion barriers. These neutral media did not alter the bulk electroneutrality condition (3.10). However, the charged membranes and tissues described in Sections 3.4 and 3.5 can greatly imbalance the internal equilibrium concentrations in order to preserve *bulk neutrality*. For example, a positively charged membrane in equilibrium with a solution of concentration c_0 such that $\bar{\rho}_m \gg Fc_0$ was found to have internal concentrations $\bar{c}_{+0}$ and $\bar{c}_{-0}$ described by the approximate bulk neutrality condition

$$\bar{\rho}_m \simeq F\bar{c}_{-0} \gg F\bar{c}_{+0} \qquad (3.126)$$

which characterizes the "Donnan exclusion" of positively charged mobile *co-ions*. We will find that the unbalancing of internal equilibrium concentrations can greatly alter the coupled transport of ions (due to diffusion or applied electric fields) from the mechanisms embodied in the coupled diffusion and charge relaxation models of Section 3.1. Therefore, we must now redevelop the model in order to account for *nonzero fixed charge density*.

Transport phenomena must still be described by the continuity, carrier flux, and electric field flux (Gauss' law) relations (3.3)–(3.5), but modified to include $\bar{\rho}_m$. These are written inside the medium for a $z:z$ electrolyte as

$$\frac{D\bar{c}_+}{Dt} = -\nabla \cdot \boldsymbol{N}_+ \tag{3.127}$$

$$\frac{D\bar{c}_-}{Dt} = -\nabla \cdot \boldsymbol{N}_- \tag{3.128}$$

$$\boldsymbol{N}_+ = \bar{u}_+ \bar{c}_+ \boldsymbol{E} - \bar{D}_+ \nabla \bar{c}_+ \tag{3.129}$$

$$\boldsymbol{N}_- = -\bar{u}_- \bar{c}_- \boldsymbol{E} - \bar{D}_- \nabla \bar{c}_- \tag{3.130}$$

$$\nabla \cdot (\epsilon \boldsymbol{E}) = \rho_e = \bar{\rho}_m + zF\left(\bar{c}_+ - \bar{c}_-\right) \tag{3.131}$$

We will first address the case of a uniformly charged material where $\bar{\rho}_m$ is independent of position. Hence, internal ion concentrations can be described in terms of equilibrium and perturbation components by

$$\bar{c}_+ = \bar{c}_{+0} + \tilde{c}_+(\boldsymbol{r}, t) \tag{3.132}$$

$$\bar{c}_- = \bar{c}_{-0} + \tilde{c}_-(\boldsymbol{r}, t) \tag{3.133}$$

where $\bar{c}_{+0}$ and $\bar{c}_{-0}$ are independent of position. These equilibrium carrier concentrations are unequal, in general, owing to the presence of fixed charge; an example is the heavily (positive) charged membrane represented by (3.126). When the perturbation concentrations are small compared with *both* majority and minority equilibrium concentrations,

$$\begin{aligned} \tilde{c}_+(\boldsymbol{r}, t) &\ll \bar{c}_{+0} \\ \tilde{c}_-(\boldsymbol{r}, t) &\ll \bar{c}_{-0} \end{aligned} \tag{3.134}$$

we will find that the governing differential equations can be linearized. (Of course, (3.134) can be quite restrictive, since Donnan exclusion of co-ions can result in a very small equilibrium minority concentration. Failure of the validity of (3.134) would therefore lead to increased mathematical complexity). Quasineutrality is assumed for characteristic dimensions much greater than $1/\kappa$, and therefore we expect that the perturbation concentrations are not very different from one another:

$$\tilde{c}_+(\boldsymbol{r}, t) = \tilde{c}_-(\boldsymbol{r}, t) \tag{3.135}$$

The quasineutrality condition is stated more strictly as

$$\left| \frac{\tilde{c}_+ - \tilde{c}_-}{\bar{c}_{i0}} \right| \ll 1 \tag{3.136}$$

where $\bar{c}_{i0}$ is the *majority* equilibrium concentration. Once again, (3.136) results when the charge relaxation time is much less than the diffusion time, $\tau_{\text{rel}} \ll \tau_{\text{diff}}$ for a given length of interest (see Problem 3.5).

3.6.1 Coupled Diffusion in Charged Porous Media

We might attempt to develop the new coupled transport model by a procedure analogous to that of Section 3.1, i.e., by incorporating (3.129) and (3.130) into (3.127) and (3.128), multiplying the continuity relations (3.127) by $\bar{D}_-$ and (3.128) by $\bar{D}_+$ and adding the equations. The result would be the coupled diffusion equation (3.21) if the term (3.20) were negligible as before. For the *binary electrolyte in neutral media*, (3.20) resulted in the inequality (comparing net migration and diffusion fluxes)

$$\frac{|\boldsymbol{E}(\bar{c}_+ - \bar{c}_-)|}{|(RT/F)(\nabla\bar{c}_+ + \nabla\bar{c}_-)|} = \left|\left(\frac{\boldsymbol{E}\delta}{2RT/F}\right)\left(\frac{\bar{c}_+ - \bar{c}_-}{\bar{c}_0}\right)\right| \ll 1 \qquad (3.137)$$

where the first equality results from *equal* equilibrium concentrations ($\bar{c}_{+0} = \bar{c}_{-0} = \bar{c}_0$). In the present case, however, where $\nabla\bar{c}_+ = \nabla\bar{c}_- = \nabla\tilde{c}_\pm$ and $\bar{c}_{+0} \neq \bar{c}_{-0}$ for finite $\bar{\rho}_m$, the equivalent ratio given by (3.20) is

$$\left|\left(\frac{\boldsymbol{E}\delta}{2RT/F}\right)\left(\frac{\bar{c}_{+0} - \bar{c}_{-0}}{\tilde{c}_\pm}\right)\right| \qquad (3.138)$$

which can be $\gg 1$ even when there is no applied field and $|\boldsymbol{E}\delta| \ll 2RT/F$. It is evident that the *net migration current* is no longer negligible with respect to the diffusion current when the equilibrium concentrations are *unbalanced*, even for distances large compared with $1/\kappa$! The net migration current in *neutral* media is proportional to $(\tilde{c}_+ - \tilde{c}_-)\boldsymbol{E}$, while that in charged media is proportional to $(\bar{c}_{+0} - \bar{c}_{-0})\boldsymbol{E}$; obviously, the latter can be much larger than the former. Since $\bar{c}_{+0} \neq \bar{c}_{-0}$, the migration flux associated with the counter-ion is much larger than that of the co-ion; furthermore the co-ion migration flux may not always be negligible compared with diffusive fluxes as was the case in Sections 3.1 and 3.2. To make this point more clearly, we compare the terms in the continuity equations (3.127) and (3.128), incorporating the flux relations (3.129) and (3.130):

$$\frac{D\bar{c}_+}{Dt} = \bar{D}_+\nabla^2\bar{c}_+ - \frac{\bar{u}_+\bar{c}_+}{\epsilon}\nabla\cdot\epsilon\boldsymbol{E} - \boldsymbol{E}\bar{u}_+\cdot\nabla\bar{c}_+ \qquad (3.139)$$

$$\frac{D\bar{c}_-}{Dt} = \bar{D}_-\nabla^2\bar{c}_- + \frac{\bar{u}_-\bar{c}_-}{\epsilon}\nabla\cdot\epsilon\boldsymbol{E} + \boldsymbol{E}\bar{u}_-\cdot\nabla\bar{c}_- \qquad (3.140)$$

where ϵ is assumed to be independent of position. It is the terms involving $\nabla\cdot\epsilon\boldsymbol{E}$ in (3.139) and (3.140) that are significant in the electric-field-coupling of the diffusion process. The ratio of the sum of the $\nabla\cdot(\epsilon\boldsymbol{E})$ terms to the sum of the $\nabla^2\bar{c}$ terms can be cast in the form

$$\left|\left(\frac{\delta^2}{(1/\kappa)^2}\right)\left(\frac{\bar{c}_+ - \bar{c}_-}{\tilde{c}_\pm}\right)\right| \qquad (3.141)$$

which can be large for any δ, even $\delta \simeq 1/\kappa$ (for which $\bar{c}_+ - \bar{c}_- > \tilde{c}_\pm$).

We proceed, therefore, by solving (3.127)–(3.131) simultaneously, without decoupling Gauss' law. This can be done by dividing (3.139) by $\bar{D}_+\bar{c}_+$ and (3.140) by $\bar{D}_-\bar{c}_-$ and adding the two equations. This will result in cancellation of the $\nabla\cdot\epsilon\boldsymbol{E}$ terms when (3.139) and (3.140) are added; however, the influence of $\nabla\cdot\epsilon\boldsymbol{E}$ is still implicit in the resulting expression:

$$\left(\frac{1}{\bar{c}_+\bar{D}_+}\frac{D\tilde{c}_+}{Dt} + \frac{1}{\bar{c}_-\bar{D}_-}\frac{D\tilde{c}_-}{Dt}\right) = \left[\frac{\nabla^2\tilde{c}_+}{\bar{c}_+} + \frac{\nabla^2\tilde{c}_-}{\bar{c}_-}\right] + \frac{\boldsymbol{E}_{TOT}}{RT/F}\cdot\left[\frac{\nabla\tilde{c}_-}{\bar{c}_-} - \frac{\nabla\tilde{c}_+}{\bar{c}_+}\right] \qquad (3.142)$$

where all time and space derivatives of $\bar{c}_\pm$ have been replaced by derivatives of $\tilde{c}_\pm$, since $\bar{c}_{+0}$ and $\bar{c}_{-0}$ are constants. Note that the total electric field in (3.142) is represented, in general, as the sum of any applied field $\boldsymbol{E}_0$ and a self-field $\boldsymbol{E}_{self}$ resulting from any small but finite space charge (i.e., separation of positive and negative mobile carrier densities due to unequal diffusion or applied fields). Therefore, $\boldsymbol{E}_{TOT} = \boldsymbol{E}_0 + \boldsymbol{E}_{self}$ in

(3.142). If the rightmost term in (3.142) is negligible compared with the first right-hand term, then (3.142) reduces to a nonlinear diffusion equation still exhibiting implicitly the electrical coupling due to the $\nabla \cdot \epsilon \boldsymbol{E}$ terms of (3.139) and (3.140). For length scales large compared with $1/\kappa$, the quasineutrality approximation (3.135) simplifies (3.142) and results in the *ambipolar diffusion equation*

$$\frac{\mathrm{D}\tilde{c}_\pm}{\mathrm{D}t} = \bar{D}_a \nabla^2 \tilde{c}_\pm \qquad (3.143)$$

where

$$\bar{D}_a = \frac{\bar{D}_+ \bar{D}_- \left(\bar{c}_+ + \bar{c}_-\right)}{\bar{D}_+ \bar{c}_+ + \bar{D}_- \bar{c}_-} \qquad (3.144)$$

Thus, (3.142) is nonlinear since the *ambipolar diffusion coefficient* $\bar{D}_a$ depends on concentrations. However, (3.143) and (3.144) can be linearized when the conditions (3.134) apply, giving

$$\bar{D}_a = \frac{\bar{D}_+ \bar{D}_- \left(\bar{c}_{+0} + \bar{c}_{-0}\right)}{\bar{D}_+ \bar{c}_{+0} + \bar{D}_- \bar{c}_{-0}} \qquad (3.145)$$

Note that the concentration dependence of $\bar{D}_a$ represents the importance of the majority carrier migration term in the self-field coupling of the diffusion process. The rightmost term in (3.142) can be neglected when

$$\left| \frac{\left(\dfrac{\boldsymbol{E}_{\mathrm{TOT}}}{RT/F}\right) \cdot \left(\dfrac{\nabla \tilde{c}_-}{\bar{c}_-} - \dfrac{\nabla \tilde{c}_+}{\bar{c}_+}\right)}{\dfrac{\nabla^2 \tilde{c}_+}{\bar{c}_+} + \dfrac{\nabla^2 \tilde{c}_-}{\bar{c}_-}} \right| = \left| \left(\frac{\boldsymbol{E}_{\mathrm{TOT}}\delta}{RT/F}\right) \left(\frac{\bar{c}_+ \tilde{c}_- - \bar{c}_- \tilde{c}_+}{\bar{c}_+ \tilde{c}_- + \bar{c}_- \tilde{c}_+}\right) \right| \ll 1 \quad (3.146)$$

For situations in which there is no applied field ($\boldsymbol{E}_0 = 0$), $\boldsymbol{E}_{\mathrm{TOT}} = \boldsymbol{E}_{\mathrm{self}}$. Since $\delta\boldsymbol{E}_{\mathrm{self}} \sim RT/F$, the concentration-dependent terms in (3.146) would have to be examined to determine whether the inequality in (3.146) was valid. If so, then the ambipolar diffusion model of (3.143) adequately represents the dynamics of electrolyte transport. For the limiting case of heavily charged media, with

$$\bar{c}_{+0} \gg \bar{c}_{-0} \text{ or } \bar{c}_{-0} \gg \bar{c}_{+0} \qquad (3.147)$$

the concentration-dependent terms in (3.146) cancel; for $\boldsymbol{E}_0 = 0$, we are left with $[\boldsymbol{E}_{\mathrm{self}}\delta/(RT/F)] \sim 1$. The inequality (3.146) is not satisfied. As an example, the steady state diffusion potential ($\sim RT/F$) induced by a concentration gradient across a heavily charged membrane would alter the intramembrane concentration profiles $\tilde{c}_\pm(x)$ to a small extent (see Problem 3.13).

For the limiting case of $\bar{\rho}_m \to 0$, (3.146) becomes

$$\left| \left(\frac{\boldsymbol{E}\delta}{2RT/F}\right) \left(\frac{\tilde{c}_+ - \tilde{c}_-}{\tilde{c}_\pm}\right) \right| \ll 1 \qquad (3.148)$$

which is exactly analogous to the neutral-media–binary-electrolyte problem (compare with (3.20) for the case where c_0 is taken to be independent of position).

Of course, there are many important applications involving *applied* electric fields $\boldsymbol{E}_0$ across tissues or membranes. For sufficiently high $\boldsymbol{E}_0$, the effect of $\boldsymbol{E}_{\mathrm{self}}$ may be negligible in comparison (although for very thin membranes, such as $\sim 50\,\text{Å}$ cell membranes, it would be very diffi-cult to apply a field as high as $V_{\text{mem}}/\delta_{\text{mem}}$), and (3.142) can be cast into the form

$$\frac{\mathrm{D}\tilde{c}_{\pm}}{\mathrm{D}t} = \bar{D}_a \nabla^2 \tilde{c}_{\pm} + \boldsymbol{E}_0 \bar{u}_a \cdot \nabla \tilde{c}_{\pm} \tag{3.149}$$

by multiplying it by $\bar{D}_+ \bar{c}_+ \bar{D}_- \bar{c}_-$ and keeping the $\boldsymbol{E}_0$ term. The ambipolar mobility $\bar{u}_a$ is defined as

$$\bar{u}_a = \frac{\bar{u}_+ \bar{u}_- \left(\bar{c}_+ - \bar{c}_- \right)}{\bar{u}_+ \bar{c}_+ + \bar{u}_- \bar{c}_-} \tag{3.150}$$

In Example 3.6.2, the case of heavily charged membranes is examined, for which $\bar{c}_{+0} \gg \bar{c}_{-0}$ and vice versa for negative and positive fixed charge densities, respectively. It is seen that the ambipolar diffusivity $\bar{D}_a$ and mobility $\bar{u}_a$ reduce to those of the *co-ion* or *minority carrier* in these lim-its. It may be surprising to examine the limiting case $\bar{\rho}_m \to 0$, for which $\bar{c}_+ \sim \bar{c}_-$ and therefore $\bar{u}_a \to 0$ in (3.150). We see that carrier migration due to $\boldsymbol{E}_0$ would tend to separate positive and negative mobile carriers, a separation opposed by charge relaxation. In charged membranes, the numerous majority carriers can readjust quite easily to such a demand, satisfying charge relaxation via conduction currents that consist pri-marily of majority carrier migration. This prevents the buildup of any significant $\nabla \cdot \epsilon \boldsymbol{E}$ and allows *co-ion transport*. In a neutral membrane, however, the positive and negative mobile carrier concentrations are essentially equal and the tendency for carrier separation due to $\boldsymbol{E}_0$ leads to self-fields that prevent such an occurrence; thus, migration effectively ceases!

It may seem puzzling, at first, that we can neglect the self-field term in (3.142) while keeping the $\boldsymbol{E}_0$ term. Note that $\boldsymbol{E}_{\text{self}}$ (i.e., internal dou-ble layer fields) may be as high as $10^7\,\text{V}\,\text{m}^{-1}$ (see Chapter 6), typically many orders of magnitude greater than $\boldsymbol{E}_0$. The point is that $\boldsymbol{E}_0$ can be significant over *macroscopic* dimensions much greater than a Debye length, while $\boldsymbol{E}_{\text{self}}$ is important only when considering dimensions on the order of $1/\kappa$. Transport equations such as (3.143) and (3.149) *only make sense on a dimensional scale much larger than $1/\kappa$, a length scale that smooths over the fine structure of the internal double layer fields.*

3.6.2 Charge Relaxation in Charged Porous Media

To complete the model for nonequilibrium transport in charged media, we derive the relevant charge relaxation equation, analogous to (3.27) and (3.30). Multiplying (3.127) by $+zF$ and (3.128) by $-zF$ and then adding the equations yields (compare with (3.25))

$$\frac{\mathrm{D}\rho_e}{\mathrm{D}t} = -\nabla \cdot \left[zF \left(\bar{u}_+ \bar{c}_+ + \bar{u}_- \bar{c}_- \right) \boldsymbol{E} \right] + \nabla \cdot \left[zF \left(\bar{D}_+ - \bar{D}_- \right) \nabla \tilde{c}_{\pm} \right] \tag{3.151}$$

The rightmost term of (3.151) can be neglected when (compare with (3.28))

$$\left| \frac{\left(\bar{D}_+ - \bar{D}_- \right) \tilde{c}_{\pm} \left(RT/zF \right)}{\left(\bar{D}_+ \bar{c}_+ + \bar{D}_- \bar{c}_- \right) |\boldsymbol{E}\delta|} \right| \ll 1 \tag{3.152}$$

or, equivalently,

$$|\boldsymbol{E}\delta| \gg \left| \frac{RT}{zF} \left[\frac{\left(\bar{D}_+ - \bar{D}_- \right) \tilde{c}_{\pm}}{\bar{D}_+ \bar{c}_+ + \bar{D}_- \bar{c}_-} \right] \right| \tag{3.153}$$

When (3.152) and (3.153) are satisfied, (3.151) reduces to the usual charge relaxation equation (analogous to (3.30))

$$\frac{\mathrm{D}\rho_e}{\mathrm{D}t} = -\nabla \cdot [\sigma(\boldsymbol{r}, t)\boldsymbol{E}] \tag{3.154}$$

where $\sigma(\boldsymbol{r}, t)$ is essentially a constant dominated by the internal majority carrier concentration. In the limit $\bar{\rho}_m \to 0$, (3.151)–(3.154) become analogous to the neutral-media–binary-electrolyte case, in which c_0 is taken to be a constant. In summary, the form of the charge relaxation relations for charged media is similar to that of neutral media, but the conductivity is basically that associated with the majority carrier. On the other hand, the self-field coupled diffusion process is seen to be quite different from that of neutral media; the effective diffusion coefficient $\bar{D}_a$ is concentration-dependent.

Example 3.6.1 Coupled Diffusion and Charge Relaxation in the Limit $\bar{\rho}_m \to 0$ For the limiting case in which $\bar{\rho}_m \to 0$, the condition of bulk electroneutrality inside the material now requires that $\bar{c}_{+0} = \bar{c}_{-0}$. Incorporation of this condition in (3.145) shows that the ambipolar diffusivity reduces to

$$\bar{D}_a \to \frac{2\bar{D}_+\bar{D}_-}{\bar{D}_+ + \bar{D}_-} \equiv \bar{D}^* \tag{3.155}$$

But this is identically the effective diffusion coefficient derived for the case of a binary electrolyte in neutral media, in Sections 3.1 and 3.2.

Example 3.6.2 Show that Diffusion in a Heavily Charged Medium is Rate-Limited or Controlled by the Co-ion Diffusivity For the case of a positively charged membrane in which $|\bar{\rho}_m/F|$ is much greater than the external bath concentrations, Donnan equilibrium predicts that $\bar{\rho}_m \simeq F\bar{c}_{-0} \gg F\bar{c}_{+0}$. In this limit, (3.145) gives

$$\bar{D}_a \to \bar{D}_+ \tag{3.156}$$

Similarly, for a heavily negatively charged membrane or tissue,

$$\bar{D}_a \to \bar{D}_- \tag{3.157}$$

Thus, we find that the ambipolar diffusion coefficient becomes, in the limit, the diffusivity of the *co-ion*.

To interpret nonequilibrium diffusion of electrolyte in heavily positively charged media, for example, (3.156) together with (3.143) describe the diffusion process:

$$\frac{\mathrm{D}\tilde{c}_+}{\mathrm{D}t} = \bar{D}_+\nabla^2\tilde{c}_+ \tag{3.158}$$

$$\frac{\mathrm{D}\tilde{c}_-}{\mathrm{D}t} = \bar{D}_+\nabla^2\tilde{c}_- \tag{3.159}$$

Equations (3.158) and (3.159) may be interpreted physically from (3.129) and (3.130). It is evident that the migration term in (3.129) is much less than that of (3.130), while the diffusion fluxes are of the same order. For $\boldsymbol{N}_+$ and $\boldsymbol{N}_-$ to be of the same order (they are equal and opposite when no net current flows), it is evident that *co-ion flux is dominated by the diffusion term* in (3.129). Being able to neglect the migration term in (3.129) essentially decouples Gauss' law (3.131) from the calculation of $\tilde{c}_\pm(\boldsymbol{r}, t)$; incorporation of (3.129) into (3.127) directly yields (3.158) and (3.159). Thus, the large background of counter-ions effectively shields the co-ions from the self-induced $\boldsymbol{E}$ field, and the co-ions flow only by diffusion. The self-$\boldsymbol{E}$ *is* important in counter-ion flow; counter-ion migration assures that $\boldsymbol{N}_+ + \boldsymbol{N}_- = 0$ for open circuit conditions. We will find in Example 3.6.3 that co-ion migration may not be negligible when there is an applied $\boldsymbol{E}$. However, the problem can often be analyzed by careful consideration of just the co-ion flux equations, rather than diving into a numerical or analytical solution of the coupled, nonlinear system of (3.127) and (3.131).

The results of this example are extremely important in the many problems of cell biology, physiology, and biomaterials science in which diffusion in *multi-ionic electrolytes* is important. Such problems involve the diffusion of certain species into or across membranes and other materials in which there is a large background of neutral-salt ions (e.g., the physiological ionic strength of about 0.15 M). The diffusing species can often be thought of as being present in perturbation amounts and being shielded from any internal electric field by the large background of ions. If this is the case, one can immediately write a diffusion equation for this species, neglecting the self-$\boldsymbol{E}$ migration term as described above. Once again, this eliminates the need to solve a complicated system of coupled partial differential equations.

Example 3.6.3 Ion Exchange [24]: Chromatography and Water Purification The ion exchange resin in Figure 3.23 can be used to remove mineral ions from water. The membrane contains fixed negative charges having concentration $[R^-]$. Ion exchangers have many important industrial applications, such as water purification and deionizing columns. Under ideal conditions, ion exchange is a *diffusion* process; i.e., the rate-limiting process is most often diffusion rather than a chemical reaction or reaction-limited diffusion. The rate that governs ion exchange depends on the particular species involved and can be found in general by solving the diffusion equation (Fick's second law). We will see in this case, however, that the effective diffusion coefficient is not a constant independent of intra-resin mobile ion concentrations.

Assumptions:

(i) $\overline{[Cl^-]} \simeq 0$ if $|z_{R^-}F[R^-]| \equiv |\bar{\rho}_m| \gg |\overline{[Cl^-]}Fz_-|$ (i.e., assuming strong Donnan exclusion). $\bar{\rho}_m$ and $\bar{D}_i$ are constants independent of concentration.

(ii) No specific chemical interaction processes occur (e.g., membrane or resin dissociation remains constant).

EXAMPLE

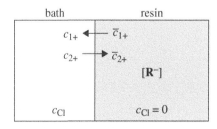

Figure 3.23 Cation exchange resin originally in the H^+ form ($c_1 = [H^+]$) being converted to Na^+ form ($c_2 = [Na^+]$); i.e., resin exchanges H^+ for Na^+.

(iii) Diffusion is rate-limited by membrane, not by interfacial (stagnant) films.

In general, there will be gradients in the concentrations of mobile species inside the membrane, $\bar{c}_{1+}$ and $\bar{c}_{2+}$ (which will be referred to from now on as $\bar{c}_1$ and $\bar{c}_2$), having respective diffusion coefficients $\bar{D}_1$ and $\bar{D}_2$, which are assumed independent of concentration. There will result an electrically coupled diffusion inside the membrane characterized by an effective diffusion coefficient $\bar{D}_{12}$, which is to be derived.

In the following development, we are concerned only with processes *inside* the membrane—the right side of Figure 3.23. We expect that if the total fluxes $\boldsymbol{N}_i$ across the interface satisfy the inequalities (3.88), then a *Donnan (quasi)equilibrium exists at the membrane/solution interface*. However, the interface is assumed to offer little or no resistance to the flow of ions across it—i.e., diffusion is rate-limited by the membrane and not by the interface ($\tau_{\text{diff}} \sim \tau_{\text{rel}}$ in the thin interfacial region). *Inside the membrane,*

$$\boldsymbol{J}_1(\boldsymbol{r}) = -z_1 F \bar{D}_1 \nabla \bar{c}_1 - z_1 F \bar{u}_1 \bar{c}_1 \nabla \Phi \tag{3.160}$$

$$\boldsymbol{J}_2(\boldsymbol{r}) = -z_2 F \bar{D}_2 \nabla \bar{c}_2 - z_2 F \bar{u}_2 \bar{c}_2 \nabla \Phi \tag{3.161}$$

$$\boldsymbol{J}_1(\boldsymbol{r}) + \boldsymbol{J}_2(\boldsymbol{r}) = 0 \tag{3.162}$$

$$\nabla \cdot \boldsymbol{J}_i = -\frac{\partial}{\partial t}(z_i F \bar{c}_i) = -\frac{\partial \rho_i}{\partial t} \tag{3.163}$$

In addition, we assume electroneutrality in the membrane under the condition of assumptions (i) and (ii) above:

$$F(z_1 \bar{c}_1 + z_2 \bar{c}_2) + \bar{\rho}_m \simeq 0 \tag{3.164}$$

which gives upon differentiation

$$z_1 \nabla \bar{c}_1 \simeq -z_2 \nabla \bar{c}_2 \tag{3.165}$$

(The terms in (3.162) are not identical, since $(\bar{D}_1 \neq \bar{D}_2) \to$ finite $\boldsymbol{E}_{\text{self}}$.) First, we add (3.160) and (3.161) under the condition (3.162), and solve for $\nabla \Phi (= -\boldsymbol{E})$:

$$\nabla \Phi = \frac{-z_1 \bar{D}_1 \nabla \bar{c}_1 - z_2 \bar{D}_2 \nabla \bar{c}_2}{z_1 \bar{u}_1 \bar{c}_1 + z_2 \bar{u}_2 \bar{c}_2} = \frac{-z_1 \bar{D}_1 + z_1 \bar{D}_2}{z_1 \bar{u}_1 \bar{c}_1 + z_2 \bar{u}_2 \bar{c}_2} \nabla \bar{c}_1 \tag{3.166}$$

where the second equality comes from (3.165). Equation (3.160) can now be rewritten in terms of $\nabla \bar{c}_1$ alone:

$$\boldsymbol{J}_1(\boldsymbol{r}) = \frac{-z_1^2 \bar{u}_1 \bar{D}_1 \bar{c}_1 - z_1 z_2 \bar{u}_2 \bar{D}_1 \bar{c}_2 + z_1^2 \bar{u}_1 \bar{D}_1 \bar{c}_1 - z_1^2 \bar{u}_1 \bar{D}_2 \bar{c}_1}{z_1 \bar{u}_1 \bar{c}_1 + z_2 \bar{u}_2 \bar{c}_2} F \nabla \bar{c}_1 \tag{3.167}$$

Using the Einstein relation, (3.167) reduces to

$$\boldsymbol{J}_1(\boldsymbol{r}) = -z_1 F \bar{D}_{12} \nabla \bar{c}_1 \tag{3.168}$$

where the coupled diffusion coefficient $\bar{D}_{12}$ (sometimes called the interdiffusion coefficient) is defined by

$$\bar{D}_{12} = \frac{\bar{D}_1 \bar{D}_2 (z_2^2 \bar{c}_2 + z_1^2 \bar{c}_1)}{z_1^2 \bar{D}_1 \bar{c}_1 + z_2^2 \bar{D}_2 \bar{c}_2} \tag{3.169}$$

Now, from (3.163), the time-dependent diffusion problem is described by

$$\frac{\partial \bar{c}_1}{\partial t} = \bar{D}_{12} \nabla^2 \bar{c}_1 \tag{3.170}$$

Remember that (3.170) is nonlinear, since $\bar{D}_{12}$ is a function of $\bar{c}_1$ and $\bar{c}_2$. Focusing on $\bar{D}_{12}$, we see from (3.169) that, for $z = 1$,

$$\bar{D}_{12} \to \bar{D}_1 \text{ for } \bar{c}_1 \ll \bar{c}_2$$

$$\tag{3.171}$$

$$\bar{D}_{12} \to \bar{D}_2 \text{ for } \bar{c}_2 \ll \bar{c}_1$$

Thus, the ion having the smaller internal concentration dominates the diffusion process. This can be seen physically from (3.160) and (3.161): the "minority" carrier will flow predominantly by diffusion, since the migration term is directly proportional to carrier concentration. Thus, for $\bar{c}_1 \ll \bar{c}_2$, for example, the migration term can be neglected in (3.160) and we can proceed *directly* to an equation of the form (3.170) *without first having to solve for the electric field*!

Example 3.6.4 Electrodiffusion: Applied and Self-Induced Electric Fields Electrodiffusion theories were initiated by Nernst and Planck, and reformulated and used by Bernstein, Teorell, Meyer, Sievers, and others to model the electrical properties of living cell membranes. These theories describe passive ion flow due to migration and diffusion, and have been summarized in standard texts and review articles (see, e.g., [25]). While the theory of electrodiffusion has not yet successfully described all of the known properties of cell membranes, such as that of the squid giant axon, it has played a key role in the understanding of transport phenomena in bulk tissues and other native and synthetic biomaterials. In this example, a linearized model is presented which clearly demonstrates the physical transport phenomena induced by the electric field.

In order to investigate electric field control of intramembrane concentration profiles, we focus on the configuration shown schematically in Figure 3.24. A membrane of thickness δ

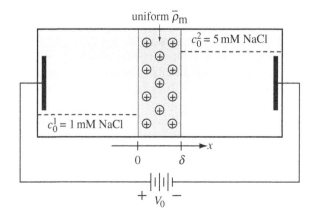

Figure 3.24 Electric field control of intramembrane concentration.

separates electrolyte baths of two different neutral salt concentrations. The membrane contains fixed, dissociated charge groups as well as mobile ions; it is assumed for now that $\bar{\rho}_m$ is independent of the intramembrane mobile ion concentrations. An electric field $\boldsymbol{E}_0$ can be applied perpendicular to the tissue membrane by means of electrodes. Inner and outer electrolytes are continually renewed (not shown in Figure 3.24) to maintain c_0^1 and c_0^2 at any desired value.

The mobile ion concentration profiles inside the membrane are determined by a balance between the competing processes of ion migration due to $\boldsymbol{E}_{\text{TOT}}$, ion diffusion due to the imposed concentration gradient, and convection of mobile species in the advent of fluid flow through the membrane. Solution of such a problem has been widely investigated with respect to fixed charge membrane models and the theory of transmembrane potentials. The additional possibility that an applied electric field can cause an electrokinetic (e.g., electrophoretic) movement of the membrane if it is nonrigid and deformable must also be included, in general, as well as the possibility of a field-induced convective flow of fluid into and across the membrane (e.g., electroosmosis). However, these latter phenomena are found to be unimportant in some cases and we will not include them in the present example. (The role of such phenomena will be described in detail in Chapter 6.)

In order to delineate the way in which an applied $\boldsymbol{E}_0$ or built-in self-$\boldsymbol{E}$ can alter the concentration profile in a moving, deformable membrane, we formulate the problem for the simple case of a uniformly charged membrane supporting a gradient in the concentration of a single mono–monovalent electrolyte such as NaCl. The outer and inner concentratrions are the bulk concentrations defined as $c_0^1 = c_{+0}^1 = c_{-0}^1$ and $c_0^2 = c_{+0}^2 = c_{-0}^2$, respectively, as shown in Figure 3.24. It is assumed for simplicity that diffusion coefficients inside the membrane, $\bar{D}_+$ and $\bar{D}_-$, as well as the ionic mobilities $\bar{u}_+$ and $\bar{u}_-$, are constant. For an incremental volume in the membrane, the continuity relations (3.139) and (3.140) for mobile positive and negative ions are ($z_+ = |z_-| = 1$)

$$\frac{\partial \bar{c}_+}{\partial t} = \bar{D}_+ \nabla^2 \bar{c}_+ - \frac{\bar{u}_+ \bar{c}_+}{\epsilon} \nabla \cdot \epsilon \boldsymbol{E} - \boldsymbol{E} \bar{u}_+ \cdot \nabla \bar{c}_+ \qquad (3.172)$$

$$\frac{\partial \bar{c}_-}{\partial t} = \bar{D}_- \nabla^2 \bar{c}_- + \frac{\bar{u}_- \bar{c}_-}{\epsilon} \nabla \cdot \epsilon \boldsymbol{E} + \boldsymbol{E} \bar{u}_- \cdot \nabla \bar{c}_- \qquad (3.173)$$

where convection has been neglected. In general, additional terms can be added to (3.172) and (3.173) to account for chemical reactions in the membrane that would lead to production or depletion of the mobile species (see Chapter 1). Here, we assume for simplicity that the monovalent electrolyte of interest does not react with or bind to any groups in the membrane. Note that $\boldsymbol{E}$ in (3.172) and (3.173) is in general the sum of an applied field, $\boldsymbol{E}_0$, and the self-field, or diffusion potential field, which exists if $\bar{D}_+ \neq \bar{D}_-$. Finally, $\boldsymbol{E}$ is related to the fixed mobile and charged species by Gauss' law:

$$\frac{\partial (\epsilon E)}{\partial x} = \rho_e = \bar{\rho}_m + F(\bar{c}_+ - \bar{c}_-) \qquad (3.174)$$

where ϵ is the dielectric constant (assumed uniform) and ρ_e is due to the membrane/tissue volume fixed charge density $\bar{\rho}_m$ and the mobile species concentrations.

For the purposes of this example, it is illuminating to examine the limiting case of a highly charged membrane, for which a simple, closed form solution is often appropriate. To assemble the system of interest, we consider a membrane whose fixed charge density is much greater than the bulk solution concentrations, which are initially equal: $\bar{\rho}_m/F \gg c_0^1 = c_0^2$. A Donnan equilibrium is assumed between membrane and bulk solutions, and for the case $\bar{\rho}_m > 0$, $\bar{\rho}_m$ is balanced by an *equal amount of uniformly distributed mobile negative ions* $\bar{c}_{-0}$ *inside the membrane*. If c_0^2 were now increased slightly above c_0^1 but remained much less than $\bar{\rho}_m/F$, and if $\boldsymbol{E}_0$ were applied across the membrane, then additional mobile ions would enter or leave in perturbational amounts due to diffusion and migration, provided $\boldsymbol{E}_0$ is much less than internal double layer fields, which are on the order of $10^8\,\mathrm{V\,m^{-1}}$.

In the bulk of the membrane phase, $\bar{c}_{-0} \gg \bar{c}_{+0}$. Therefore, from the results of Example 3.6.2, we expect that the minority (co-ion) positive ion will not be significantly influenced by the $\nabla \cdot \epsilon \boldsymbol{E}$ term in (3.172). The majority carriers ($\bar{c}_{-0}$) effectively shield perturbation of minority carriers so that no unbalanced space charge exists in the membrane that could give rise to $\nabla \cdot \boldsymbol{E}$. An applied field *will not disturb electroneutrality* as long as $|\boldsymbol{E}_0|$ is much less than the internal double layer field strengths. However, $\boldsymbol{E} = \boldsymbol{E}_0 + \boldsymbol{E}_{\mathrm{self}}$ *can* still influence the transport of minority co-ions. Therefore, the rightmost term in (3.172) cannot be neglected, as was the case in Example 3.6.2, for which $\boldsymbol{E} = 0$. Therefore, we can focus on the co-ion continuity relation and write

$$\frac{\partial \bar{c}_+}{\partial t} = \bar{D}_+ \frac{\partial^2 \bar{c}_+}{\partial x^2} - \bar{u}_+ E \frac{\partial \bar{c}_+}{\partial x} \qquad (3.175)$$

Inherent to (3.175) and the above discussion is the assumption of quasineutrality in the *bulk* of the uniformly charged membrane, where we may write

$$\bar{\rho}_m + F\left(\bar{c}_+(x) - \bar{c}_{-0} - \tilde{c}_-(x)\right) \simeq 0 \qquad (3.176)$$

which is valid for $\delta \gg 1/\kappa$.

The boundary conditions on the concentrations at $x = 0^+$ and $x = \delta^-$, $\bar{c}_+(0^+)$ and $\bar{c}_+(\delta^-)$, can be calculated approximately from the Donnan equilibrium, where 0^+ and δ^- are several Debye lengths from the edges of the membrane:

$$\bar{c}_+(0^+) \simeq (c_0^1)^2/\bar{c}_- \qquad (3.177)$$

$$\bar{c}_+(\delta^-) \simeq (c_0^2)^2/\bar{c}_- \qquad (3.178)$$

Equations (3.177) and (3.178) are reasonable if the net fluxes of ions across the membrane satisfy (3.88); i.e., they are each less than the individual migration and diffusion fluxes associated with the interfacial regions 0^+ and δ^- (note that these interfacial regions contain significant space charge, where, (3.176) does not apply).

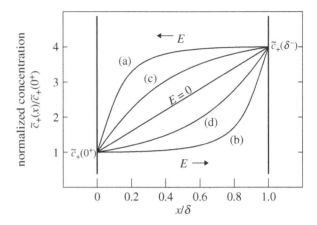

Figure 3.25 Calculated normalized perturbation concentration inside a membrane of thickness δ for the following values of the applied electric field E and $\rho_m > 0$: (a) $E\delta = -200$ mV; (b) $E\delta = +200$ mV; (c) $E\delta = -50$ mV; (d) $E\delta = +50$ mV.

Equation (3.175) is a modified diffusion equation; the steady state $(\partial/\partial t \to 0)$ solution for the cases $E \neq 0$ and $E = 0$ are, respectively

$$\bar{c}_+(x) = \frac{\left[\bar{c}_+(\delta^-) - \bar{c}_+(0^+)\right] e^{+E(x-\delta)/V_T} + \bar{c}_+(0^+) - \bar{c}_+(\delta^-)e^{-E\delta/V_T}}{1 - e^{-E\delta/V_T}}$$

(3.179)

$$\bar{c}_+(x) = \bar{c}_+(0^+) + \left[\bar{c}_+(\delta^-) - \bar{c}_+(0^+)\right] x/\delta$$

(3.180)

where $\bar{c}_+(0^+)$ and $\bar{c}_+(\delta^-)$ are assumed to be known constants and $V_T = RT/F$ is the equivalent thermal voltage $\simeq 25.7$ mV. As $\tilde{c}_-(x) \simeq \tilde{c}_+(x)$ in the bulk, the total intramembrane concentration $\bar{c}_-(x)$ can be easily represented as $\bar{c}_{-0} + \tilde{c}_+(x)$.

Figure 3.25 shows the calculated, normalized concentration profile $\tilde{c}_+(x)/\tilde{c}_+(0^+)$ corresponding to (3.179) and (3.180) for the case $E\delta = 0$, $\pm 2V_T$, $\pm 8V_T$, with boundary concentrations $\bar{c}_+(\delta^-) = 4\bar{c}_+(0^+)$. We note that reversing the polarity of E in (3.179) has the effect of reversing the curvature of $\bar{c}_+(x)$. For a negatively charged membrane, the curvatures change sign for each of the respective cases above.

Example 3.6.5 Transient Electrodiffusion: Find the Equivalent Electrodiffusion Time Constant that Characterizes the Response $c(x, t)$ to an Applied Step in Electric Field, $E_x = E_0 u_{-1}(t)$, for the problem of Example 3.6.4 The intramembrane minority concentration is described by the diffusion equation derived in Example 3.6.4:

$$\frac{\partial \bar{c}_+}{\partial t} = \bar{D}_+ \frac{\partial^2 \bar{c}_+}{\partial x^2} - \bar{u}_+ E_x \frac{\partial \bar{c}_+}{\partial x}$$

(3.181)

which can be used to establish the initial conditions in the membrane. To find the transient solution due to the applied field, we may write (3.181) as

$$\frac{\partial \bar{c}_+}{\partial t} = \bar{D}_+ \frac{\partial^2 \bar{c}_+}{\partial x^2} - \bar{u}_+ E_0 \frac{\partial \bar{c}_+}{\partial x}$$

(3.182)

We assume a solution that is the superposition of two terms:

$$\bar{c}_+(x, t) = \bar{c}_+(x, t = \infty) - \hat{c}(x, t)$$

(3.183)

EXAMPLE

where $\bar{c}_+(x, t = \infty)$ is the steady state solution represented by (3.179) in Example 3.6.4. (Hence, $\bar{c}_+(x, t = \infty)$ automatically satisfies (3.181).) To find $\hat{c}(x, t)$, we may proceed by recognizing that the substitution of variables

$$\hat{c}(x, t) = e^{-(-ax + a^2\alpha^2 t)} m(x, t) \qquad (3.184)$$

will transform (3.181) into a standard diffusion equation

$$\frac{\partial^2 m}{\partial x^2} = \frac{1}{\alpha^2} \frac{\partial m}{\partial t} \qquad (3.185)$$

Alternatively, we may recognize that a variable-separable solution of the form

$$\hat{c}(x, t) = \left(e^{\lambda x} \sin kx\right) e^{-t/\tau} \qquad (3.186)$$

satisfies the original modified diffusion equation (3.181). Substitution of (3.186) into (3.181) yields the relation

$$0 = +\frac{1}{\tau}(e^{\lambda x} \sin kx)e^{-t/\tau}$$
$$+ e^{-t/\tau}\left[\bar{D}_+\left(-k^2 e^{\lambda x} \sin kx + 2\lambda k e^{\lambda x} \cos kx + \lambda^2 e^{\lambda x} \sin kx\right)\right.$$
$$\left. - E_0 \bar{u}_+ \left(k e^{\lambda x} \cos kx + \lambda e^{\lambda x} \sin kx\right)\right] \qquad (3.187)$$

But (3.187) can only be satisfied if the $\sin kx$ and $\cos kx$ terms individually sum to zero. These conditions yield the two relations

$$\frac{1}{\tau} = (k^2 - \lambda^2)\bar{D}_+ + E_0 \bar{u}_+ \lambda \qquad (3.188)$$

$$2\lambda \bar{D}_+ = E_0 \bar{u}_+ \qquad (3.189)$$

where the Einstein relation can be used to transform (3.189) to

$$\lambda = \frac{E_0}{2RT/F} \qquad (3.190)$$

It can be seen that (3.186) will satisfy the initial and boundary conditions only if $\hat{c}(x)$ is represented as a superposition of eigenfunctions $\sin k_n(x)$ with eigenvalues $k_n = n\pi/\delta$ (δ being the membrane thickness). Therefore, with $\tau \to \tau_n$ and $k \to k_n$, (3.188) will give the desired time constants τ_n. Substitution of (3.190) into (3.188) and division of all terms by $\bar{D}_+$ yields

$$\frac{1}{\tau_n \bar{D}_+} = \frac{n^2 \pi^2}{\delta} + \frac{E_0^2}{4(RT/F)^2} \qquad (3.191)$$

which becomes

$$\tau_n = \frac{\delta^2}{\pi^2 \bar{D}_+}\left(\frac{1}{n^2 + [E_0\delta/(2\pi RT/F)]^2}\right) \qquad (3.192)$$

The effect of the magnitude of the applied E_0 can be seen from (3.192). In terms of the $n = 1$ term of the Fourier sine series ($n = 1, 2, 3, \ldots$), the limit $|E_0\delta| \ll RT/F$ reduces (3.192) to a simple diffusion time. However, for $|E_0\delta| \gg RT/F$, the *electrodiffusion time* can be significantly shorter than τ_{diff}.

3.7 PROBLEMS

Problem 3.1 This problem considers a mathematically simplified model for the microchannel flow-measuring technique pictured in Figure 3.2. However, we will now compare two alternative approaches, that of injecting net charge ρ_e (not pictured in Figure 3.2) versus injecting a concentrated but electroneutral "pulse" of electrolyte solution. The flow channel is treated here as two-dimensional (see Figure 3.26), and the fluid (e.g., blood) is modeled as a homogeneous, incompressible fluid, with *uniform* conductivity σ and permittivity ϵ. The fluid moves to the right in Figure 3.26 with a velocity profile $v = v_x(y)\boldsymbol{i}_x$ as shown. (Note that v satisfies $\nabla \cdot v = 0$).

Method I At $t = 0$, an initial free charge distribution $\rho_e(x, y, t = 0) = \rho_0 e^{-(x/d)^2}$ is placed in the fluid stream as shown in Figure 3.27. (ρ_e fills the channel in the y direction and is Gaussian in the x direction.) An electrode is placed 5 cm downstream in the channel wall so as to measure any charge ρ_e that leaks to the wall and thereby measure a signal proportional to flow velocity. We wish to find the way in which the free charge relaxes as the fluid moves, and whether or not such a velocity measuring scheme is feasible ($\sigma \simeq 1\,\mathrm{S\,m^{-1}}$ assumed).

(a) Assume that the ohmic model of Section 3.1 applies, so that free charge in the fluid is governed by (3.30). Thus, charge relaxation occurs in a manner similar to that described by (2.80) in Chapter 2, except that relaxation now proceeds with respect to a *moving* fluid. Hence, we look for a solution to (3.30) with x replaced by $x' = x - v_x(y)t$ in $\rho_e(x, y, t)$.

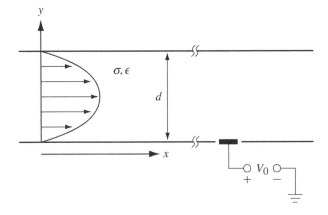

Figure 3.26 Parabolic fluid velocity profile $v = v_x(y)\,\boldsymbol{i}_x = \left[(Uy/d)(1 - y/d)\right]\boldsymbol{i}_x$, $U = 4\,\mathrm{cm\,s^{-1}}$ in a channel of width d.

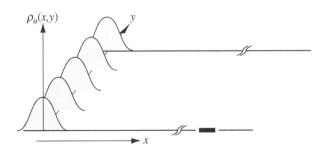

Figure 3.27 An initial free charge distribution $\rho_e(x, y, t = 0) = \rho_0 e^{-(x/d)^2}$ is placed in the fluid stream at $t = 0$, filling the channel in the y direction.

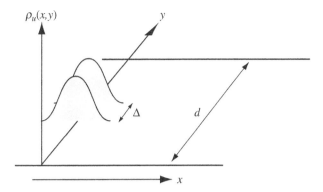

Find a solution to (3.30) (you can guess a solution by the suggested technique and then substitute it into (3.30) to prove that it is a valid solution).

(**b**) Sketch $\rho_e(x, y)$ at $t = \Delta t$ and $t = 2\Delta t$ on the same graph as Figure 3.27 showing the redistribution of ρ_e as the fluid moves, where Δt is a relatively short time interval.

(**c**) Repeat parts (a) and (b) for the case where ρ_e does *not* fill the channel in the y direction, but rather is a pulse of thickness Δ located in the center of the channel as in Figure 3.28.

(**d**) Discuss the feasibility of the flow measurment schemes suggested in parts (a)–(c).

Method II A pulse of concentrated NaCl having exactly the same shape as that of Figure 3.27 is injected into the channel rather than the pulse of ρ_e.

(**e**) Sketch qualitatively $c_{\text{NaCl}}(x, y)$ for $t = \Delta t$ and $t = 2\Delta t$ as the fluid moves. Discuss the feasibility of this technique and the kinds of downstream measurements that could be used to find the fluid velocity (i.e., concentration, conductivity, etc.). The mathematical solution to this problem was derived by GI Taylor [1] and will be discussed in more detail in Chapter 5 and Problem 5.7. Variations of this technique are still used today.

Problem 3.2 Electromagnetic Blood Flow Meter The electromagnetic flowmeter has been optimized for noninvasive blood flow measurement in *intact* vessels. It can be used for vessels as large as the aorta and as small as 1 mm diameter, with ±5% accuracy on larger vessels. Furthermore, we will show below that output voltage is proportional to the mean flow velocity, independent of the exact fluid velocity *profile*! You will be able to compare the measurement theory with that of the prototype electroquasistatic flowmeters of Problem 3.1.

Figure 3.29(a) shows an example of an electromagnetic flow probe; Figure 3.29(b) is a schematic of the magnetic field and measurement configuration.

(**a**) Assume that the fluid velocity profile is uniform across the vessel cross-section (a good approximation for turbulent flow). Use (2.92) to calculate the measured voltage V_0 (sign and magnitude) in terms of B_0 and flow velocity U. Assume that the vessel wall has approximately the same

Figure 3.29 Electromagnetic flow probes.

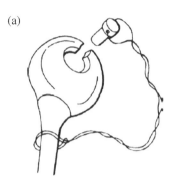

(a)

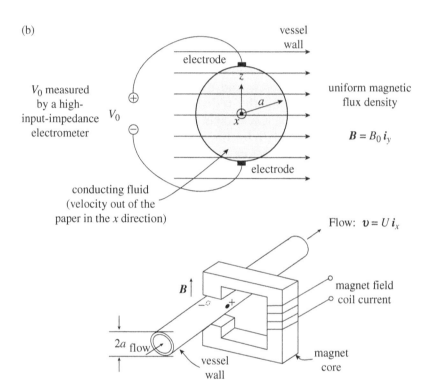

(b)

V_0 measured by a high-input-impedance electrometer

V_0

conducting fluid (velocity out of the paper in the x direction)

electrode

vessel wall

a

electrode

uniform magnetic flux density

$\boldsymbol{B} = B_0\, \boldsymbol{i}_y$

Flow: $\boldsymbol{v} = U\boldsymbol{i}_x$

$\boldsymbol{B}$

magnet field coil current

$2a$ flow

vessel wall

magnet core

conductivity as blood, and that there is no saline around the outside of the vessel to shunt V_0 (see Figure 3.29(a) and (c)). (The input impedance of the meter is very high.)

(b) To examine the dependence of V_0 on the fluid velocity profile, we examine the two-dimensional flow geometry of Figure 3.30. For mathematical simplicity, assume that the fluid velocity does not vary in the y direction ($w \gg a$):

$$v = \frac{3}{2}U\left(1 - \frac{z^2}{a^2}\right)\boldsymbol{i}_x \qquad (3.193)$$

(Note that the parabolic profile (3.193) can be derived in a straightforward manner, as will be shown in Chapter 5. What is the mean velocity?) This simplifies the calculation of the electric field. Find $\boldsymbol{E}$ inside the "vessel" and the resulting V_0; compare with part (a). In practice, the vessel wall conductivity $\neq \sigma_{\text{blood}}$ and correction factors (i.e., calibration) must be included.

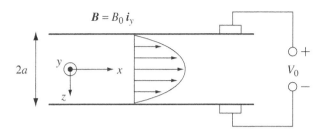

Figure 3.30 Parabolic flow.

The theory and extensive clinical use of these meters are discussed in detail in the pioneering publications by Shercliff [26] and Cobbold [27]. Recent nonmedical applications have also been discussed in the literature [28].

Problem 3.3 Coupled Diffusion Step Response: Characteristic Diffusion Time In Example 3.2.1, coupled diffusion of a binary electrolyte across a neutral membrane was found to occur with an effective diffusion coefficient

$$\bar{D}^* = \frac{2\bar{D}_+\bar{D}_-}{\bar{D}_+ + \bar{D}_-} \tag{3.194}$$

(a) Starting with the flux equation (3.1), use the results of Section 3.2 relating the concentration and potential profiles within a membrane (of thickness $\delta \gg 1/\kappa$) to show that

$$\bar{N}_+ = -\bar{D}^*\frac{d\bar{c}}{dx}; \text{ given } \bar{c}_+ \simeq \bar{c}_- \equiv \bar{c} \tag{3.195}$$

and similarly for $\bar{N}_-$.

We now solve the *transient diffusion problem* where the concentration takes the form of a step, with

$$c'_+ = c'_- = (c'_0 - c''_0)u_{-1}(t) + c''_0. \quad \begin{cases} c'_0 \text{ and } c''_0 \text{ for } t > 0 \text{ are} \\ \text{maintained constant by} \\ \text{continuous recirculation} \end{cases}$$

$$c''_+ = c''_- = c''_0 \text{ for all time}$$

$$\tag{3.196}$$

where c' is the concentration to the left ($x \le 0$) and c'' that to the right of the membrane ($x \ge \delta$); see Figure 3.31.

Figure 3.31 Coupled diffusion across a membrane of thickness δ.

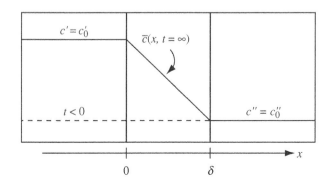

PROBLEM

(**b**) Show that the differential equation that describes the one-dimensional transient diffusion of a binary electrolyte in the region $0 \leq x \leq \delta$ is

$$\frac{\partial \bar{c}(x, t)}{\partial t} = \bar{D}^* \frac{\partial^2 \bar{c}(x, t)}{\partial x^2} \qquad (3.197)$$

where $\bar{c}$ is the concentration of electrolyte inside the membrane and $\bar{D}^*$ is the effective diffusion coefficient inside the membrane. Assume quasineutrality and neglect convection.

(**c**) Write the boundary conditions for $\bar{c}(x, t)$ valid at times $t > 0$, and write the initial condition $\bar{c}(x, t = 0^+)$.

(**d**) Express $\bar{c}(x, t)$ in the form

$$\bar{c}(x, t) = \hat{c}(x, t) + \bar{c}_0(x) \qquad (3.198)$$

where $\bar{c}_0(x)$ is the concentration at infinite time, $\bar{c}(x, t = \infty)$, and $\hat{c}(x, t)$ is a transient term that approaches zero as $t \to \infty$. Find $\bar{c}_0(x)$. Using the results of parts (b) and (c), find a governing equation with boundary and initial conditions for $\hat{c}(x, t)$. Solve for $\hat{c}(x, t)$. (Separation of variables is one possible method of solution. To match the boundary and initial conditions, $\hat{c}(x, t)$ will be a Fourier series expansion in terms of $\bar{c}_0(x)$.

Find $\bar{c}(x, t)$ by finding a consistent set of eigenfunctions, diffusion times τ_n, and Fourier coefficients for $\hat{c}(x, t)$.

(**e**) Evaluate the longest diffusion time $\tau_{n=1}$ for $\delta = 10$ nm and $\delta = 0.5$ mm, assuming a value of $\bar{D}^*$ corresponding to that in water (e.g., dilute NaCl).

(**f**) Find $\Delta\Phi_{\text{diff}}(t) = \Phi'' - \Phi'$ for $t > 0^+$. Carefully consider the diffusion transient phase after many charge relaxation times ϵ/σ. (Note that $t = 0^+$ means $t \gg \epsilon/\sigma$, i.e., after many relaxation times).

(**g**) Find the total amount of electrolyte gained by the membrane (per unit membrane area) at time t.

Problem 3.4 Membrane Electrodes In parts (a) and (b) below, specify all relevant properties of a desired membrane (e.g., neutral or heavily charged, dimensions, etc.), and all other materials needed, for the respective designs.

(**a**) Design an "electrode" capable of measuring an unknown concentration of a known single binary electrolyte. The "electrode" must consist of a membrane along with any other materials to be specified, and should be used in conjunction with a "reference electrode" (specify), both of which are to be dipped simultaneously into the unknown solution.

Write an expression for the unknown concentration in terms of the measured "electrode" potential.

(**b**) Design a simple "electrode" that will measure single-ion concentrations (activities) of a given species. The species can be one of many in a multi-ionic solution, but it is known to be the *only one present having a given sign* (+ or −); all other ions present are of opposite sign. This form of "ion-selective electrode" will again consist of a membrane (specify properties) and other materials, including a reference electrode (specify). Write an expression for the unknown ion concentration (activity) in terms of the measured electrode potential.

For parts (a) and (b), briefly discuss the limitations of your design (e.g., the effect of unknown contaminating ions or salts, specificity of designed "electrode" to the measurement of a single species or binary salt, etc.) The thought models of parts (a) and (b) are the "forerunners" of modern ion-exchange specific-ion electrodes.

Problem 3.5 The Validity of the Quasineutrality Assumption Write the potential profile $\Phi(x)$ of (3.40) in terms of ionic transport numbers instead of diffusivities. Then:

(**a**) Find the field $E_x(x)$ and the space charge $\rho_e(x)$ corresponding to the potential $\Phi(x)$.

(**b**) Write an analytical expression for the ratio $|\rho_e(x)/\rho_0| \equiv |(c_+ - c_-)/c|$ at $x = 0$ and at $x = \delta$. Show that your answer can be expressed in two different forms, first as a ratio that is proportional to $|(1/\kappa)/\delta|^2$ and second as a ratio proportional to

$$\frac{\tau_{\text{charge relaxation}}}{\tau_{\text{diffusion}}}$$

(**c**) Calculate the ratio obtained in part (b) using typical physiological values for parameters and show that for $\delta = 100\,\mu\text{m}$,

$$\left|\frac{\rho_e(\delta)}{\rho_0}\right| \ll 1 \tag{3.199}$$

What about $\delta = 10\,\text{nm}$?

(**d**) Discuss the validity of the quasineutrality assumption in terms of a ratio of lengths and a ratio of time constants (refer to part (b)).

(**e**) Show that an expression for the Debye length can be written in the form

$$\frac{1}{\kappa} \propto \sqrt{D\tau_{\text{rel}}}, \quad \text{where } \tau_{\text{rel}} \sim \frac{\epsilon}{\sigma} \tag{3.200}$$

Interpret this result in terms of a balance between migration and diffusion.

Problem 3.6 Theoretical and Measured $V_0(t)$ for Transient Electrolyte Diffusion Across a Neutral Membrane, with an Initially Imposed Concentration Gradient In Sections 3.2 and 3.3, we developed a theoretical expression for the steady state diffusion potential resulting from *binary*

Figure 3.32 Measurement of transient diffusion potential with reversible Ag/AgCl electrodes. The cell volume is V on both sides of the membrane.

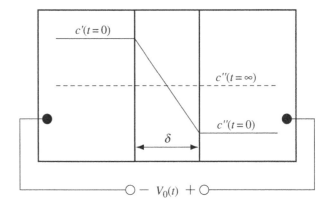

electrolyte diffusion across a neutral membrane. It was assumed that the concentrations c' and c'' were constant in time even in the face of irreversible carrier flow from left to right. This assumption is plausible if (1) the solutions on both sides are continually renewed, or (2) the cell volumes are so large that finite carrier flow will hardly perturb c' and c'' during the time needed to measure V_0, even without renewal of the solution. The validity of the latter approach can be tested by *finding the rate of decay of an initially imposed concentration gradient resulting from irreversible diffusion.*

We will find that the measured decay rate can provide additional useful information concerning several other assumptions made in Section 3.2. First, the assumption that $\bar{D}^* = D$ (the salt diffusion coefficient inside the membrane is identical with that in pure water) is valid only in a few instances, and certainly *not* for the case of a charged membrane. The measurements of the decay rate can be considered an experimental measurement of $\bar{D}^*$, provided that the membrane thickness δ is known. Conversely, δ can be predicted from the decay rate if $\bar{D}^*$ is known. This is particularly important when inadequate stirring of the solutions leads to the formation of stagnant electrolyte films at each membrane interface. Such films result in an uncontrollable increase in the "effective" membrane thickness—a problem that plagues all steady state membrane diffusion problems. This problem relates back to the assumption that convection is unimportant.

With respect to the experimental configuration of Figure 3.32, the chambers on both sides of the membrane have equal volumes V and cross-sectional areas A equal to that of the membrane. The concentrations $c'(t)$ and $c''(t)$ are uniform throughout the bulk on both sides of the membrane owing to efficient stirring. A linear concentration profile is assumed for electrolyte in the membrane (see (3.39)). The flux of *salt*, N, from left to right is (see Example 3.2.1)

$$N = -\bar{D}^* \left(\frac{c'' - c'}{\delta} \right) \tag{3.201}$$

Assume that $c'(0^+) = c'(0^-)$ and $c''(\delta^+) = c''(\delta^-)$ for all time. Assume also that $\bar{c}_0(x, t)$ is linear at all times. Neglect stagnant films at membrane surfaces.

(**a**) Derive a differential equation that represents the time rate of change of the uniform concentration $c'(t)$ as salt diffuses from left to right. This can be done by relating the concentration $c'(t + \Delta t)$ to the concentration at time t, $c'(t)$, and the amount of salt leaving the left side in time Δt due to salt diffusion. You will need one other relation between $c''(t)$ and $c'(t)$ to obtain the differential equation; namely, show first that

$$c''(t) + c'(t) = \text{ constant } = K \qquad (3.202)$$

(**b**) Obtain the solution to the differential equation. Show that the solution satisfies the expected infinite-time response, $c'(t = \infty) = K/2$. Your solution should be representable in terms of an exponentially decaying response, with time constant τ. Compare τ with the diffusion time $\tau_{\text{diff}} \sim \delta^2/2D$ of Problem 3.3. Clarify the validity of the "steady state" diffusion potential measurement. *Discuss the assumptions stated after* (3.201).

(**c**) Show that the total measured potential $V_0(t)$, using Ag/AgCl electrodes takes the form

$$V_0(t) = \frac{2RT}{F}(1 - t^- + t^+) \tanh^{-1}\left[e^{-t/\tau}\left(1 - \frac{2c''(t=0)}{K} \right) \right]$$

where τ is the same as that in part (b) above.

(**d**) Now assume that the Ag/AgCl electrodes are connected to the electrolyte baths of Figure 3.32 via saturated KCl salt bridges. Find $V_0(t)$.

Problem 3.7 Ion Exchange Membranes as Models of Biological Membranes In Example 3.2.2, the limiting case of the diffusion potential for an ideal semipermeable (e.g., heavily charged) membrane is discussed. In this limit, diffusion of electrolyte across the membrane is completely stopped owing to the constraint of bulk electroneutrality.

Some specialized biological membranes are believed to be "mosaics" composed of patches of membrane permeable to either anions or cations alone (i.e., "anion- and cation-selective" patches, respectively) and arranged in a quilt-like fashion along the plane of the membrane. Such membranes are heterogeneous in their permeability properties, unlike those we have been studying so far. One such membrane is shown schematically in Figure 3.33(a) (+ + + represents fixed positive charge corresponding to a patch permeable only to *negative* ions, etc.). Early investigators puzzled over the possibility of electrolyte diffusion across such a model mosaic membrane, keeping in mind the actual biological analog. (Remember, with a pure cation- or anion-selective membrane, no diffusion can occur.)

To answer the question, experiments were performed with model systems such as those of Figure 3.33(b) and (c), composed of pure cation and anion selective membranes.

(**a**) In Figure 3.33(b), calculate the potential of the upper left-hand solution with respect to the upper right-hand solution

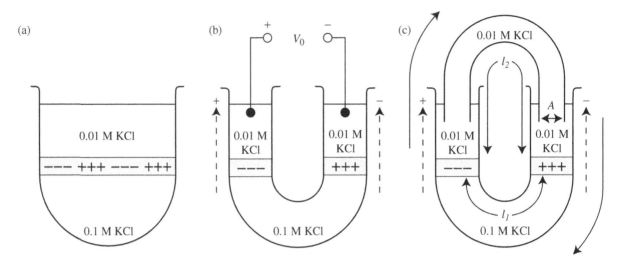

Figure 3.33 Pictoral development of the theory of electrolyte permeability of mosaic membranes that are composed of ideally anion-selective and ideally cation-selective parts. (a) Mosaic membrane with adjacent cation-selective and anion-selective parts. (b) Spatial and electrical separation of the cation-selective and anion-selective parts of the membrane. (c) The spatially separated cation-selective and anion-selective parts joined electrically by a bridge of solution in an all-electrolyte circuit.

that would be measured using Ag/AgCl electrodes *with* salt bridges in conjunction with a high-input-impedance electrometer. The electrolytes (and their concentrations) on both sides of the two membranes are shown in the figure. (*Neglect* activity coefficients.)

(b) The upper solutions are now connected via a 0.01 M KCl salt bridge as shown in Figure 3.33(c). Give a brief description of the events in the overall system that occur after the salt bridge connection is made.

(c) Find the direction and magnitude of the total current I immediately after the salt bridge connection is made in Figure 3.33(c). Plug in numbers! To make the calculations easy, assume that the cross-sectional areas A of all tubular sections are identical and equal to $1\,cm^2$. Assume $l_1 = 10\,cm$ and $l_2 = 20\,cm$; neglect the thickness of the membranes.

(d) What are the qualitative implications of your analysis in parts (a)–(c) for transport and potential drops in the system of Figure 3.33(a)?

Historical literature related to this problem includes the article by Sollner [29].

Problem 3.8 Goldman Equation At the heart of the Hodgkin and Katz model for the resting potential of biological membranes is the "Goldman equation," which expresses the resting potential in terms of membrane permeability to the three ions, Na^+, K^+, and Cl^-

$$\Phi'' - \Phi' = \frac{RT}{F} \ln\left(\frac{P_K c_K' + P_{Na} c_{Na}' + P_{Cl} c_{Cl}''}{P_K c_K'' + P_{Na} c_{Na}'' + P_{Cl} c_{Cl}'}\right) \tag{3.203}$$

PROBLEM

The permeability coefficients P_i are empirical constants for any given membrane, which are related to c_i', $\bar{c}_i'$, $\bar{c}_i''$, and c_i'' according to the definitions

$$\bar{c}_i' = \beta_i c_i' \quad \text{at} \quad x = 0$$
$$\bar{c}_i'' = \beta_i c_i'' \quad \text{at} \quad x = \delta \tag{3.204}$$

$$P_i = \frac{u_i \beta_i RT}{F\delta} = \frac{D_i \beta_i}{\delta} \tag{3.205}$$

In essence, all of the microscopic details of the effect of membrane *structure* and *charge* on the transport of ions are lumped into the interfacial constants P_i or β_i, which are determined empirically. The membrane itself is taken to be neutral. (Note that $\bar{c}_i'$ and $\bar{c}_i''$ are the concentrations just inside the membrane at $x = 0^+$ and $x = \delta^-$ respectively). β_i is sometimes called the *partition coefficient*.

In this problem, we wish to compare the Goldman equation with the model for diffusion potentials presented thus far. The basic assumption in the Goldman treatment is that the field across the membrane is constant:

$$-E_x(x) = \frac{\Phi'' - \Phi'}{\delta} \tag{3.206}$$

(a) Write the flux equation for Na$^+$, K$^+$, and Cl$^-$ ions *inside* the membrane including the constant-field assumption. Using one of the three equations, find the concentration profile $\bar{c}_i(x)$ and potential profile $\Phi(x)$ inside the membrane. Express your answer in terms of the coefficients J_i, $\bar{c}_i$, Φ'', and Φ'.

(b) Compare your results for $\bar{c}_i(x)$ and $\Phi(x)$ with those derived in Section 3.2 by sketching the two different solutions directly on Figure 3.34 for the ith species. Under what limiting conditions do the two different approaches yield the same results?

(c) Finish the derivation of the Goldman equation by writing out J_{Na}, J_K, and J_{Cl}. Use (3.204) and (3.205) to replace $\bar{c}_i$ with c_i and P_i. Equation (3.203) results by using the open circuit condition $\Sigma_i J_i = 0$.

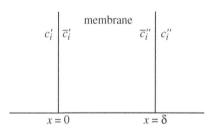

Figure 3.34 Internal and external concentrations.

Problem 3.9 Donnan Equilibrium We first treat the membrane system (Figure 3.11) originally described by Donnan (parts (a)–(c)). The results are then applied, by analogy, at each of the interfaces of a charged membrane system like that of Figure 3.18 (part (d)).

(a) With respect to Figure 3.11, write statements of equilibrium in terms of the electrochemical potentials $\bar{\mu}_i$ of all appropriate species (including the *solvent*) on both sides of the membrane. Use your results to find the Donnan potential drop $\Delta\Phi_D$ across the membrane in terms of the bulk concentrations (on both sides of the membrane) of all appropriate species. (Attention should be paid to the use of the pressure–volume term in writing the electrochemical

PROBLEM

potential; see any good physical chemistry text as a reference for parts (a) and (b).)

(**b**) Find the osmotic pressure, $\Delta \Pi^{os} = P' - P''$ (i.e., left side − right side), by using the equilibrium relation for the solvent written in part (a). Express your answer in terms of ion concentrations on both sides.

(**c**) Write statements of electroneutrality for the bulk phases on *both* sides of the membrane. Use your results to express the Donnan ratio r in terms of $\bar{\rho}_m$ and the *bulk electrolyte concentration* c_0, on the *right* side of Figure 3.11, where

$$r \equiv \frac{\bar{c}_{1+}}{c_{1+}} = \frac{\bar{c}_{2+}}{c_{2+}} = \cdots = \frac{\sum \bar{c}_{i+}}{\sum c_{i+}} = \frac{-(\bar{\rho}_m/F)}{2 \sum_i c_{i0}} + \left[\left(\frac{\bar{\rho}_m/F}{2 \sum_i c_{i0}} \right)^2 + 1 \right]^{1/2}$$
(3.207)

$$r \equiv \frac{c_{1-}}{\bar{c}_{1-}} = \frac{c_{2-}}{\bar{c}_{2-}} = \cdots = \frac{\sum c_{j-}}{\sum \bar{c}_{j-}}$$
(3.208)

Write the Donnan potential in terms of r.

(**d**) Calculate the total measured potential across the charged membrane of Figure 3.18 for the case of AgCl electrodes using the TMS theory, given the following values:

electrolyte (left side): 0.2 M NaCl + 0.001 M HCl

electrolyte (right side): 0.05 M NaCl + 0.001 M HCl

$\bar{\rho}_m = 0.1$ M (this is the membrane fixed charge known to correspond to pH = 3.)

Make reasonable approximations and assumptions; neglect activity coefficients.

Problem 3.10 Prediction of Intramembrane Ion Concentrations $\bar{c}_i$ Using the Donnan Equilibrium Model In Problem 3.9, the Donnan equilibrium model was developed for the case of a charged membrane supporting gradients in single, *binary electrolyte* concentration. In this problem, we extend the analysis to the *multi-ion* case, which is more commonly found in practice.

In Example 3.4.1, it was shown that the Donnan relation between intramembrane concentrations and external bath concentrations holds for each and every ion in equilibrium. If one particular intramembrane ion concentration $\bar{c}_i$ was known, then all other $\bar{c}_j$ could be found from relations such as (3.95)–(3.97). However, we may *not* know any of the $\bar{c}_i$. Instead, biochemical analysis may tell us the value of $\overline{\rho}_m$ for the membrane. (We assume a *spatially uniform* $\bar{\rho}_m$ throughout this problem.)

Given the system of Figure 3.14, a membrane with positive $\bar{\rho}_m$ is in equilibrium with a bath having known external concentrations c_{Na^+}, c_{Cl^-}, and c_{H^+}. Find the average internal concentrations $\bar{c}_{Na^+}$, $\bar{c}_{H^+}$, and $\bar{c}_{Cl^-}$ in terms of $\bar{\rho}_m$ and the external concentrations. Assume that *quasineutrality* holds *inside the membrane*. (Express your answer as analytical expressions.)

Problem 3.11 Electrical Conductivity as a Measure of Charge Density $\bar{\rho}_m$ and of the Isoelectric Point (pI) of the Material

Background: A measurement of bulk *electrical conductivity* σ of a "plug" of ribonuclease has been made as a function of pH. Can the data be interpreted so as to predict (i) the isoelectric point (the pH value at which there is zero *net* macromolecular *fixed* charge), and (ii) the charge $\bar{\rho}_m$ as a function of pH?

Experimental Procedure

(1) The RNase plug is equilibrated in a bathing solution containing only a known concentration c_0 of NaCl at a given value of pH (Figure 3.35). The pH is adjusted by proper amounts of HCl or NaOH. (Assume that $c_0 \gg c_{HCl}$ or c_{NaOH} needed to adjust pH.)

(2) The RNase plug is *removed*, surface water is padded dry, and the plug is placed between electrodes (Figure 3.36). Its low-frequency conductivity σ is measured (assume that electrode interfacial impedances, etc., have been accounted for).

(3) The procedure is repeated over a broad pH range ($3 \lesssim$ pH $\lesssim 11$).

(a) Find the conductivity $\bar{\sigma}$ of the *interstitial water* of the RNase plug in terms of (1) $\bar{\rho}_m$ (the known, averaged macroscopic fixed charge density of the plug, in coulombs per liter of interstitial water, (2) "intra-plug" ion mobilities $\bar{u}_{Na}$ and $\bar{u}_{Cl}$, (3) c_0 (external [NaCl]), and (4) other relevant constants. For simplicity, HCl and/or NaOH concentrations may be neglected in the quasineutrality assumption, since they are so small.

(b) Invert the answer to part (a) to find $\bar{\rho}_m$ as a function of $\bar{\sigma}$, c_0, $\bar{u}_\pm$, and other constants.

(c) An independently measured titration curve of ribonuclease is shown in Figure 3.37. Also included is the amino acid

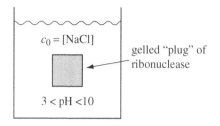

Figure 3.35 Ribonuclease gel plug in equilibration solution.

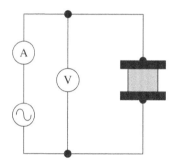

Figure 3.36 Conductivity measurement.

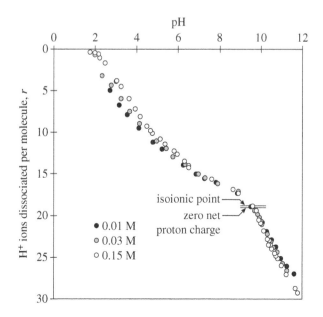

Figure 3.37 Titration data for ribonuclease at 25°C and ionic strengths 0.01 M, 0.03 M, and 0.15 M [30]. The plot shows the point of zero net charge (the isoelectric point, corresponding to $r = 19$) and also the isoionic pH at the particular concentration that was used.

Table 3.1 Titratable groups of ribonuclease [30].

	pK_{int} expected from data on small molecules	Number of sites from amino acid analysis	Data obtained from titration curve Number of sites	pK_{int}
α-COOH	3.75	1	(1)	—
Side-chain COOH	4.6	10	11 { 10	4.7
Imidazole	7.0	4	4	6.5
α-NH$_2$	7.8	1	5 { (1)	7.8
Phenolic	9.6	6	6	9.95
Side-chain NH$_2$	10.2	10	16 { 10	10.2
Guanidyl	>12	4	4	>12

Numbers in parentheses were assumed correct because of the overwhelming evidence that ribonuclease consists of a single polypeptide chain.

analysis of the same RNase and the pK values of its various positive and negative charge groups (Table 3.1). From these data, the *isoelectric point* of RNase is independently estimated to be at pH $\simeq$ 9.5 (Figure 3.37).

How do you expect the *isoelectric point* to manifest itself in the measured curve of σ versus pH? Sketch approximately the expected variation of measured σ with pH (just trends, no conductivity values).

(**d**) Comment on the difference between the bulk, measured σ and the calculated $\bar{\sigma}$ of interstitial water (part (a)). How might you correct the measured value of σ so as to predict $\bar{\sigma}$?

Problem 3.12 pH Electrode Figure 3.8(a) shows one configuration of a glass pH electrode. Figure 3.38 is a schematic for modeling the potential drops associated with the total electrode potential which is translated into a pH reading. The total potential drop (Figure 3.38) between the inner buffer and the solution of unknown pH can be modeled as the sum of *four* contributions: a Nernstian equilibrium potential between each glass surface (inner and outer) and its adjacent solution phase, and two "ion-exchange diffusion potentials" within the glass phase.

Figure 3.38 Schematic for potential drops associated with the "responsive" pH glass element. (Modern pH electrodes incorporate a "combination" of glass, inner and outer reference within a single body.)

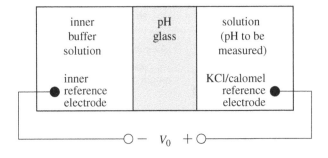

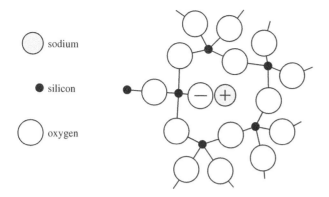

Figure 3.39 Structure of sodium silicate glass showing "fixed charge" (oxygen) and mobile Na^+ ions within the silica matrix.

The origin of these latter diffusion potentials is thought to be as follows. The pH glass can be modeled as a charged membrane having fixed charge (due to ionized oxygen in the silica matrix: Figure 3.39) and mobile counter-ions (e.g., Na^+ ions in sodium silicate glass). When the pH glass is brought into contact with a solution containing H^+ ions, these H^+ ions are preferentially exchanged with the sodium ions (see Example 3.6.3). This ion exchange equilibrium is defined by the reaction

$$[H^+]^{soln} + [\overline{Na^+}]^{glass} \rightleftharpoons [\overline{H^+}]^{glass} + [Na^+]^{soln} \tag{3.209}$$

$$K \equiv \frac{[\overline{H^+}]^{glass}[Na^+]^{soln}}{[H^+]^{soln}[\overline{Na^+}]^{glass}} \tag{3.210}$$

Because the exchange involves only a *small* fraction of ions, the concentration of mobile Na^+ ions within the inner regions of the glass can be assumed to remain constant ($\equiv \bar{c}_0$) over the useful lifetime of the electrode. Hence, the ion exchange processes occur within thin regions on each side of the glass, and the inner glass material remains unaffected. We may also assume that, within the glass, electroneutrality is satisfied:

$$\bar{c}_H(x) + \bar{c}_{Na}(x) = \text{const} = \bar{c}_0 \tag{3.211}$$

where the last equality assumes that $\bar{c}_H = 0$ in the middle of the glass throughout its useful lifetime.

(**a**) Since the inner buffer solution (Figures 3.8(a) and 3.38) is never changed, justify the assumption that V_0 can be written in the form

$$V_0 = \text{const} - \Delta\Phi_{Donnan} - \Delta\Phi_{diff} \tag{3.212}$$

(**b**) Assume that the interface between the glass surface and the unknown solution (Figure 3.40) reaches a quasiequilibrium (why is this possible?), which can be modeled by equations analogous to that of the AgCl electrode: (3.65) and (3.67). Write an analytical expression for $\Delta\Phi''$ (interfacial) in terms of the concentration of Na^+ ions in the bulk solution, $\bar{c}_{Na}$,

Figure 3.40 pH glass interface.

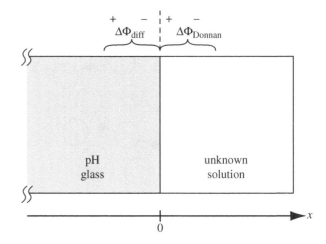

and that in the surface layers of the glass, $\bar{c}''_{Na}$. Thus, assume that the activity of Na$^+$ ions in the hydrated glass phase is equal to their concentration, unlike the assumption that led to (3.68). Write a similar expression for $\Delta\Phi''$ in terms of $\bar{c}_H$ and $\bar{c}'_H$. Note that the standard potential (i.e., the constant $\Delta\Phi^\circ$) is different for H$^+$ and Na$^+$.

(c) Use the method of Example 3.6.3 to find $\Delta\Phi''_{diff}$ in terms of $\bar{c}_0$, $\bar{c}''_{Na}$, $\bar{D}_{Na}$, and $\bar{D}_H$. Does the value of this diffusion potential depend on the profile of ions within the glass?

(d) Use (3.210) and (3.211) to express V_0 in terms of K, c_H, and c_{Na} (i.e., bulk solution concentrations) and the $\bar{D}_i$. Your answer shows the c_H dependence of V_0 and the so-called *sodium error* of the electrode, as manifested by K and c_{Na}. Thus, the sodium error is pH-dependent—an empirically well-known phenomenon!

(e) Examine the limiting V_0 for the cases $c_{Na} \gg Kc_H$ and $c_{Na} \ll Kc_H$.

Problem 3.13 Steady State Electrodiffusion Carry out a detailed derivation of the steps in Example 3.6.4, and then consider electroosmotic convection and intramembrane conductivity by referring to Figure 3.25 and carrying out the following steps:

(a) To formulate the electrodiffusion problem, write the complete (time-varying) continuity and flux equations for $\bar{c}_+(x,t)$ and $\bar{c}_-(x,t)$ inside the membrane, along with Gauss' law (neglect convection). Together, these relations will enable us to solve for intramembrane concentration profiles of mobile ions.

(b) Assuming that $|\bar{\rho}_m/F| \gg c'_0, c''_0$, then, in the bulk of the positively charged membrane, we may represent the co-ion and counter-ion concentrations, respectively, as

$$\bar{c}_+ \simeq \tilde{c}_+(x), (\bar{c}_{+0} \simeq 0) \qquad (3.213)$$

$$\bar{c}_- \simeq \bar{c}_{-0} + \tilde{c}_-(x) \qquad (3.214)$$

where

$$\tilde{c}_+(x) \simeq \tilde{c}_-(x) \qquad (3.215)$$

is assumed to be valid in the membrane bulk since $\delta \gg 1/\kappa$. Find $\bar{c}_{-0}$. Your answer is a good approximation *even in the face of steady diffusion* across the membrane.

Now, find an approximate quasiequilibrium boundary condition relating the minority carrier or co-ion boundary concentrations $\bar{c}_+(0^+)$ and $\bar{c}_+(\delta^-)$ to c'_0, c''_0, and $\bar{\rho}_m$; make (and state) reasonable approximations and state the conditions for which your answer is valid, even in the face of nonequilibrium diffusion across the membrane.

(c) At $t = 0$, a step in electric field $E_x = E_0 u_{-1}(t)$ is applied across the membrane. Find an expression for the *steady state* $(t \to \infty)$ co-ion concentrations $\bar{c}_+(x) \simeq \tilde{c}_+(x)$ in terms of known constants from your answer to part (b) and other relevant constants. State all assumptions and approximations.

(d) Find $\tilde{c}_\pm(x)$ in the limit $E_0 \to 0$ from your answer to part (c). Sketch your solution to part (c) for $E_0 > 0$ and $E_0 < 0$, where $|E_0|$ is a few $(RT/F)/\delta$. Also sketch the limiting solution for $E_0 = 0$. Compare with Figure 3.25.

(e) In Chapter 6, we will find that an applied E_0 can induce bulk "electroosmotic" fluid flow through the membrane. Under certain conditions, we can model this flow by a continuum, phenomenological relation, $v_x = -L_{12}V_0 = -(L_{12}\delta)E_0$, where v_x is the average (mass-centered) fluid velocity with respect to the membrane and L_{12} is the electroosmotic transduction coefficient (a constant). (The negative sign comes from the definition of L_{12}.) *Modify* your answers to parts (a) and (c) above to *include convective flow* induced by E_0.

(f) Calculate the change in conductance of the membrane, $\Delta G/G(E_0 = 0)$, due to $(E_0\delta) = +4RT/F$. (In this case, $\Delta G/G$ is small because $\Delta\tilde{c} \ll \bar{c}_{-0}$; when c'_0, $c''_0 \sim \bar{\rho}_m$, much larger $\Delta G/G$ ratios are possible for $V_0 \sim 100\,\mathrm{mV}$.)

(g) From your answer to parts (c) and (d), find $\Delta\tilde{c}$ for the limiting cases when (i) $E_0\delta \gg +V_T$, (ii) $E_0\delta \ll -V_T$, and (iii) $|E_0\delta| \ll V_T$. Express your answer in terms of $\tilde{c}(\delta^-)$ and $\tilde{c}(0^+)$. (Note that $V_T = RT/F$.)

Problem 3.14 Electrodiffusion (Transient)

PROBLEM

(a) Carry out a detailed derivation of the steps of Example 3.6.5 and show that the electrodiffusion time constant is

$$\tau_{\mathrm{ed}} = \frac{\delta^2}{\pi^2 \bar{D}_+} \left(\frac{1}{1 + \left(\dfrac{|E_0\delta|}{2\pi RT/F} \right)^2} \right) \qquad (3.216)$$

(b) Interpret τ_{ed} in the limits of large and small $|E_0\delta|$.

(c) Complete the solution to the transient problem by finding the coefficients.

Problem 3.15 For parts (a)–(c), the membrane in Figure 3.41 is uncharged ($L_{12} = L_{21} = 0$) and separates KCl solutions of two different concentrations, $c_0^2 > c_0^1$. Assume $D_+ = \bar{D}_+ = \bar{D}_- = D_-$ and $W \gg \delta$; c_0^1 and c_0^2 are maintained constant right up to the membrane surfaces by vigorous stirring. L_{11} is independent of concentration. At time $t = T$ shortly after the concentration gradient is established (where $T \gg \delta^2/2D$ and $T \ll \delta W/2D$), a pressure drop $\Delta P u_{-1}(t = T)$ is applied across the membrane by filling the left-hand funnel to a desired height. (Δh is found to remain almost constant for times of interest; flow through the membrane is found to occur at a constant rate as measured via the right-hand capillary tube.) The rigid porous backing at $x = \delta^+$ prevents bulk membrane motion or deformation, but does not affect electrical, chemical, or fluid mechanical parameters in any other way.

(a) Derive an expression and appropriate boundary conditions that can be used to find the approximate time and space dependences of the $K^+(x, t)$ and $Cl^-(x, t)$ ion profiles inside the membrane for times such that c_0^1 and c_0^2 have not changed appreciably.

(b) Find an expression for and graph the "steady state" profiles $K^+(x)$ and $Cl^-(x)$ for the times of interest in part (a).

(c) Find the "steady state" V_0 as measured by *silver/silver chloride* electrodes placed in the bulk solutions on both sides of the membrane.

For parts (d)–(f), the membrane separates two different concentrations of NaCl, where $D_- > D_+$. Further, the membrane is positively, uniformly charged such that $\bar{\rho}_m \gg Fc_0^1$ and $\bar{\rho}_m \gg Fc_0^2$. The L_{ij} can be considered, approximately, as constants independent of space. (Justify this approximation if you use it in any questions below.) $\bar{D}_i$ and $\bar{u}_i$ are approximately constant throughout the membrane. At time $t = T$ after the concentration gradient is established, a pressure drop $\Delta P u_{-1}(t = T)$ is applied across the membrane.

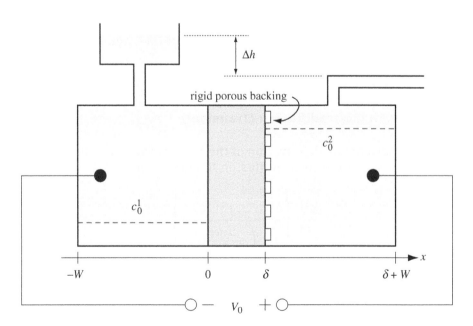

Figure 3.41 Membrane parameters L_{11}, L_{22}, and $L_{12} = L_{21}$ are independent of ΔP, V_0, and concentration of ions. $\delta \gg 1/\kappa$. Electrometer input impedance is very high.

(**d**) Derive an expression and appropriate boundary conditions that can be used to find the approximate time and space dependences of the minority (co-ion) carrier concentration inside the membrane valid for reasonably small internal concentration gradients and for times such that c_0^1 and c_0^2 have not changed appreciably.

(**e**) Find an expression for the "steady state" minority carrier concentration profile inside the membrane for the times of interest in part (d).

(**f**) Find an expression for the potential V_0 measured at $t = T^-$ (just before application of ΔP) and at $t = T^+$. Is the "steady state" V_0 necessarily the same or different than that measured at $t = T^+$? (Neglect changes in c_0^1 and c_0^2.)

(**g**) For the conditions of parts (a)–(c), derive an expression for the time constant associated when with changes in c_0^1 and c_0^2 after application of $\Delta P u_{-1}(T)$.

3.8 REFERENCES

[1] Taylor GI (1953) Dispersion of salts injected into large pipes or the blood vessels of animals. *Appl. Mech. Rev.* **6**, 265–267. Taylor dispersion is encountered in many transport examples in biological systems and in microelectromechanical systems and micro- and nanofluidic devices.

[2] Goldman DE (1943) Potential, impedance and rectification in membranes. *J. Gen. Physiol.* **27**, 37–60.

[3] Ives DJG and Janz GJ (eds) (1961) *Reference Electrodes, Theory and Practice.* Academic Press, New York.

[4] Plonsey R (1969) *Bioelectric Phenomena.* McGraw-Hill, New York.

[5] Eisenman G (ed.) (1967) *Glass Electrodes for Hydrogen and other Cations: Principles and Practice.* Marcel Dekker, New York.

[6] Spiegler KS and Wyllie MRJ (1956) Electrical potential differences. In *Physical Techniques in Biological Research*, 2nd ed., Volume II (Oster G and Pollister AW, eds). Academic Press, New York, pp.301–392.

[7] Geddes LA (1972) *Electrodes and the Measurement of Bioelectric Events.* Wiley-Interscience, New York.

[8] Donnan FG (1911) The theory of membrane equilibrium in the presence of a non-dialyzable electrolyte. *Z. Elektrochem.* **17**, 572–581.

[9] Bartlett JH and Kromhout RA (1952) The Donnan equilibrium. *Bull. Math. Biophys.* **14**, 385–391.

[10] Mauro A (1962) Space charge regions in fixed charge membranes and the associated property of capacitance. *Biophys. J.* **2**, 179–198.

[11] Freeman WDS and Maroudas A (1975) Charged group behaviour in cartilage proteoglycans in relation to pH. *Ann. Rheum. Dis.* **35**(Suppl 2), 44–45.

[12] Tanford C (1961) *Physical Chemistry of Macromolecules.* Wiley, New York, Chapters 7 and 8.

[13] Timasheff SN (1970) Polyelectrolyte properties of globular proteins. In *Biological Polyelectrolytes* (Veis A, ed.). Marcel Dekker, New York, pp. 1–64.

[14] Venn MF and Maroudas A (1977) Chemical composition and swelling of normal and osteoarthrotic femoral head cartilage. II. Swelling. *Ann. Rheum. Dis.* **36**, 399–406.

[15] Teorell T (1953) Transport processes and electrical phenomena in ionic membranes. *Prog. Biophys. Biophys. Chem.* **3**, 305–369.

[16] Plonsey [4], Chapters 2 and 3.

[17] Plonsey [4], Section 2.8.

[18] Scatchard G (1953) Ion exchanger electrodes. *J. Am. Chem. Soc.* **75**, 2883–2887.

[19] Kirkwood JG (1954) Transport of ions through biological membranes from the standpoint of irreversible thermodynamics. In *Ion Transport Across Membranes* (Clark HT, ed.). Academic Press, New York, pp. 119–127.

[20] Picheny MA and Grodzinsky AJ (1976) Methods for measurement of charge in collagen and polyelectrolyte composite materials. *Biopolymers* **15**, 1845–1851.

[21] Yannas IV, Burke JF, Huang C, and Gordon PL (1975) Suppression of *in vivo* degradability and of immunogenicity of collagen by reaction with glycosaminoglycans. *Polymer Preprints* **16**, 209–214.

[22] Nernst W (1889) Eine electromotorische Wirksamkert der Ionen. *Z. Physik. Chem.* **4**, 129–181.

[23] Planck M (1890) Ueber die Potential-differenz zwischen zwei verdunnten Losungen binarer Elektrolyte. *Ann. Physik. Chem.* **40**, 461–576.

[24] Helfferich F (1962) *Ion Exchange.* McGraw-Hill, New York, (reprinted by Dover Publications, New York, 1995). This book gives detailed theoretical analyses and describes many practical applications.

[25] Cole KS (1965) Electrodiffusion models for the membrane of squid giant axon. *Physiol. Rev.* **45**, 340–379.

[26] Shercliff JA (1962) *The Theory of Electromagnetic Flow-Measurement.* Cambridge University Press, Cambridge.

[27] Cobbold RSC (1974) *Transducers for Biomedical Measurements: Principles and Applications.* Wiley, New York.

[28] Zhou Z, Li LB, Qin Y, et al. (2010) Failure modes analysis of electromagnetic flowmeter based on grey relational analysis. In *Advances in Functional Manufacturing Technologies* (Zuo D, Guo H, Xu X, et al., eds). Applied Mechanics and Materials, Volume 33. Trans Tech Publications, Switzerland, pp. 322–326.

[29] Sollner J (1969) The electrochemistry of porous membranes, with particular reference to ion exchange membranes and their use in model studies of biophysical interest. *J. Macromol. Sci., Part A: Chem.* **3**, 1–86.

[30] Tanford [12], p. 555.

Electrical Interaction Forces: From Intramolecular to Macroscopic

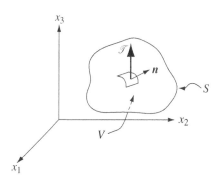

4.1 INTRODUCTION

Electrical interactions are critically important along and between molecules, at cell–matrix and cell-surface interfaces, and within tissues. At the nanoscale (e.g., molecular and interfacial regimes), these interactions are associated with electrical dipole or "double layer" charge configurations. In order to examine such interactions in detail, we first introduce the general concept of stresses and force densities in Section 4.2, and the Maxwell (electrical) stress tensor in Section 4.3. The force density associated with polarization in dielectric media is introduced in Section 4.4, with applications including cell dielectrophoresis. With this as background, the electrical double layer is described in Section 4.5 and the relation between the double layer and charge groups along molecules and surfaces is summarized in Section 4.6. Electromechanical coupling and double layer repulsion forces are treated in Sections 4.7–4.9. These repulsive interactions are central to the classical DLVO (Derjaguin, Landau, Verwey, Overbeek) theory describing the balance of electrical and nonelectrical forces applied to biological and physicochemical systems.

4.2 FORCE, STRESS, TRACTION, AND THE FORCE DENSITY

In this section, we first consider the concept of a generalized stress function $\mathscr{T}$ without being specific as to its particular physical origins. The starting point is a seemingly abstract question: Under what circumstances can the total force $\boldsymbol{f}$ on the material within a volume V that is enclosed by a surface S be found by integrating a force per unit area $\mathscr{T}$ over that surface?

$$f_i = \oint_S \mathscr{T}_i \, da, \quad \boldsymbol{f} = \oint_S \mathscr{T} \, da \tag{4.1}$$

where the integrals in (4.1) are expressed in index and vector notation so that familiarity with these two ways of making statements can be gained. The ith component of the force equation is represented by the index notation. The subscript i can be 1, 2, or 3, in which case it represents the force in the x_1, x_2, or x_3 directions, respectively. The force per unit area, $\mathscr{T}$, is called the *traction*. The geometry of the surface S upon which the traction acts is specified by the unit normal $\boldsymbol{n}$. As illustrated by Figure 4.1, $\mathscr{T}$ may act in any direction at the surface with respect to $\boldsymbol{n}$.

When the section of surface being considered has a normal in one of the axis directions, the components of $\mathscr{T}$ are the components of the stress tensor σ_{ij}. The cubical volume shown in Figure 4.2 has surfaces with normals in the direction of one of the cartesian coordinates. Each surface has a set of three stress components acting on it. σ_{ij} is the stress acting in the jth direction on a surface whose normal points in the ith

Figure 4.1 Material within volume V enclosed by surface S acted on by force $\boldsymbol{f}$ found by integrating the traction $\mathscr{T}$ over S.

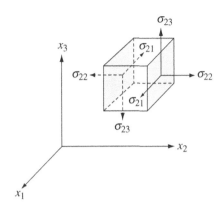

Figure 4.2 Cubical volume showing components of stress acting on the "2" surfaces.

direction. When the stress acts normally to a surface ($i = j$), it is called a "normal stress." Stresses acting tangentially to a surface are "shear stresses," σ_{ij}, where $i \neq j$. Thus, σ_{i2} is a force per unit area acting in the 2 direction on the ith surface. By convention, these components act on both the "front" and "back" surfaces of the cube (Figure 4.2).

With these definitions, what is the relationship between $\mathscr{T}$ and the stress tensor $\underline{\underline{\sigma}}$ for a section of surface that has arbitrary orientation denoted by an arbitrary $\boldsymbol{n}$? To answer this question, consider the infinitesimal triangular shaped area da of Figure 4.3. By definition, $\boldsymbol{n}$ is normal to this elemental surface, but $\mathscr{T}$ has an arbitrary direction relative to $\boldsymbol{n}$.

As a matter of geometry, note that the surfaces da_i are

$$da_i = n_i\, da \tag{4.2}$$

where n_i is the component of $\boldsymbol{n}$ in the ith direction. This can be seen by writing Gauss' theorem

$$\oint_S \boldsymbol{A} \cdot \boldsymbol{n}\, da = \int_V \nabla \cdot \boldsymbol{A}\, dV \tag{4.3}$$

with $\boldsymbol{A} = \boldsymbol{i}_j$, where $\boldsymbol{i}_j$ is the unit vector in the jth direction and therefore a constant. Hence, $\nabla \cdot \boldsymbol{i}_j = 0$, and the surface integral must vanish. This integration over the infinitesimal surfaces of Figure 4.3 gives (4.2).

Now consider the force due to $\mathscr{T}$ acting in the jth direction. Using the first surface element of Figure 4.3, we write $\mathscr{T}_j\, da$, and this must be equal to the sum of the forces acting on the equivalent surface elements of the second infinitesimal surface element:

$$\mathscr{T}_j\, da = \sigma_{1j}\, da_1 + \sigma_{2j}\, da_2 + \sigma_{3j}\, da_3 = \left(\sigma_{1j} n_1 + \sigma_{2j} n_2 + \sigma_{3j} n_3\right) da \tag{4.4}$$

We use the Einstein summation convention: *if an index k appears twice in one term, this implies a sum of three terms, k = 1, 2, 3.* Thus, (4.4) gives the general relationship between the traction and the stress:

$$\mathscr{T}_j = \sigma_{ij} n_i, \quad \mathscr{T} = \underline{\underline{\sigma}} \cdot \boldsymbol{n} \tag{4.5}$$

Equation (4.5) can now be used to write the total force in (4.1) as

$$f_j = \oint_S \sigma_{ij} n_i\, da, \quad \boldsymbol{f} = \oint_S \underline{\underline{\sigma}} \cdot \boldsymbol{n}\, da \tag{4.6}$$

Therefore, one method for obtaining the total force on a material surrounded by a surface S is to integrate the stress tensor over that surface as in (4.6). As an alternative, we can also integrate the force per unit volume, i.e., the force density $\boldsymbol{F}$, over the volume enclosed by S. What

Figure 4.3 Infinitesimal area element da of Figure 4.1, showing equivalent area elements da_i with normals in the x_i directions.

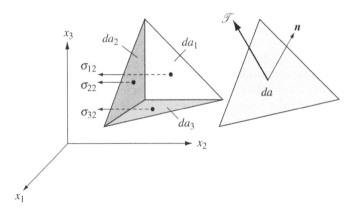

is the relationship between $\underline{\underline{\sigma}}$ and $\boldsymbol{F}$? There are two ways of answering this question, each lending insights into the nature of the stress tensor.

First, using a mathematical approach, we extend Gauss' theorem, as given by (4.3), to convert the surface integration called for in (4.6) to a volume integration. To extend the vector form of Gauss' theorem for use with tensors, think of i as a given; then f_i is a scalar and σ_{ij} represents three numbers that can be thought of as the components of a vector $\boldsymbol{A}$: $A_j = (\sigma_i)_j$. Equation (4.6) then becomes

$$f_j = \int_V \frac{\partial \sigma_{ij}}{\partial x_i} \, dV, \quad \boldsymbol{f} = \int_V \nabla \cdot \underline{\underline{\sigma}} \, dV \qquad (4.7)$$

The quantity that is integrated over the volume to find the jth component of the total force is naturally taken as the force per unit volume, i.e., the force density $\boldsymbol{F}$:

$$\boxed{F_j = \frac{\partial \sigma_{ij}}{\partial x_i}, \quad \boldsymbol{F} = \nabla \cdot \underline{\underline{\sigma}}} \qquad (4.8)$$

The force density is the divergence of the associated stress tensor, with the tensor divergence defined by the summation convention and (4.8). Note that for any given component of $\boldsymbol{F}$, j is a predetermined integer. Hence, F_j is the sum of three terms.

A more physical approach to the derivation of (4.8) involves a simple force balance. Consider the volume of Figure 4.2 to be an infinitesimal one with sides having dimensions Δx_1, Δx_2, and Δx_3, volume $(\Delta x_1 \Delta x_2 \Delta x_3)$, and center at the general coordinate (x_1, x_2, x_3) (Figure 4.4). The force density is defined as being the force per unit volume acting on the elemental volume in the limit where the volume vanishes. Hence, for the component of the force density in the 2 direction, for example,

$$F_2 = \lim_{\Delta x_1, \Delta x_2, \Delta x_3 \to 0} \frac{1}{\Delta x_1 \Delta x_2 \Delta x_3}$$

$$\times \left\{ \left[\sigma_{12} \left(x_1 + \frac{\Delta x_1}{2}, x_2, x_3 \right) - \sigma_{12} \left(x_1 - \frac{\Delta x_1}{2}, x_2, x_3 \right) \right] \Delta x_2 \Delta x_3 \right.$$

$$+ \left[\sigma_{22} \left(x_1, x_2 + \frac{\Delta x_2}{2}, x_3 \right) - \sigma_{22} \left(x_1, x_2 - \frac{\Delta x_2}{2}, x_3 \right) \right] \Delta x_1 \Delta x_3$$

$$\left. + \left[\sigma_{32} \left(x_1, x_2, x_3 + \frac{\Delta x_3}{2} \right) - \sigma_{32} \left(x_1, x_2, x_3 - \frac{\Delta x_3}{2} \right) \right] \Delta x_1 \Delta x_2 \right\} \quad (4.9)$$

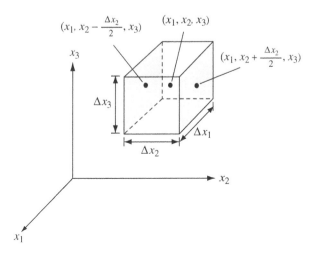

Figure 4.4 Volume of Figure 4.2 taken as elemental.

where terms that vanish in the limit are not included. By the definition of a partial derivative, (4.9) becomes (4.8) in the special case where $i = 2$. Of course, the other two components of F could also be written out in the form of (4.9), and the other two components of (4.8) deduced by this argument. So, we arrive at the same relationship between the force density and the stress tensor whether we start from the tensor form of Gauss' theorem or from consideration of forces on an elemental volume.

Finally, we recall that the stress components acting on the "back" surfaces in Figure 4.2 are taken as acting in the negative axes directions. This was built into (4.9). On these surfaces, the normal vector, which by definition points out of the volume V, is the negative of the unit vectors i directed in the axis directions. This picture is consistent with the relationship between stress and traction given by (4.5).

4.3 FORCE DENSITY AND THE MAXWELL STRESS TENSOR

If a force density can be written as the divergence of a stress tensor, then we know that the total force can be found by integrating the traction over the enclosing surface S. In this section, the force density acting on a medium supporting a net free charge density ρ_e is taken as an important illustration.

We start with the jth component of the force density,

$$F_j = \rho_e E_j \tag{4.10}$$

where ρ_e in biological systems is often associated with the ionizable charge groups of nucleic acids, the polysaccharides of proteoglycans and glycoproteins, and the amino acid residues of protein peptides, along with mobile ions (see Section 1.4). Since these charged biomacromolecules are embedded within a physiological saline environment, the fluid phase will often have a net charge having a density equal and opposite to the density of the "fixed" (immobile) charge groups of the solid macromolecular phase.

In the presence of a local electric field due to either matrix charge or an applied electric field, the electrical force acting on these charge groups will be transmitted to the medium as a whole. Positive charges would give rise to a force density acting in the direction of E, while negative charges would result in a negative force to the surrounding medium. Therefore, it is the *net* charge that must appear in (4.10).

To write (4.10) in the form of a divergence of stress tensor $\underline{\sigma}$, the charge density is first represented in terms of E by using Gauss' law:

$$F_j = \frac{\partial(\epsilon E_i)}{\partial x_i} E_j \tag{4.11}$$

where we assume that the medium can be modeled as homogeneous and isotropic, having a constant (linear) permittivity ϵ. The following steps are typical of the "art" of deriving a stress tensor from a force density. The object is to write (4.11) in the form of (4.8).

We can automatically generate a term in the desired form by taking E_j inside the derivative in (4.11), but then the additional term that is generated by making the derivative of the product must be subtracted:

$$F_j = \frac{\partial}{\partial x_i}(\epsilon E_i E_j) - \epsilon E_i \frac{\partial E_j}{\partial x_i} \tag{4.12}$$

The fact that $\boldsymbol{E}$ is irrotational ($\nabla \times \boldsymbol{E} = 0$) means that, regardless of i and j,

$$\frac{\partial E_j}{\partial x_i} = \frac{\partial E_i}{\partial x_j} \tag{4.13}$$

so the indices can be reversed in the last term of (4.12) and, because of the product rule of differentiation, E_i can be taken inside the derivative:

$$\epsilon E_i \frac{\partial E_j}{\partial x_i} = \epsilon E_i \frac{\partial E_i}{\partial x_j} = \frac{\partial}{\partial x_j}(\tfrac{1}{2}\epsilon E_k E_k) \tag{4.14}$$

The second equality simply recognizes that j is a summation variable and, to avoid confusion in the next step, can just as well be replaced by k. The last term in (4.14) can be written as the gradient of a scalar. To make this term fit into our stress tensor formalization, we define the Kronecker delta function

$$\delta_{ij} = \begin{cases} 1 & (i = j) \\ 0 & (i \neq j) \end{cases}$$

so that the last term in (4.12), multiplied by δ_{ij}, becomes the desired derivative with respect to x_j. Hence, (4.12) becomes

$$F_j = \frac{\partial}{\partial x_i}(\epsilon E_i E_j - \tfrac{1}{2}\delta_{ij}\epsilon E_k E_k) \tag{4.15}$$

Comparison of this expression with (4.8) identifies the stress tensor associated with the force density of (4.10) as

$$\boxed{\sigma_{ij}^{\mathrm{el}} = \epsilon E_i E_j - \tfrac{1}{2}\delta_{ij}\epsilon E_k E_k} \tag{4.16}$$

where $\underline{\underline{\sigma}}^{\mathrm{el}}$ is the Maxwell stress tensor.

Example 4.3.1 Use of the Maxwell Stress Tensor   The example of Figure 4.5 illustrates how the stress tensor can be used to find the total force on an object by integrating over a surface enclosing the object upon which the force acts. The inner cylindrical electrode has radius b, and is at a potential V_0 relative to the outer electrode, which has radius a. These electrodes are highly conducting and so can be regarded as equipotentials. On the top of the inner cylinder is a highly conducting "blob" of material, perhaps water, that has an arbitrary shape. The object is to find the electrical force in the z direction on this "blob."

A surface chosen to exploit the geometry and enclose only the material upon which the force is to be computed is shown as the dashed surface in Figure 4.5. In terms of the stress tensor given by (4.16), the total force is

$$f_z = \oint \sigma_{iz} n_i \, da \tag{4.17}$$

Contributions to the surface integral are considered as the sum of contributions acting on the surfaces I–V in the figure:

- On I, the surface is sufficiently removed from the field region that $\boldsymbol{E} \to 0$ and hence $\sigma_{ij} \to 0$.
- On II, $\boldsymbol{n} = \boldsymbol{i}_r$, $\sigma_{zr} = \epsilon_0 E_z E_r$, and, because $E_z = 0$ on this equipotential surface, $\sigma_{zr} = 0$.

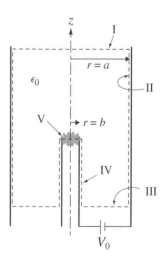

Figure 4.5 Cylindrical coaxial electrodes at $r = a$ and $r = b$ support a "blob" of perfectly conducting material of arbitrary shape, identified as suface V.

- On III, $\boldsymbol{n} = -\boldsymbol{i}_z$, $\sigma_{zz} = \frac{1}{2}\epsilon_0(E_z^2 - E_r^2) = -\frac{1}{2}\epsilon_0 E_r^2$ (the surface is well below the "blob"). Specifically,

$$E_r = \frac{V}{r\ln(a/b)}, \quad \sigma_{zz} = -\frac{1}{2}\frac{\epsilon_0 V^2}{[\ln(a/b)]^2 r^2} \tag{4.18}$$

- On IV, the situation is as on II, except $\boldsymbol{n} = -\boldsymbol{i}_r$.
- On V, there is no field, $\sigma_{zz} = 0$.

Thus, the only contribution to the surface integral comes from an integration over surface III:

$$f_z = \int_{S_{\mathrm{III}}} \frac{1}{2}\frac{\epsilon_0 V^2 2\pi r\,dr}{[r\ln(a/b)]^2} = \frac{\pi\epsilon_0 V^2}{\ln(a/b)} \tag{4.19}$$

That the force is independent of the specific shape can also be argued by an energy method (virtual work). Note that the force is upward, as would be expected from the fact that the electric field lines tend to diverge upward and outward from the top of the inner cylinder toward the outside cylinder. The positive surface charges on the "blob" would therefore be pulled in a generally upward direction. The voltage needed for lift-off can be computed by equating the electrical force (4.19) to the weight of the "blob," mg.

4.4 POLARIZATION FORCE DENSITY

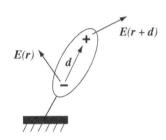

Figure 4.6 A nonuniform field imparts a net force on dipoles, and this force is transferred to the surrounding medium.

As ions migrate through a liquid under the influence of an electric field, they impart a net force density $\rho_e\boldsymbol{E}$ to the liquid. This is a consequence of the collisions between the individual ions and the neutral liquid molecules.

An additional force density of electrical origin is associated with the electrical force exerted on dipoles (pairs of charged particles) or molecular structures that have dipole moments. The picture shown in Figure 4.6 emphasizes that a dipole will only experience a net electrical force when the electric field is nonuniform.

The net electrical force on the dipole is

$$\boldsymbol{f}_{\mathrm{dipole}} = q\left[\boldsymbol{E}(\boldsymbol{r} + \boldsymbol{d}) - \boldsymbol{E}(\boldsymbol{r})\right] \tag{4.20}$$

In the limit of small $\boldsymbol{d}$,

$$\boldsymbol{f}_{\mathrm{dipole}} = q\left[\boldsymbol{E}(\boldsymbol{r}) + \boldsymbol{d}\cdot\nabla\boldsymbol{E} - \boldsymbol{E}(\boldsymbol{r})\right] = (q\boldsymbol{d})\cdot\nabla\boldsymbol{E} \tag{4.21}$$

Now, if a sum is made over all dipoles within a unit volume δV, and the polarization density $\boldsymbol{P}$ is defined as

$$\boldsymbol{P} = \sum(q_i\boldsymbol{d}_i)/\delta V \tag{4.22}$$

it follows that the force per unit volume is

$$\boxed{\boldsymbol{F}_{\mathrm{dipole}} = \boldsymbol{P}\cdot\nabla\boldsymbol{E}} \tag{4.23}$$

This polarization force density can be superimposed on the free charge force density to obtain a total force density of electrical origin on materials that support both charged particles (fixed or mobile) and polarized molecules. Called the *Kelvin force density*, $\mathbf{P} \cdot \nabla \mathbf{E}$ has the advantage of being supported by a simple physical model that does not depend for its derivation on a specific relation between $\mathbf{E}$ and $\mathbf{P}$. A disadvantage is that evaluation of total forces by integration of this force density over a volume is not always straightforward.

In (4.23), however, what is meant by $\mathbf{E}$? Is this the macroscopic electric field in terms of which the theory of electromagnetics is framed? From the derivation, it is actually the microscopic electric field experienced by each particle. So, the force density (4.23) is strictly correct only if the dipoles are dilute enough that one dipole does not appreciably distort the electric field of its nearest neighbor. The gradient in electric field called for in (4.23) must be based on a dimension large compared with the distance between dipoles or the expression is clearly not correct.

It is asking a lot to derive a force density that takes into account the average effects of interactions (which are electrical in origin) between particles. Therefore, it is often more reasonable to start with certain laboratory measurements summarized in the form of constitutive laws and then deduce a force density in terms of this empirical information. This approach is familiar from the energy method applied to the determination of electrical forces in discrete systems, as illustrated in the following example.

Example 4.4.1 Evaluation of Force Using an Energy Method

The total electrical force is to be found on the dielectric slab shown in Figure 4.7. This material, which is allowed to have only the single degree of freedom ξ, supports no free charge. Hence, whatever the force, its origins must be in the polarizability of the material. This is summarized in the measurement of the voltage as a function of the charge on the capacitor plate connected to the positive terminal, and the displacement ξ of the material: $v(q, \xi)$. Given this constitutive law, the force is deduced by considering an isolated "thermodynamic subsystem." Considering only energy storage in the electric field, the change in this energy, δW, is the result either of imparting the increment of electrical energy $v\, \delta q$ through the electrical terminals or by doing the incremental work $-f\, \delta \xi$ on the electrical system. (Hence, f is defined as the electrical force acting *on* the external world.) With the understanding that the isolated system is conservative, the energy W is found by first putting the system together mechanically; with the electrical excitation set to zero, the work performed in placing the slab at its final position is zero. The remaining energy storage as the electrical variables are raised is then found from

$$W = \int v(q, \xi)\, \delta q \qquad (4.24)$$

This is an integral that can be carried out once $v(q, \xi)$ is known.

E X A M P L E

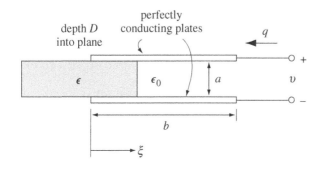

depth D
into plane

perfectly
conducting plates

q

ϵ

ϵ_0

a

v

$+$

$-$

b

ξ

Figure 4.7 A dielectric slab free to slip between capacitor plates.

To isolate the system from the external world, the electrical terminals can now be open-circuited: the charge q is maintained constant, and the slab is given a virtual displacement. The resulting change in energy is

$$\delta W + f\,\delta\xi = 0 = \left(\left.\frac{\partial W}{\partial \xi}\right|_{q=\text{const}} + f\right)\delta\xi \qquad (4.25)$$

and it follows that the required force is

$$f = -\frac{\partial W}{\partial \xi} \qquad (4.26)$$

To evaluate this expression, the constitutive law for $v(q,\xi)$ is used to evaluate (4.24), and then the required derivative is taken in (4.26).

The *Korteweg–Helmholtz (KH) force density* is derived in a manner that represents a generalization of the method used in Example 4.4.1 (for details, see [1]):

$$\boldsymbol{F} = \nabla\left(\tfrac{1}{2}E^2\rho\frac{\partial\epsilon}{\partial\rho}\right) - \tfrac{1}{2}E^2\nabla\epsilon + \rho_e\boldsymbol{E} \qquad (4.27)$$

The first term is called the electrostriction force density because of its origins in the effect of a change in specific volume on the polarization, while the second term is caused by variations in the dielectric constant. For example, it is responsible for the net force on the slab of Figure 4.7.

For the case of dilute dipoles, the polarization is proportional to the mass density, so the force density is specialized further:

$$\epsilon - \epsilon_0 = k\rho \qquad (4.28)$$

where k is a constant. Then, (4.27) becomes

$$\boldsymbol{F} = \nabla(\tfrac{1}{2}E^2 k\rho) - \tfrac{1}{2}E^2 k\nabla\rho = k\rho\nabla(\tfrac{1}{2}E^2) = \boldsymbol{P}\cdot\nabla\boldsymbol{E} + \rho_e\boldsymbol{E} \qquad (4.29)$$

which is again the Kelvin force density plus the force density associated with freely mobile charge. For noninteracting dipoles, the Kelvin and KH force densities are in agreement.

In general, the Kelvin and KH force densities differ by the gradient of a pressure function (a scalar). To see this, observe that

$$\boldsymbol{D}\cdot\nabla\boldsymbol{E} = (\epsilon - \epsilon_0)\boldsymbol{E}\cdot\nabla\boldsymbol{E} + \epsilon_0\boldsymbol{E}\cdot\nabla\boldsymbol{E} = \boldsymbol{P}\cdot\nabla\boldsymbol{E} + \nabla(\tfrac{1}{2}\epsilon_0\boldsymbol{E}\cdot\boldsymbol{E}) \qquad (4.30)$$

Example 4.4.2 Use of Polarization Forces to Separate Living and Dead Cells: Dielectrophoresis (For examples of pioneering work in this area, see [2]. Recent studies relevant to cell biology and biophysics include [3, 4].) The experimental apparatus in Figure 4.8 is to be used to separate living from dead cells. The polarization force on living biological cells differs from that of dead cells owing to differences in material properties. We are to *find an expression for the force* on any given cell in a very dilute suspension of these cells.

As a first approximation, we will model all the cells under investigation to be insulating dielectric spheres ($\sigma_1 = 0$, ϵ_1). They are suspended in a fluid that is also a linear dielectric

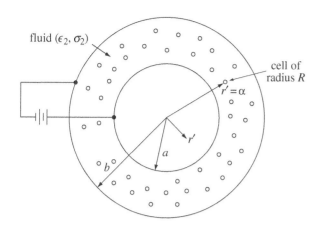

Figure 4.8 Concentric cylindrical electrode configuration of depth L into plane filled with fluid of conductivity σ_2 and dielectric constant ϵ_2. Typical dimensions are $R = 10\,\mu m$, $a = 5\,cm$, $b = 10\,cm$, and $L = 20\,cm$.

material having permittivity ϵ_2 and conductivity $\sigma_2 = 0$. In parts (a)–(c) below, the polarization force on the cells is to be modeled under the assumption that there is *no free charge* anywhere in the system. The assumptions and models described above lead to an electric field solution that is also applicable for lossy materials ($\sigma \neq 0$), provided that the applied voltage is at a high enough frequency. This will be shown in part (d).

(**a**) Consider one isolated cell at the position $r' = \alpha$ in Figure 4.8 far away from any other cell. Given that the cell radius $R \ll b - a$ and $R \ll a$, the electric field in the region of the cell can be approximated as that due to the presence of a dielectric sphere in an essentially uniform applied field.

We first find an expression for the electric field **E** *inside and outside the cell*, in terms of the approximately uniform applied field that would exist at $r' = \alpha$ in the absence of the cell (call the latter E_0). This is justified if the concentration of cells is reasonably dilute. Assuming that there is no free charge anywhere in the system, Maxwell's equations in the electroquasistatic limit are

$$\nabla \cdot \epsilon \boldsymbol{E} = 0 \tag{4.31}$$

$$\nabla \times \boldsymbol{E} = 0 \tag{4.32}$$

Hence, the fields inside and outside the cell must satisfy Laplace's equation, $\nabla^2 \Phi = 0$. We assume that Laplacian potentials inside and outside the cell take the form (Table B.7)

$$\Phi^{\mathrm{out}} = -E_0 r \cos\theta + \frac{A}{r^2} \cos\theta \tag{4.33}$$

$$\Phi^{\mathrm{in}} = B r \cos\theta \tag{4.34}$$

The boundary condition far from the cell is

$$\lim_{r \to \infty} \Phi^{\mathrm{out}} \to -E_0 r \cos\theta \tag{4.35}$$

(Note that this corresponds to the field "far" from the $\sim 10\,\mu m$ diameter cell, but still approximately in the region $r' \simeq \alpha$.) At $r = R$,

$$\boldsymbol{n} \cdot (\epsilon_2 \boldsymbol{E}^{\mathrm{out}} - \epsilon_1 \boldsymbol{E}^{\mathrm{in}}) = \sigma_s = 0 \tag{4.36}$$

$$\Phi^{\mathrm{out}}(r = R) = \Phi^{\mathrm{in}}(r = R) \tag{4.37}$$

The assumed form of Φ^{out} already satisfies the boundary condition (4.35). Incorporating the assumed forms of Φ^{out} and Φ^{in} with the boundary conditions (4.36) and (4.37) gives

$$A = E_0 R^3 \left(\frac{\epsilon_1 - \epsilon_2}{\epsilon_1 + 2\epsilon_2} \right), \quad B = -E_0 \left(\frac{3\epsilon_2}{\epsilon_1 + 2\epsilon_2} \right) \qquad (4.38)$$

$$\Phi^{\text{out}} = -E_0 r \cos\theta + E_0 \left(\frac{R^3}{r^2} \right) \left(\frac{\epsilon_1 - \epsilon_2}{\epsilon_1 + 2\epsilon_2} \right) \cos\theta \qquad (4.39)$$

$$\Phi^{\text{in}} = -E_0 \left(\frac{3\epsilon_2}{\epsilon_1 + 2\epsilon_2} \right) r \cos\theta \qquad (4.40)$$

With $\boldsymbol{E} = -\nabla\Phi$, the electric field inside and outside (in the neighborhood of) the cell is

$$\boldsymbol{E}^{\text{out}} = E_0(\boldsymbol{i}_r \cos\theta - \boldsymbol{i}_\theta \sin\theta)$$

$$+ E_0 \frac{R^3}{r^3} \left(\frac{\epsilon_1 - \epsilon_2}{\epsilon_1 + 2\epsilon_2} \right) (\boldsymbol{i}_r 2 \cos\theta + \boldsymbol{i}_\theta \sin\theta) \qquad (4.41)$$

$$\boldsymbol{E}^{\text{in}} = +E_0 \left(\frac{3\epsilon_2}{\epsilon_1 + 2\epsilon_2} \right) (\boldsymbol{i}_r \cos\theta - \boldsymbol{i}_\theta \sin\theta) \qquad (4.42)$$

We can sketch the field for the case $\epsilon_2 > \epsilon_1$, by noting that $|E^{\text{in}}| > E_0$. In addition, knowledge of the induced polarization surface charge σ_p at $r = R$ will aid in our sketch:

$$\sigma_p = -\boldsymbol{n} \cdot (\boldsymbol{P}^{\text{out}} - \boldsymbol{P}^{\text{in}}) = \boldsymbol{n} \cdot \epsilon(\boldsymbol{E}^{\text{out}} - \boldsymbol{E}^{\text{in}}) \propto \left(\frac{\epsilon_1 - \epsilon_2}{2\epsilon_2 + \epsilon_1} \right) \cos\theta \qquad (4.43)$$

Therefore, for $\epsilon_2 > \epsilon_1$, σ_p has the *sign* shown in the sketch of Figure 4.9.

(b) We now find an expression for the equivalent dipole moment of the cell that represents the field induced outside the spherical cell by polarization of the sphere itself. In general, the potential of a point charge dipole having dipole moment $\boldsymbol{p} \equiv q\boldsymbol{d}$ has the form in spherical coordinates

$$\Phi_{\text{dipole}} = \frac{qd}{4\pi\epsilon_2 r^2} \cos\theta \qquad (4.44)$$

Figure 4.9 A cell in a uniform electric field.

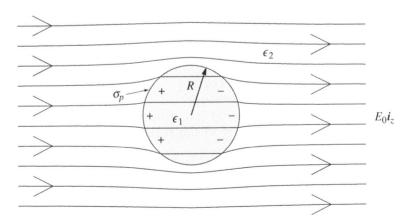

for a dipole in an infinite medium of permittivity ϵ_2. The electric field $\boldsymbol{E} = -\nabla\Phi$ of the dipole, corresponding to (4.44), is

$$\boldsymbol{E}_{\text{dipole}} = \frac{qd}{4\pi\epsilon_2 r^3}(\boldsymbol{i}_r 2\cos\theta + \boldsymbol{i}_\theta \sin\theta) \qquad (4.45)$$

By comparing (4.45) with the dipole term of $\boldsymbol{E}^{\text{out}}$ in (4.41), we find that the equivalent induced dipole moment of the cell is

$$\boldsymbol{p} = 4\pi\epsilon_2 R^3 \boldsymbol{E}_0 \left(\frac{\epsilon_1 - \epsilon_2}{\epsilon_1 + 2\epsilon_2}\right) \qquad (4.46)$$

(**c**) To find an expression for the polarization force on the cell at $r' = \alpha$ that results from the fact that the applied field is really nonuniform, we evaluate the force

$$\boldsymbol{f} \simeq \boldsymbol{p} \cdot \nabla\boldsymbol{E} = 4\pi\epsilon_2 R^3 \left(\frac{\epsilon_1 - \epsilon_2}{\epsilon_1 + 2\epsilon_2}\right) \boldsymbol{E}_0 \cdot \nabla\boldsymbol{E}_0 \qquad (4.47)$$

where

$$\boldsymbol{E}_0 = \boldsymbol{i}_{r'} \left(\frac{V}{r'\ln(b/a)}\right) \qquad (4.48)$$

In (4.47), we have assumed that $\boldsymbol{p}$ corresponds to that induced by the uniform field problem of Figure 4.9, while $\boldsymbol{E}_0$ has the form (4.48); if $\boldsymbol{E}_0$ were really uniform, $\nabla \cdot \boldsymbol{E}_0 = 0$ and hence $\boldsymbol{f} = 0$. Therefore,

$$\boldsymbol{E}_0 \cdot \nabla\boldsymbol{E}_0 = E_0 \frac{\partial E_0}{\partial r'} = \frac{-V^2}{(r')^3 \left[\ln(b/a)\right]^2} \qquad (4.49)$$

$$\boldsymbol{f}\,|_{r'=\alpha} = \boldsymbol{p} \cdot \nabla\boldsymbol{E}\,|_{r'=\alpha} = -4\pi\epsilon_2 R^3 \left(\frac{\epsilon_1 - \epsilon_2}{\epsilon_1 + 2\epsilon_2}\right) \left(\frac{V^2}{\alpha^3 \left[\ln(b/a)\right]^2}\right) \boldsymbol{i}_{r'} \qquad (4.50)$$

Note that for $\epsilon_2 > \epsilon_1$, $\boldsymbol{f}$ is in the direction of decreasing electric field strength. $\boldsymbol{f} = 0$ for $\epsilon_1 = \epsilon_2$.

(**d**) The experiment is performed with an outer fluid bath having $\epsilon_2 \sim 100\epsilon_0$ and $\sigma_2 = 10^{-3}\,\text{S}\,\text{m}^{-1}$. The results show that the accumulation of free charge at the surface of the cell leads to anomalous behavior. A suggestion is made to use an ac ($V = V_0 \cos\omega t$) rather than dc voltage applied to the electrodes. Calculate the range of frequency that should be used to suppress the behavior associated with free charge. Does this change the polarization force?

To answer these questions, we first realize that if the cell membrane is reasonably insulating, we may still assume that $\sigma_1 \equiv 0$ as far as the solution of the electric field problem is concerned. Now, with $\sigma_s \neq 0$ at $r = R$, the boundary conditions at the cell surface take the form

$$\boldsymbol{n} \cdot (\boldsymbol{J}^{\text{out}} - \boldsymbol{J}^{\text{in}}) = -\frac{\partial \sigma_s}{\partial t} \qquad (4.51)$$

$$\boldsymbol{n} \cdot (\epsilon_2 \boldsymbol{E}^{\text{out}} - \epsilon_1 \boldsymbol{E}^{\text{in}}) = \sigma_s \qquad (4.52)$$

In the sinusoidal steady state, with $\partial/\partial t \to j\omega$, (4.51) and (4.52) can be combined to give

$$\boldsymbol{n} \cdot (j\omega\epsilon_2 + \sigma_2)\boldsymbol{E}^{\text{out}} = \boldsymbol{n} \cdot j\omega\epsilon_1 \boldsymbol{E}^{\text{in}} \qquad (4.53)$$

or

$$\boldsymbol{n} \cdot \left(j\frac{\omega\epsilon_2}{\sigma_2} + 1 \right) \boldsymbol{E}^{\text{out}} = \boldsymbol{n} \cdot j\frac{\omega\epsilon_1}{\sigma_2} \boldsymbol{E}^{\text{in}} \qquad (4.54)$$

If a frequency is used such that $\omega\epsilon_2/\sigma_2 \gg 1$, then this problem reduces to the same boundary value problem as in parts (a)–(c). Physically, this is the frequency range for which free charge does *not* have time to relax to the surface of the cell (see Chapter 2):

$$f \gg \left(\frac{1}{2\pi} \right) \frac{\sigma_2}{\epsilon_2} = \frac{10^{-3}\,\text{S}\,\text{m}^{-1}}{2\pi \times 10^{-9}\,\text{F}\,\text{m}^{-1}} = \frac{10^6}{2\pi}\,\text{Hz} \qquad (4.55)$$

Note that $\lambda = c/f = 2\pi \times 3 \times 10^8/10^6 = 600\pi$ m $\gg$ dimensions of interest. Thus, there is a usable frequency range still corresponding to the electroquasistatic analysis.

Note that $\boldsymbol{p} \cdot \nabla\boldsymbol{E} \propto V^2$; $\langle \boldsymbol{f} \rangle_{\text{time}} \propto \frac{1}{2}V_0^2$. We will see in Chapter 6 that the frequency dependence of *electrophoretic* forces is quite different than that of "dielectrophoretic" forces. In electrophoresis, the force is proportional to E_0, not E_0^2.

4.5 THE DIFFUSE DOUBLE LAYER: SITE OF INTRA- AND INTERMOLECULAR ELECTRICAL INTERACTIONS

The concept of the diffuse or space charge layer and the analysis first proposed by Gouy [5] has found a tremendously wide range of applications. These include the ionic atmosphere theory of Debye and Huckel [6], the space charge distribution at semiconductor junctions and semiconductor/electrolyte interfaces, and the distribution of ions around hydrophobic colloidal particles in electrolyte solutions. We present here a derivation of the potential and space charge distribution in the electrolyte phase for the simple case of a plane parallel model, as illustrated in Figure 4.10. This derivation is based on the

Figure 4.10 Planar diffuse double layer.

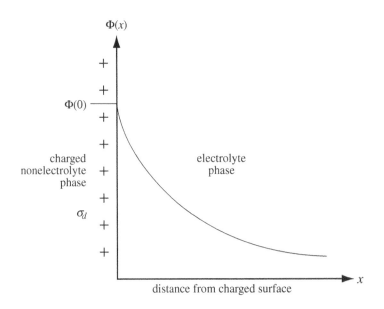

distance from charged surface

Poisson–Boltzmann theory, the most widely used approach to modeling electrostatic interactions in biological and physicochemical systems.

The figure depicts an interface between two phases—one an electrolyte having bulk concentration c_0 and the other a phase whose surface is known to be charged. The latter may be a metal, an insulating solid, a synthetic polyelectrolyte, or (of particular interest) a biological macromolecule or cell surface. This surface has a fixed charge distribution due to ionizable charge groups (e.g., on amino acid residues; see Section 1.4) or caused by adsorption of charges to surfaces. We now find the self-consistent *equilibrium* distribution of space charge (i.e., positive and negative ions), and potential in the electrolyte phase adjacent to the surface.

The relevant parameters associated with Figure 4.10 are

$c_i(r) \equiv$ concentration of the ith ionic species in the electrolyte (mol m^{-3})

$c_{io} \equiv$ bulk concentration of the ith species (ideally the concentration for $r \to \infty$, but practically, just far enough away from the space region—several Debye lengths) (mol m^{-3})

$z_i \equiv$ valence of the ith ion

$\Phi(r) \equiv$ electrical potential

The reference, or zero potential, is chosen to be in the bulk electrolyte phase $\Phi(x \to \infty) \equiv \Phi_{bulk} = 0$

$R \equiv$ universal gas constant = 8.314 J K^{-1} mol^{-1}

$T \equiv$ temperature (K)

$F \equiv$ Faraday constant (96,500 C (mol of electronic charges)$^{-1}$)

In the analysis, we make the following assumptions:

(1) The surface charge of the left-hand (nonelectrolyte) phase can be represented by a smoothed surface charge density of magnitude σ_d; we neglect any discreteness-of-charge effects, so that our continuum formulation can be employed.

(2) Hydrated ions in the solution phase will be treated as *point charges* in the sense that they are present right up to the interface.

(3) The dielectric permittivity ϵ of the space charge region is that of the bulk solution, and is independent of field strength.

(4) There are no other charged species or impurities in the system.

We note that assumptions (2)–(4) are consistent with the so-called Poisson–Boltzmann model for the electrical potential, as derived below. Thus, the potential and charge distribution in the space charge region are related by Poisson's equation

$$\nabla^2 \Phi = \frac{-\rho_e}{\epsilon} \tag{4.56}$$

For mathematical simplicity, we assume that c_i and Φ vary only in the x direction. Therefore, in the one-dimensional case,

$$\frac{d^2 \Phi(x)}{dx^2} = \frac{-\rho_e(x)}{\epsilon} = -\frac{1}{\epsilon} \sum_i z_i F c_i(x) \tag{4.57}$$

The volume charge density $\rho_e(x)$ is expressed in terms of ion concentrations in the space charge region, and we make the additional assumption that the probability of finding a given ion of species i and valence z_i at the position x is proportional to the Boltzmann factor

$\exp[-z_i F\Phi(x)/RT]$. Thus, we assume that ionic concentrations can adequately be described by Boltzmann statistics, which accounts for the opposing tendencies of electrical migration and chemical diffusion in the vicinity of the charged surface. Therefore, we can write

$$c_i(x) = c_{i0} \exp\left[-\frac{z_i F\Phi(x)}{RT}\right] \tag{4.58}$$

Physically, we see that for the case of Figure 4.10, a positively charged surface at $x = 0$ implies a positive surface potential ($\Phi(0)$), and therefore a positive potential $\Phi(x)$ for $x > 0$ (assuming $\Phi_{bulk} = 0$ is chosen as the reference potential). Further, the concentration of negatively charged ions $z_i < 0$, the counter-ions in this case, has a higher value in the space charge region compared with that of the bulk, as predicted by (4.58). The converse is true for the positive ions, or "co-ions." Combining (4.57) and (4.58), we arrive at the Poisson–Boltzmann equation:

$$\frac{d^2\Phi(x)}{dx^2} = -\frac{1}{\epsilon}\sum_i z_i F c_{i0} \exp\left[\frac{-z_i F\Phi(x)}{RT}\right] \tag{4.59}$$

where $\rho_e(x)$ has been expressed in terms of ionic concentrations. Implicit to writing the Poisson–Boltzmann equation (4.59) is the so-called "mean field approximation" that the potential Φ in (4.58) (i.e., the potential of the mean force acting on an ion in the electrolyte) is the same as the mean potential in Poisson's equation (4.57). This foundational approximation is discussed in many texts on thermodynamics and statistical mechanics (see, e.g., [7]).

To find $\Phi(x)$, (4.59) must be integrated twice. We present the simple case of a single symmetrical electrolyte ($z_+ = -z_-$, $c_{+0} = c_{-0}$); more complicated situations have been analyzed numerically. This simplifies the summation in (4.59), which becomes

$$\frac{d^2\Phi(x)}{dx^2} = \frac{2zFc_{i0}}{\epsilon} \sinh\left[\frac{zF\Phi(x)}{RT}\right] \tag{4.60}$$

For conditions such that $zF\Phi(x) \ll RT$, (4.59) and (4.60) both reduce to

$$\frac{d^2\Phi(x)}{dx^2} = \kappa^2 \Phi(x) \tag{4.61}$$

where

$$\frac{1}{\kappa} = \sqrt{\frac{\epsilon RT}{2z^2F^2 c_{i0}}} \equiv \text{Debye length} \tag{4.62}$$

Equation (4.61) is the so-called "linearized Debye–Hückel approximation" having the solution

$$\Phi(x) = \Phi(0)e^{-\kappa x} \tag{4.63}$$

The Debye length is the distance over which the electric field and potential decay to $1/e$ of their values at $x = 0$. (Note that (4.61) and (4.63) do not apply to asymmetrical electrolytes, such as $CaCl_2$.) For the general case, the summation in (4.59) must be carried through the analysis. Direct integration of (4.60) leads to the transcendental relation

$$\tanh\left[\frac{zF\Phi(x)}{4RT}\right] = \tanh\left[\frac{zF\Phi(0)}{4RT}\right] e^{-\kappa x} \tag{4.64}$$

which reduces to (4.63) for a small argument of the hyperbolic tangent.

Having found the potential distribution, we can also obtain a relation between the surface charge σ_d and the surface potential $\Phi(0)$, which in this model is equivalent to the entire potential drop across the diffuse layer. This can be obtained by one integration of (4.60), since we know from the boundary condition on Gauss' law that

$$\sigma_d = -\epsilon \left. \frac{\partial \Phi(x)}{\partial x} \right|_{x=0} \tag{4.65}$$

The result is

$$\sigma_d = (8\epsilon R T c_{i0})^{1/2} \sinh\left[\frac{z F \Phi(0)}{2RT}\right] \tag{4.66}$$

which, for $zF\Phi(0) \ll RT$, reduces to

$$\sigma_d = \frac{\epsilon}{1/\kappa}\Phi(0) = \epsilon\sqrt{\frac{2z^2 F^2 c_{i0}}{\epsilon RT}}\Phi(0) \tag{4.67}$$

Equation (4.67) suggests a simple parallel-plate capacitor model for the double layer, with plate spacing equal to the Debye length, $1/\kappa$.

The equations for potential and charge, (4.63)–(4.67), are expressed in terms of the parameters σ_d and $\Phi(0)$, which may or may not be known in actual experiments. The other parameters in these equations are essentially those that comprise the Debye length. In later chapters, experiments will be discussed in which it will be very important to distinguish between systems in which either the surface charge or the surface potential is specified, and to determine whether the constraint of constant double layer capacitance (i.e., constant ionic strength) is relevant. The change in potential distribution $\Phi(x)$ resulting from changes in the *chemical*, *electrical*, or *mechanical* environment will certainly depend on which of the above constraints is known to be imposed. For example, varying the electrolyte bulk concentration and/or valence will cause a change in the Debye length, as seen from (4.62). If the surface potential is held constant, then an increase in c_{i0} (or z_i) will lead to a decrease in the Debye length, and thus an increase in the spatial decay rate of $\Phi(x)$ as sketched in Figure 4.11(a). In Figure 4.11(b), the surface charge σ_d is held constant—i.e., there is a constant slope of $\Phi(x)$ at $x = 0$. Thus, from (4.67), an increase in c_{i0} or z_i results in a decrease in $\Phi(0)$. We will see that it is possible to duplicate these categories of constraints experimentally, by controlling the chemical environment of the material.

With respect to electrokinetic (e.g., microfluidic) experiments, it is usually not the total potential drop across the double layer that is

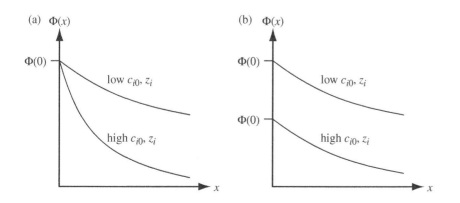

Figure 4.11 Diffuse double layer potential distribution at different ionic strengths: (a) $\Phi(0)$ constant; (b) σ_d constant.

Figure 4.12 Diffuse double-layer potential distribution at different ionic strengths. Variation of ζ potential for conditions of (a) constant $\Phi(0)$, (b) constant σ_d.

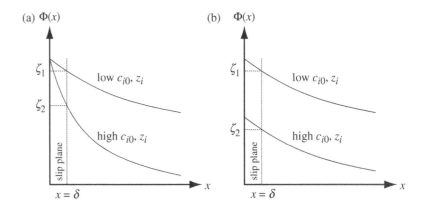

important. A tight or compact inner region of the diffuse double layer composed of rigidly bound solvent molecules exists at surfaces. A fluid mechanical slip plane is thought to occur at a distance of about two or three molecular diameters from the interface, defining the beginning of the mobile portion of the diffuse layer. As electrical and viscous shears must equilibrate over this *mobile* region, it is the potential drop across the mobile region that assumes importance in electrokinetic experiments. Commonly referred to as the ζ (zeta) potential, this may be close in magnitude to $\Phi(0)$, or drastically different if adsorption of specific ions or dipoles occurs (see Section 6.3).

Figure 4.11 is redrawn in Figure 4.12 to include the changes in ζ found by altering c_{i0} and z_i. (Note that Figures 4.11 and 4.12 do not include complications due to specific adsorption, as shown in Figure 6.1(c,d).) A simple capacitor-like model for the mobile portion of the diffuse layer relates the effective electrokinetic surface charge density σ_{ek} to ζ:

$$\zeta \equiv \int_{x=\infty}^{\mathrm{slip\ plane}} \frac{d\Phi(x)}{dx}\, dx \tag{4.68}$$

$$\sigma_{\mathrm{ek}} \equiv \int_{\mathrm{slip\ plane}}^{x=\infty} [-\rho_e(x)]\, dx = \int_{\mathrm{slip\ plane}}^{x=\infty} \epsilon \frac{d^2\Phi(x)}{dx^2}\, dx \tag{4.69}$$

where $\Phi(x)$ is given by (4.63) or (4.64) and σ_{ek} is an equivalent surface charge equal in magnitude and opposite in sign to the net charge in the *mobile* fluidic region. The results of (4.68) and (4.69) can then be expressed as

$$\sigma_{\mathrm{ek}} = C_{\mathrm{ek}}\zeta. \tag{4.70}$$

For the case where $\zeta = \Phi(0)$, σ_{ek} and C_{ek} take on definitions in terms of the total surface charge and capacitance per unit area (cf. (4.67)).

At this point, it is worth noting that there are several qualifications involving many of the basic assumptions upon which the model of the diffuse double layer is based. Errors have been computed for many types of discrepancies, amounting to several percent and often canceling each other (within certain confines for the total potential drop across the double layer). However, the basic form of the model remains valid over a wide range of constraints. A discussion of these problems can be found in many standard texts and review articles: for a historical perspective, see [8] and for a recent review, see [9].

One of the first experimental verifications of the applicability of the Poisson–Boltzmann-based diffuse model was due to Grahame [10]. We mention it here since his experimental technique and double layer model are very germane to this discussion. Grahame's measurements involved the double layer at a mercury/NaF solution interface, the electrolyte being chosen because specific adsorption effects were found to be absent with NaF. While this system is clearly nonphysiological, it proved to be electrochemically "clean" and could thereby be used to quantify terms in the double layer model. The model chosen to represent the double layer is sketched in Figure 4.13, corresponding to a series capacitance model as shown in the inset. The first capacitance corresponds to the compact or Helmholtz layer, while the second corresponds to the usual diffuse layer as defined here by (4.67). The Helmholtz layer incorporates corrections to the previous assumptions, (4.57) and (4.59), upon which our diffuse model is based. First, the fact that hydrated electrolyte ions are not infinitely small point charges, but have finite size, leads to the postulation of a plane or locus of closest approach at $x = \delta$; there are no ions in the region $(0 < x < \delta)$. Second, it is known that the dielectric constant of the solvent in this region is less than that further out in the diffuse and bulk regions, owing to high-field-strength dielectric saturation. Thus, with the mercury surface charge forming one plate of the Helmholtz capacitor, Grahame assumed a capacitance per unit area

$$C_H = \frac{\epsilon'}{\delta} \tag{4.71}$$

where ϵ' is the dielectric constant of the compact zone. Then the diffuse layer capacitance per unit area is

$$C_d = \frac{\sigma_d}{\psi_d} \tag{4.72}$$

and therefore the total series capacitance per unit area is

$$C_T = \frac{\sigma_d}{\psi(0)} \tag{4.73}$$

where ψ_d is the potential drop across the diffuse layer and $\psi(0)$ is the surface potential, or, equivalently, the total double layer potential drop. As there are no ions in the compact layer, it is assumed that C_H is only a function of mercury surface charge, and independent of electrolyte concentration.

To differentiate between these capacitances (which are actually treated as differential capacitances), Grahame measured C_T at very high

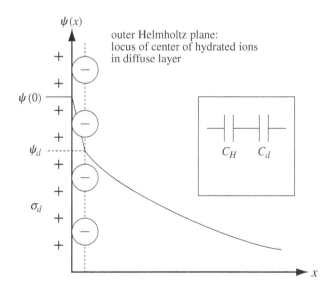

Figure 4.13 Grahame's model for a double layer without specific adsorption (also applicable to a polyelectrolyte/electrolyte interface): potential distribution showing Helmholtz plane. The inset shows the equivalent capacitive circuit model where the total series capacitance is

$$\frac{1}{C_T} = \frac{1}{C_H} + \frac{1}{C_d}$$

and

$$C_T = \frac{\sigma_d}{\psi(0)}, \quad C_H = \frac{\epsilon'}{\delta}, \quad C_d = \frac{\sigma_d}{\psi_d}$$

NaF concentration (about 1 M), where

$$\frac{1}{C_T} = \frac{1}{C_H} + \frac{1}{C_d} \simeq \frac{1}{C_H} \tag{4.74}$$

This follows from (4.62), since increasing the concentration decreases $1/\kappa$ and finally makes C_d much larger than C_H. The resulting value of C_H is then applicable for all (lower) electrolyte concentrations, including those for which C_d is expected to dominate. In this manner, C_H could effectively be subtracted out of the experiment, and the remaining C_d could be compared with a mathematical model such as that of (4.66). Excellent agreement with such a model was found.

We should note here that the model of Figure 4.13 may also be appropriate for a *polyelectrolyte/electrolyte interface.*

4.6 THE DOUBLE LAYER IN RELATION TO NATIVE AND SYNTHETIC BIOLOGICAL POLYELECTROLYTES: A HISTORICAL PERSPECTIVE

We consider two states in which a polyelectrolyte may be found: as an isolated macromolecule (rodlike or coiled), or as a condensed macroscopic assembly of many macromolecules, whether in the form of a fiber, membrane, or tissue. We expect that each of these aggregates can be charged in a roughly similar fashion. The ionizable groups in their dissociated state comprise the polyelectrolyte primary charge, whose magnitude and sign depends on the pH and ionic strength of the bathing electrolyte. Most physiological polyelectrolytes are surrounded by saline solution in their normal environment. Thus, electrical double layers are inherent to all such systems, since electrolyte counter-ions will naturally be attracted by the presence of polyelectrolyte net charge. At first, we will consider some examples in terms of the simple diffuse model as pictured in Figure 4.10.

The ionic atmosphere surrounding a charged polyelectrolyte is as important for electromechanical processes as the primary molecular charge. The theory of this atmosphere is similar to the Debye–Hückel theory for ionic solutions. Fuoss [11] provided experimental evidence for the existence of such an atmosphere, using an electric field applied across a solution of labeled sodium polyacrylate. Upon dissociation into polyanions and $^{22}Na^+$, the application of the field resulted in transport of the polyanions along with some of the sodium (the latter moving *against* the field). This was interpreted in terms of an electrostatic association of $^{22}Na^+$ with the polyanion, with resulting formation of an atmosphere of counter-ions. In addition, a dynamic equilibrium was established between "free" and associated Na^+. By means of experiments of this kind, it could be shown that the charge on the polyelectrolyte and the diffuse layer of counter-ions in solution form an electrical double layer.

Since the conformation of a molecule in solution, or the precise state of aggregation of the many such molecules in a larger assembly, determines the spatial distribution of the primary charge, so does it determine the geometry of the double layer. This is emphasized because it is an all-pervading theme behind all experiments and theoretical models. By way of example, Morawetz [12] and Oosawa [13] treated the case of a coiled macromolecule in terms of a spherically symmetric charge distribution accompanied by a spherical atmosphere of mobile ions and spherical double layer potential. Oosawa [13] and Fuoss, Katchalsky, and Lifson [14] calculated the electrical (double layer) potential at an ideal, extended, rod-like, polyelectrolyte molecule in solution with its counter-ions.

For a polyelectrolyte *membrane*, one may look at double layer charge and potential at several structural levels. First, net charge might be described in terms of an average surface charge per unit area in the plane of the membrane (neglecting discrete charge effects). This would imply a planar double layer model at the membrane interfaces. Looking further into the fine structure of a porous membrane, one may be interested in the presence of charge at the inner surface of the pores. Gliozzi et al. [15] studied extruded collagen films having a dry thickness of 25 μm. Here, "pores" are spaces between the interwoven fibrils, and might be modeled as cylinders having a diameter equal to the mean fibril spacing (e.g., Figure 6.4(c)). Such a charged pore model is important in the study of membrane transport properties, and in understanding the effect of membrane deformations on such transport properties.

Any model, such as a cylindrical charged pore, must be motivated by the actual microscopic/molecular polyelectrolyte structure as being the best suited model for the case at hand. For example, if the known spacing between charge groups on a membrane surface is much greater than the Debye length (as determined by adjacent electrolyte concentration), then a planar double layer model based on an average surface charge density would *not* make sense. Rather, one might use a model involving "clumps" of charge, each with its associated double layers. Such charge clumps might be treated in terms of equivalent charged pores. With regard to collagen, for example, the basic structural units are the filament-like macromolecules. As the latter may aggregate in many different forms (e.g., as oriented parallel fibrils in tendon fibers, or as a matrix in skin and cornea), different continuum models could accompany each respective aggregation of charge even though the most basic unit of structure is the same for all cases.

A Survey of Electromechanical Effects Involving Polyelectrolyte Macromolecules in Solution

We begin by considering the shape assumed by polyelectrolyte molecules in solution. This shape is thought to be determined by a combination of double layer repulsion, van der Waals attraction forces, and Brownian motion: a state of dynamic equilibrium. Thus, whenever the polyelectrolyte molecule has a net charge along its length, electrostatic repulsion usually counteracts intramolecular thermal coiling (or any intramolecular attractive forces that might be present, such as intramolecular hydrogen bonding in the non-ionized state). As a result, increasing the magnitude of net charge causes the coiled molecule to extend or unfold. This "electrical shaping" effect is independent of the sign of the charge, and can be detected in dilute solutions by an increase in viscosity, due to an effective increase in the hydrodynamic radius. At the so-called *isoelectric point*—the electrolyte pH at which there is no net primary charge on the polyelectrolyte—the molecule assumes its most contracted form.

Although it is qualitatively correct to think in terms of charge repulsion, Mysels [16] stressed the importance of a more quantitative model in terms of double layer repulsion, involving molecular charge and the ionic atmosphere. According to this approach, it is the increase in double layer thickness that leads to increased solution viscosity by forcing the molecule to uncoil. Uncoiling is a result of mutual repulsion of double layers surrounding different sections of the same molecule. This corresponds to a classification of electromechanical effects due to interaction through *internal* electric fields. (More recently, folding of proteins and nucleic acids have challenged the limits of the Poisson–Boltzmann theory and its application to this field [9].)

It has also been noted that neutral salt concentration, valence, and salt type affect the extension of the macromolecule. Katchalsky [17] found that, at a fixed nonisoelectric pH, an increase in salt concentration decreased the extension of polyelectrolyte molecules in solution. This is explained qualitatively in terms of the screening of molecular charge, and of a corresponding decrease in intramolecular repulsion due to the presence of more neutral salt ions. More quantitatively, an increase in concentration predicts a decrease in intramolecular repulsion due to the presence of the double layer along the length of the molecule. (Note that (4.62) applies to a plane, diffuse, double layer model, but may be applicable to curved interfaces if the radius of curvature is much greater than the double layer thickness. If not, the field problem must be solved for the specific geometry at hand, but the form of (4.62) will be similar.) Equation (4.62) also predicts that neutral salts of higher valence will lead to greater suppression of double layer repulsion, again resulting in the polyelectrolyte contraction. Such an effect was seen by Katchalsky and Zwick [18], who compared Na^+ and Ba^{2+} as counter-ions. Specific salt effects are not accounted for in the diffuse double layer model represented by (4.67), but manifest themselves in more complex corrections to this model.

Polyelectrolyte Fibers and Membranes

Polyelectrolyte biomaterials can swell in an ionic medium as a result of a change in pH, in both the presence and absence of neutral salts. Swelling experiments have been performed with many connective tissues and muscle, and show that swelling increases with increasing net charge and is minimal at the isoelectric pH. One interpretation of such swelling is based on mutual repulsion of double layers surrounding each macromolecule. Elliott [19] applied an interacting double-layer picture to interpret the equilibrium fibril separation in muscle. Adamson [20] and Mysels [16] stressed that an interpretation in terms of interacting double layers is equivalent to one based on osmotic pressure considerations. However, it is conceptually advantageous to utilize a double layer model rather than one based on osmotic pressure if it is desired to deal with coupling to the existing internal electric fields, whether by electrical, chemical, or mechanical means. The effect of adding neutral salts at low concentrations (up to a few tenths molar) is to decrease the swelling, since (4.62) again suggests a screening effect due to the decrease in double layer thickness.

Anderson and Eriksson [21] experimented with hydrated and dry collagen fibers, as well as with wet and dry bone. A mechanical stimulus in the form of a stretching force was found to give rise to an electrical signal. For the case of the wet fiber, the electrical output (measured by implanted Ag/AgCl electrodes) went to zero at the independently measured isoelectric pH of the collagen fibers, and increased at higher and lower pH levels. As a result, the authors concluded that the effect in this case was due to a streaming potential—a pH-dependent electrokinetic effect—rather than a piezoelectric phenomenon. They concluded that stretching of the fiber caused a streaming of fluid past the fibrils.

4.7 POTENTIAL PROFILE FOR INTERACTING PLANE PARALLEL DOUBLE LAYERS

In the next sections, the concepts of electrical forces and stress tensors are used in the context of several classical examples in biology, biophysics, and colloid physical chemistry. When electrical double layers are brought in contact with one another such that their separation distance is on the order of a Debye length, the resulting redistribution of the electric field leads to very large forces of electrical origin. Such forces are important to the understanding of interactions within and between charged macromolecules, fibers, and membranes. As the concentration of neutral salt ions must also be redistributed consistent with the double layer electrical potential, osmotic pressure forces will also come into play.

The theory of interacting double layers was developed by Dutch (Verwey and Overbeek) and Russian (Derjagin and Landau) groups in the second quarter of the last century. The "DLVO" theory [22] has been applied to proteins, nucleic acids, macromolecules, and many other biological systems, although its origins are in classical problems of colloid chemistry. We will examine some of the consequences of the interactions of charged particles, paying attention to the way in which surface potential or surface charge can be experimentally controlled. The *rate-limiting processes* associated with changes in surface potential or charge are of great importance in the understanding of the dynamics of electromechanical and electromechanochemical interactions which are mediated by electrical double layer phenomena.

In Figure 4.14, we picture two positively charged, planar particles suspended in an electrolyte medium. The particles are surrounded by their electrical double layers, considered to be infinite in extent in the y and z directions for the purposes of a one-dimensional analysis. We first calculate the potential profile $\Phi(x)$ as sketched in Figure 4.14.

For the case of a symmetrical electrolyte, where

$$|z_+| = |z_-| = z \tag{4.75}$$

$$c_{+0} = c_{-0} = c_0 \tag{4.76}$$

the Poisson–Boltzmann equation takes the form

$$\nabla^2 \Phi = \frac{+2zFc_0}{\epsilon} \sinh\left(\frac{zF\Phi}{RT}\right) \tag{4.77}$$

Note that (4.77) has not been linearized—no assumption has yet been made concerning the magnitude of $zF\Phi(x)/RT$. We now solve for $\Phi(x)$ in

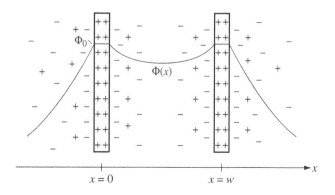

Figure 4.14 The potential profile for plane parallel interacting double layers. (Sketched for the case of constant surface potential $\Phi(0)$.)

the region ($0 \leq x \leq w$) subject to the boundary conditions

$$\Phi = \Phi_0 \quad \text{at} \quad x = 0, \quad x = w \tag{4.78}$$

$$\left.\begin{array}{l} \dfrac{\partial \Phi}{\partial x} = 0 \\[2mm] \Phi \equiv \Phi_m \end{array}\right\} \text{ at } x = \dfrac{w}{2} \tag{4.79}$$

where Φ_0 and Φ_m are constants (either Φ_0 or the surface charge σ_d may be known or controlled experimentally, while Φ_m simply represents the unknown potential at the mid-plane, $x = w/2$). Multiplying both sides of (4.77) by zF/RT, the Poisson–Boltzmann equation is rewritten in normalized form

$$\frac{d^2 \phi(x)}{dx^2} = \kappa^2 \sinh \phi(x) \tag{4.80}$$

where

$$\phi(x) = \frac{zF\Phi(x)}{RT} \tag{4.81}$$

The nonlinear differential equation (4.80) can be integrated once by first multiplying both sides by $d\phi(x)/dx$ and then integrating:

$$\frac{d\phi(x)}{dx} \frac{d^2\phi(x)}{dx^2} = \frac{d}{dx}\left[\frac{1}{2}\left(\frac{d\phi}{dx}\right)^2\right] = \kappa^2 \sinh \phi(x) \frac{d\phi}{dx} \tag{4.82}$$

$$\boxed{\frac{d\phi}{dx} = \pm\kappa \left[2\cosh\phi(x) - 2\cosh\phi_m\right]^{1/2}} \tag{4.83}$$

The constant of integration going from (4.82) to (4.83) was found from the boundary condition (4.79). In the region $0 \leq x \leq w/2$, the negative sign in (4.83) is chosen to be consistent with physical reasoning (Figure 4.14); the opposite sign applies for ($w/2 \leq x \leq w$).

Before considering the second integration of (4.80), we note that (4.83) already provides much useful information, since the electrical stress σ_{xx}^e can be written directly in terms of $d\phi/dx$. Further, we can immediately find the relation between surface potential and surface charge σ_d by evaluating (4.83) at $x = 0$:

$$\sigma_d = \boldsymbol{n} \cdot \epsilon \boldsymbol{E}\bigg|_{x=0} = -\epsilon \frac{\partial \Phi}{\partial x}\bigg|_{x=0} = +\frac{RT\epsilon\kappa}{zF}\left[2\cosh\phi_0 - 2\cosh\phi_m\right]^{1/2} \tag{4.84}$$

(σ_d in (4.84) is that amount of charge on the particle surface equal and opposite to the volume integral of $\rho_e(x)$ from $x = 0$ to $x = w/2$). Once again, (4.84) applies to the nonlinearized case. If σ_d rather than Φ_0 is in fact the physical variable under control, then we can exchange Φ_0 for σ_d using (4.84) at the end of our analysis.

Integration of (4.83) must be done numerically (it can be expressed in the form of elliptic integrals). For the purposes at hand, we find $\phi(x)$ by linearizing (4.83). For $\phi(x) \ll 1$, (4.83) becomes, in the region $0 \leq x \leq w/2$,

$$\frac{d\phi(x)}{dx} \simeq -\kappa\sqrt{\phi^2(x) - \phi_m^2} \tag{4.85}$$

Integrating (4.85) to find $\phi(x)$,

$$\int_{\phi_0}^{\phi(x)} \frac{d\phi(x)}{\sqrt{\phi^2(x) - \phi_m^2}} = \int_0^x -\kappa\,dx = \cosh^{-1}\left[\frac{\phi(x)}{\phi_m}\right] + C = -\kappa x \tag{4.86}$$

At $x = 0$, $\phi = \phi_0$, and therefore $C = -\cosh^{-1}(\phi_0/\phi_m)$:

$$\frac{\phi(x)}{\phi_m} = \cosh\left[\cosh^{-1}\left(\frac{\phi_0}{\phi_m}\right) - \kappa x\right] \tag{4.87}$$

Evaluating the integral (4.86) between the limits $x = 0$ to $x = w/2$ and ϕ_0 to ϕ_m, we obtain a relation for ϕ_m:

$$\phi_m = \frac{\phi_0}{\cosh[\kappa w/2]} \tag{4.88}$$

Substituting this result into (4.87) we have the simpler expression

$$\phi(x) = \phi_0 \frac{\cosh[\kappa(w/2 - x)]}{\cosh(\kappa w/2)} \tag{4.89}$$

Equation (4.89) thus represents the potential profile for the case of two interacting plane parallel double layers for the limiting case of *small surface potential*, $\phi_0 = zF\Phi_0/RT \ll 1$ (sketched in Figure 4.14).

4.8 FORCE EQUILIBRIUM WITH INTERACTING PLANE PARALLEL DOUBLE LAYERS

In equilibrium, the repulsive and attractive forces between the particles of Figure 4.15 balance, resulting in an equilibrium separating distance w. Both electrical and osmotic forces lead to repulsion between plates. For the case of biological tissues, equilibrium spacing is often the result of repulsive forces working against structural mechanical linkages such as the intermolecular and interfibrillar crosslinks of connective tissues or the more complex structure of muscle fibers. Typical of colloid stability problems is an attractive interaction based on London–van der Waals forces.

In Figure 4.15, a force per unit area σ^m is imagined to represent the attractive forces holding the left-hand plate or particle in equilibrium. Thus, σ^m might be an external or internal force (examples of the latter being van der Waals forces or chemical crosslinkages). σ^m is assumed here to be constant. We will find that the repulsive forces, both electrical and osmotic, may be modified by changing the chemical environment,

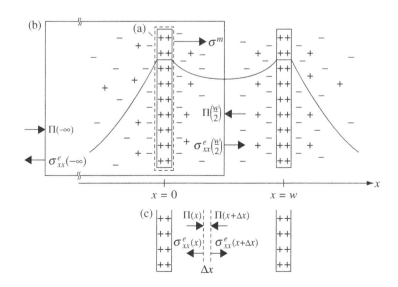

Figure 4.15 Sketch of $\Phi(x)$, osmotic pressure, and electrical stresses of interacting plane parallel double layers with surface potential Φ_0 constant. (a) Stress on a control surface surrounding unit area of one plate. (b) The control surface includes electrolyte in addition to the plate. (c) Stresses on a slab of liquid (Δx) between the plates.

such as the pH and/or ionic strength of the bath. Such a change in repulsive force given fixed σ^m will result in a new equilibrium spacing between the two plates. An example of this kind of interaction is the swelling and deswelling of gels and other polyelectrolyte tissues, such as muscle and tendon, upon changes in bath concentrations (see Chapter 7).

In order to predict the equilibrium spacing as well as the physical characteristics of the repulsive forces, we now examine the spatial dependences of the various forces on the left-hand plate. We then evaluate the total force per unit area on this plate in the x direction using first a stress tensor "box" that just surrounds the plate (Figure 4.15(a)) and then a box that includes some adjacent electrolyte on both sides of the plate (Figure 4.15(b)). Stress equilibrium for an incremental thickness of liquid between the plates is then examined (Figure 4.15(c)).

From Section 4.2, the total force on the left-hand plate can be found by evaluating the integral

$$f_x = \oint_S \sigma_{ix} n_i \, da \tag{4.90}$$

where σ_{ix} includes *all stresses acting on the plate* (i.e., electrical, osmotic, σ^m, etc.). We first consider a surface S as in Figure 4.15(a), i.e., a surface that just surrounds the plate. The force/unit area σ_x in the x direction is then (with $\boldsymbol{n} = \boldsymbol{i}_x$)

$$\sigma_x^e = \sigma_{xx}^e(0^+) - \sigma_{xx}^e(0^-) \tag{4.91}$$

In evaluating the *electrical* contribution to (4.91) we use the Maxwell stress tensor corresponding to the force density on free charge in a dielectric media of constant permittivity ϵ. The first two terms of the force density of (4.27) (including the electrostriction term) are absent in this limiting case, and the stress tensor is

$$\sigma_{ij}^e = \epsilon E_i E_j - \delta_{ij} \frac{\epsilon}{2} E_k E_k \tag{4.92}$$

Equation (4.91) then reduces to

$$\sigma_x^e = \tfrac{1}{2}\epsilon \left[E_x^2(0^+) - E_x^2(0^-) \right] \tag{4.93}$$

For our one-dimensional model, $E_x(0^+)$ is found from (4.83), while $E_x(x < 0)$ corresponds to the potential decay of a single, noninteracting double layer, where (see Figure 4.15)

$$\left. \begin{array}{c} d\Phi/dx \to 0 \\ \Phi \to 0 \end{array} \right\} \text{ as } \quad x \to -\infty \tag{4.94}$$

$$E_x(x < 0) = -\frac{d\Phi}{dx} = -\frac{RT}{zF}\sqrt{2\kappa^2[\cosh \Phi(x) - 1]} \tag{4.95}$$

(Equation (4.95) results from the integration of (4.82) subject to the boundary conditions (4.94) rather than those used to obtain (4.83).) The sign of $E_x(x < 0)$ is consistent with the positive charge of the plate.

With $\phi(0^+) = \phi(0^-) = \phi_0 = zF\Phi_0/RT$, (4.93) becomes

$$\sigma_x^e = -\epsilon \left(\frac{RT}{zF} \right)^2 \kappa^2 \left[\cosh\left(\frac{zF\phi_m}{RT} \right) - 1 \right] \tag{4.96}$$

$$\boxed{\sigma_x^e = -2RTc_0 \left[\cosh\left(\frac{zF\phi_m}{RT} \right) - 1 \right]}$$

Note that the electrical force on the plate is in the negative (replusive) x direction, as indicated by the sign of (4.96).

In writing an overall force balance for the plate, we must also account for hydrostatic pressure forces due to the surrounding liquid on both sides of the plate. We will see in Chapter 5 that hydrostatic pressure is a normal surface force and can be represented as such by a traction

$$\mathscr{T}^p = -p\,\boldsymbol{n} \tag{4.97}$$

with the corresponding stress tensor

$$\sigma_{ij}^p = -p\,\delta_{ij} \tag{4.98}$$

For a stationary liquid with a single, vertically oriented, *uncharged* plate, the pressure forces on the sides of the plate in the x direction would cancel. When the plate is charged, the potential $\Phi(x)$ in the liquid on both sides of the plate will lead to nonuniform concentrations of electrolyte ions: $c_i(x) = c_0 e^{-z_i F \Phi(x)/RT}$. Nonuniform electrolyte concentrations in *equilibrium* (Section 3.3) will lead to osmotic pressures that vary with position in the liquid (see Problem 3.9). The pressure p thus becomes the osmotic pressure Π, which is related to the concentrations of solutes (charged and uncharged) by the chemical equation of state (van't Hoff's law):

$$\Pi(x_1) - \Pi(x_2) = RT\left[\sum c_i(x_1) - \sum c_i(x_2)\right] \tag{4.99}$$

In Figure 4.15, $\Phi(0^+) = \Phi(0^-) = \Phi_0$. Therefore, $c_i(0^+) = c_i(0^-)$, with the result that $\Pi(0^+) = \Pi(0^-)$; the osmotic pressure cancels on both sides of the plate right at the surfaces ($x = 0^+, 0^-$).

The final surface stress balance of (4.91) can then be written as

$$\sigma_x^{\text{TOT}} = \sigma^m + \sigma_x^e - [\Pi(0^+) - \Pi(0^-)] = \boxed{\sigma^m - 2RTc_0\left[\cosh\left(\frac{zF\Phi_m}{RT}\right) - 1\right]} \equiv 0 \tag{4.100}$$

where $\sigma_x^{\text{TOT}} \to 0$ follows from the assumption of an equilibrium balance between attractive and repulsive forces between the plates. The negative sign in front of the osmotic pressure term results from the sign convention of (4.97).

If we chose to place the stress tensor box in the position of Figure 4.15(b), then the surface integral of the stresses around the box would give the force on all constituents within the box (including the fluid to the right and left of the plate), along with the unit area of plate itself. With the right-hand edge of the box situated at $x = w/2$, the electrical stress tensor component $\sigma_{xx}^e(x = w/2) = 0$ since $E_x \sim d\Phi/dx = 0$ there; this is also true for σ_{xx} evaluated at $x \to -\infty$. However, the osmotic pressure $\Pi(w/2)$ is now greater than that in the bulk $\Pi(-\infty) \equiv \Pi_0$; for this choice of positioning of the box, the sum of the forces per unit area in the x direction is

$$\sigma_x^{\text{TOT}} = \sigma_{xx}^e(w/2) - \sigma_{xx}^e(-\infty) - [\Pi(w/2) - \Pi_0] + \sigma^{\text{attr}} \tag{4.101}$$

For now, we assume that the σ^{attr} is identical with σ^m as shown in Figure 4.15(a), i.e., the attractive force is either an externally or internally imposed constant "pressure" that can be represented as acting on the particles in a direction to bring them together. From (4.99), the osmotic pressure term in (4.101) is

$$\Pi(w/2) - \Pi_0 = RT\left(c_0 e^{zF\Phi_m/RT} + c_0 e^{-zF\Phi_m/RT}\right) - RT(2c_0) \tag{4.102}$$

$$\sigma_x^{\text{TOT}} = \sigma^m - \Pi(w/2) + \Pi_0 = \boxed{\sigma^m - 2RTc_0\left[\cosh\left(\frac{zF\Phi_m}{RT}\right) - 1\right]} \equiv 0$$
$$\tag{4.103}$$

We note that the result (4.103) is identical to (4.100), even though the stresses were evaluated over different surfaces in the two cases. The key to the situation lies in the fact that the system of Figure 4.15 is in equilibrium, and the solute ions distribute themselves in a manner consistent with the equilibrium potential $\Phi(x)$. Thus, what appears to be an entirely electrical repulsive force in Figure 4.15(a) becomes entirely a repulsive "*osmotic swelling pressure*" in Figure 4.15(b) when the right-hand edge of the box is located at $x = w/2$. We can see a more general result with the right-hand stresses evaluated at arbitrary x, where, for the region ($0 \le x \le w/2$),

$$\sigma_{xx}^e(x) = \frac{1}{2}\epsilon\left[\frac{d\Phi(x)}{dx}\right]^2 = \left(\frac{RT}{zF}\right)^2 \epsilon\kappa^2[\cosh\phi(x) - \cosh\phi_m] \tag{4.104}$$

$$\Pi(x) = 2RTc_0\cosh\phi(x) \tag{4.105}$$

where (4.104) follows from (4.83). The stress balance now becomes

$$\sigma_x^{\text{TOT}} = \sigma^m + \sigma_{xx}^e(x) - \sigma_{xx}^e(-\infty) - [\Pi(x) - \Pi(-\infty)] \equiv 0 \tag{4.106}$$

Substituting (4.104) and (4.105) into (4.106), we see that all the x-dependent terms cancel, leaving, as before, (4.103):

$$\sigma_x^{\text{TOT}} = \boxed{\sigma^m - 2RTc_0(\cosh\phi_m - 1)} \equiv 0 \tag{4.107}$$

The equivalence of (4.107), (4.103), and (4.100) suggests a play-off between osmotic and electrical stresses on the fluid between the plates. (Once again, the electrical forces "on the fluid" are actually exerted on the solute ions; the latter transfer their force to the fluid as discussed in Section 4.3.) This situation can easily be seen by evaluating the electrical and osmotic forces on a slice of liquid between the plates having thickness Δx, in the limit $\Delta x \to 0$ (Figure 4.15(c)). (Since σ^m acts on the particle and not the intervening fluid, it is not evaluated in the force balance below.) Since the liquid is assumed to be in a stationary equilibrium,

$$\sigma_{xx}^e(x + \Delta x) - \sigma_{xx}^e(x) - [\Pi(x + \Delta x) - \Pi(x)] \equiv 0 \tag{4.108}$$

$$\frac{\partial}{\partial x}\left[\sigma_{xx}^e - \Pi(x)\right] = 0 \tag{4.109}$$

$$\frac{\partial}{\partial x}\left(\sigma_{xx}^e + \sigma_{xx}^p\right) = 0 \tag{4.110}$$

where (4.110) has incorporated the definition (4.98). But (4.110) is simply a statement that the total force on the slice of liquid is zero; in the limit $\Delta x \to 0$, $F_x = \partial(\sigma_{xj}^e + \sigma_{xj}^p)/\partial x_j = 0$ for Figure 4.15(a). That (4.110) is true follows directly from (4.104) and (4.105). We finally evaluate the total repulsive force between the plates, defined as Π^{rep}, in terms of ϕ_0

rather than ϕ_m, in the limit $\phi_m \ll 1$:

$$\Pi^{\text{rep}} = 2RTc_0(\cosh\phi_m - 1) \tag{4.111}$$

$$\Pi^{\text{rep}} \simeq \left(\frac{RT}{zF}\right)^2 \epsilon\kappa^2 \frac{\phi_m^2}{2} = \frac{1}{2}\epsilon\kappa^2\Phi_m^2 \tag{4.112}$$

Using (4.88) in (4.112) gives

$$\Pi^{\text{rep}} \simeq \frac{1}{2}\epsilon\kappa^2\Phi_0^2\left[\frac{1}{\cosh(\kappa w/2)}\right]^2 \tag{4.113}$$

If we further ask for the limiting case of $\kappa w/2 \gg 1$ (i.e., interparticle spacing so large that only weak, but finite, interaction between plates results), then

$$\Pi^{\text{rep}} \simeq 2\epsilon\kappa^2\Phi_0^2 e^{-\kappa w} \tag{4.114}$$

Thus, we have expressed the repulsive force in terms of the surface potential Φ_0, which was assumed to be constant in Figure 4.15. If the surface charge is experimentally controllable, or if a physical process is known to occur at constant charge, then we can rewrite (4.113) in terms of σ_d rather than Φ_0 by using (4.84). However, to be consistent with (4.114), which applies in the limit $\phi_0 \ll 1$, we need the linearized limit of (4.84):

$$\sigma_d = \epsilon\left(\frac{RT}{zF}\right)\kappa\sqrt{\phi_0^2 - \phi_m^2} = \epsilon\kappa\Phi_0\left[1 - \frac{1}{\cosh^2(\kappa w/2)}\right]^{1/2} = \epsilon\kappa\Phi_0\tanh(\kappa w/2) \tag{4.115}$$

where (4.88) and the identity $\cosh^2 x - \sinh^2 x = 1$ have been used. Then (4.113) becomes

$$\Pi^{\text{rep}} = \frac{1}{2}\frac{\sigma_d^2}{\epsilon}\left[\frac{1}{\sinh(\kappa w/2)}\right]^2 \simeq 2\frac{\sigma_d^2}{\epsilon}e^{-\kappa w} \tag{4.116}$$

where the second equality in (4.116) applies for large $\kappa w/2$. Figure 4.16 pictures an interaction process showing the change in potential profile as two charged plates are brought together. The case where the surface charge σ_d is constrained to be constant is compared with a constant-Φ_0 model.

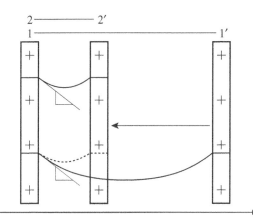

Figure 4.16 Two charged plates originally at positions 1–1′ are brought together to a new position 2–2′. Solid curve: the double layer surface charge σ_d is held constant; dotted curve: Φ_0 is held constant.

4.9 INTERACTING DOUBLE LAYERS: RATE PROCESSES AND ELECTRICAL TERMINAL CONSTRAINTS

The interacting double layer model outlined in Sections 4.7 and 4.8 has been applied to many systems of charged particles and molecules. Several examples are presented here, concerning widely differing classes of particles. Typical of phenomena involving solutions of colloidal particles (i.e., particles in the 1 nm–1 μm range) is the classic problem of equilibrium separation between tungstic acid crystals shown in Figure 4.17. These flat "platelets" tend to sediment in suspension owing to gravity, but reach an equilibrium spacing owing to double layer repulsion forces. The separation distance is fairly regular and can be estimated by interpreting the optical interference colors seen upon shining a light beam through the suspension.

Figure 4.18 shows the modification of lipid bilayer separation due to changes in aqueous electrolyte concentration [23]. The distance W results from a balance between repulsive and attractive forces, the former being modified by valence and ionic strength as we will presently explore. In Figure 4.19, we see that both the ionic strength and the pH of the bath can modify the interparticle separation distance [23]. Models of the equilibrium spacing of muscle fibers have been proposed that are based on a double layer interaction theory [19, 24]. We will examine such forces with respect to another fibrous protein—collagen—after investigating the role of electrolyte ionic strength and pH with respect to particle charge, potential, and double layer capacitance per unit area ($\epsilon\kappa$).

The data shown in Figures 4.18 and 4.19 suggest that a change in equilibrium particle spacing results from a change in the double layer repulsion force. This conclusion is based on the fact that changes in electrolyte pH and ionic strength commonly lead to changes in σ_d and Φ_0, which in turn affect Π^{rep} (see (4.113) and (4.116)). Whether Φ_0 or σ_d is in fact an externally controllable variable depends on the type of particle involved. For example, metal plates are simply amenable to the control of Φ_0 if contact can be made to a battery. The Φ_0 of colloidal particles typical of Figure 4.17 can often be controlled by fixing the concentration of a "potential-determining ion." For example, the Φ_0 of

Figure 4.17 Vertical stacking reveals a sedimentation equilibrium resulting from the balance of gravitational and double layer forces. (a) WO₃ crystal. (b) Colloidal suspension of tungstic acid (WO₃) crystals.

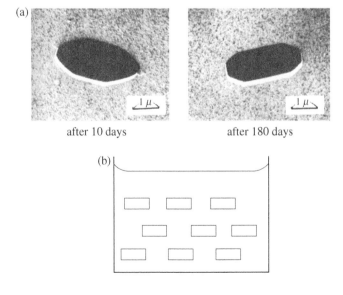

(a)

1 μ   1 μ

after 10 days   after 180 days

(b)

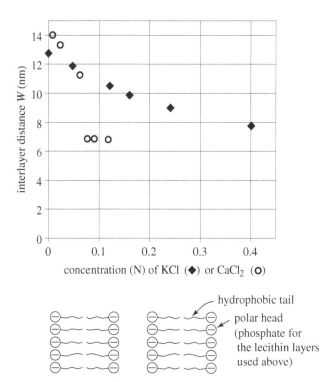

Figure 4.18 Interlayer distance in lipid layers, separated by aqueous solution containing varying amounts of electrolyte as determined by KJ Palmer and FO Schmitt [23,25].

a AgI crystal is determined by the concentration of I⁻ ions in solution according to the Nernstian relation

$$\Phi_0^{\mathrm{AgI}} - \Phi_{\mathrm{bulk}} = -\frac{RT}{F}\ln\!\left(\frac{c_{\mathrm{I}^-}}{\mathrm{const}}\right) \qquad (4.117)$$

This relation was found empirically, and applies to many other types of crystal/electrolyte interfaces.

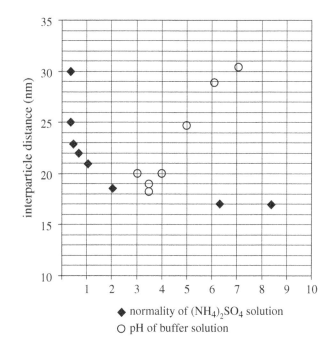

Figure 4.19 Interparticle distance in a suspension of tobacco mosaic virus protein (TMVP), determined from X-ray diffraction, as a function of pH and of electrolyte content [23].

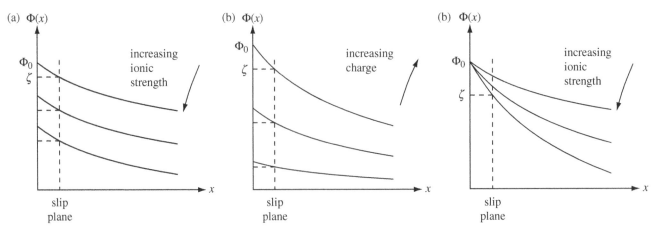

Figure 4.20 Plots of $\Phi(x)$ (the double layer potential as a function of the distance x from a charged polyelectrolyte surface), showing (a) changes in Φ_0 (the surface potential) and ζ (the zeta-potential difference between the slip plane and $x = \infty$; see Chapter 6) at constant charge (constant $|\partial\Phi/\partial x|_{x=0}$); (b) changes in Φ_0, ζ, and charge at constant capacitance (ionic strength); and (c) changes in charge and capacitance at constant Φ_0.

Most biological molecules have charge groups whose ionization state is determined by a dissociation reaction in an electrolyte bath. Thus, the surface charge σ_d can often be controlled, and Φ_0 adjusts itself to be consistent with σ_d and the double layer capacitance. This is the situation found with proteins. A titration curve of collagen is shown in Figure 1.12 in Chapter 1, where σ_d is determined primarily by bath pH, but secondarily by total ionic strength (the latter affects the dissociation reactions owing to modification of $1/\kappa$).

Figure 4.20 depicts graphically the relation between σ_d, Φ_0, and double layer capacitance per unit area $\epsilon\kappa$ for processes occurring at (a) constant charge, (b) constant $\epsilon\kappa$, and (c) constant Φ_0. It will be important to have those pictures in mind when considering the *dynamics* of double layer interactions. If one were to model equilibrium swelling of tendon by means of a double layer approach, the tensile force would be pictured to result from changes in fibril double layer repulsion forces. Such changes could occur by a change in either fibril Φ_0 or fibril σ_d, for example. However, the rate processes associated with changes in Φ_0 and σ_d are very different. The chemical reaction times for changes in σ_d are often much longer than the electrolyte redistribution times associated with changes in Φ_0.

4.10 PROBLEMS

Problem 4.1   A vector $\boldsymbol{F}$ is often defined by the way in which its components transform from one orthogonal coordinate system to another. Thus, $\boldsymbol{F}$ has components F_j in the coordinate system (x_1, x_2, x_3), which are related to the components F_i' in the coordinate system (x_1', x_2', x_3') by the vector transformation law

$$F_i' = a_{ij}F_j \qquad (4.118)$$

where a_{ij} are the direction cosines (e.g., a_{12} is the cosine of the angle between the x_1' and x_2 axes as shown in Figure 4.21). We also note that the coordinate systems are related by

$$x'_k = a_{kl}x_l \qquad (4.119)$$

(**a**) Show that the stress tensor transforms according to the law

$$\sigma'_{ik} = a_{ij}a_{kl}\sigma_{jl} \qquad (4.120)$$

by representing F'_i and F_j as the divergence of a stress tensor (e.g., $F'_i = \partial\sigma'_{ki}/\partial x'_k$) and using the chain rule of differentiation. A tensor is often defined by showing that it obeys a transformation law such as (4.120), just as a vector is defined by (4.118).

(**b**) A tensor σ_{mn} in the (x_1, x_2, x_3) frame of Figure 4.21 has elements $\sigma_{11} = \sigma_0$, $\sigma_{22} = -\sigma_0$. All other σ_{ij} are zero. Find the a_{ij} corresponding to Figure 4.21 and represent the a_{ij} in matrix form. Find the stress elements in the (x'_1, x'_2, x'_3) frame by using (4.120).

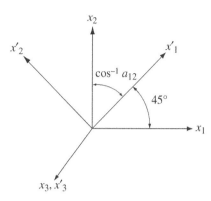

Figure 4.21 Coordinate system transform.

Problem 4.2 A pair of parallel insulating sheets is shown in Figure 4.22. The sheet at $y = d$ supports a surface charge density $-\sigma_e$, whereas the sheet at $y = 0$ supports the image surface charge density σ_e. Hence the electric field between the plates *due to the charges* is $(\sigma_e/\epsilon)\boldsymbol{i}_y$. External electrodes are used to impose an additional uniform electric field given everywhere by $\boldsymbol{E} = E_0\boldsymbol{i}_x + E_0\boldsymbol{i}_y$, where E_0 is a constant.

(**a**) Write the components of the Maxwell stress tensor at points A and B in terms of σ_e and E_0.

(**b**) Use the Maxwell stress tensor to find the total electric force in each of the coordinate directions on the section of the lower sheet between $x = a$ and $x = b$ having depth D in the z direction.

    Show that your answers to part (b) agree with forces found by multiplying the surface charge density σ_e by the appropriate averaged electric field intensity and the appropriate area.

Problem 4.3 Find the stress tensor σ_{ij} consistent with the Kelvin force density (including free charge)

$$\boldsymbol{F} = \rho_e\boldsymbol{E} + \boldsymbol{P}\cdot\nabla\boldsymbol{E}$$

Compare your answer with the σ_{ij} corresponding to a $\rho_e\boldsymbol{E}$ force density in a medium with constant permittivity ϵ.

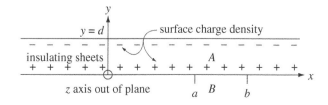

Figure 4.22 Parallel insulating sheets.

Problem 4.4 What is the boundary condition used to relate σ_d to Φ_0 in (4.115)? Discuss the validity of this equation for plate-like particles (e.g., WO_3) and membranes, each having front and back surfaces with identical electrical properties; do the same for rod-like fibrils with radial symmetry. (The discussion should concern not only the surface versus volume charge modeling, but also the validity of the boundary condition used to obtain (4.115) and the geometry of interest.)

Problem 4.5 The asymptotic expression for Π^{rep} for two flat double layers with constant surface charge σ_d, in the limit $\kappa w/2 \to 0$, *cannot* be found from (4.116). In this limit, ϕ_0 becomes very large, violating the conditions for which (4.116) was derived.

Use simplifying arguments based on van't Hoff's law and electroneutrality to find Π^{rep} for $\kappa w/2 \to 0$ in terms of σ_d, w, and the thermal voltage RT/zF.

This result can also be derived in terms of diffuse double layer theory.

Problem 4.6 With (4.98), we make all of the diagonal terms in σ_{ij}^p equal to $-p$ because there must be no shear stress on a surface having normal $\boldsymbol{n}$, regardless of the orientation of that surface. Assume that the diagonal components are not equal in the (x_1, x_2, x_3) frame of reference, and use the tensor transformation of (4.120) in Problem 4.1 to find the implied stress in the frame (x_1', x_2', x_3'). Show that, with the diagonal components equal, there is no shear stress, but otherwise there is.

Problem 4.7 As early as 1940, Zocher proposed a double layer repulsion model to explain the equilibrium separation distance between anisodimensional particles found to stack in suspension. An example of this phenomenon is the stacking of WO_3 crystals as pictured in Figure 4.17. The model supposes a balance between double layer repulsion and gravitational forces causing sedimentation.

(a) For the configuration of Figure 4.17, consider two of the plate-like particles having thickness $d = 70\,nm$ and assume $\Phi_0 = 25\,mV$. The solution contains a neutral electrolyte ($z_+ = |z_-| = 1$) having concentration $10^{-3}\,M$ in addition to the WO_3 crystals. Take the density of WO_3 to be $7\,g\,cm^{-3}$ and $T = 25°C$. *Find* the equilibrium spacing w between platelets, *assuming* $\kappa w/2 \gg 1$. Does your final answer justify this assumption? Justify all other assumptions used.

(b) Is the "equilibrium" spacing stable? Discuss stability in terms of a graph of force versus separation w.

(c) Find expressions for the energy functions associated with the gravitational and electrical forces (again for $\kappa w/2 \gg 1$). Discuss stability in terms of a potential well diagram.

Problem 4.8

PROBLEM

(**a**) Find an analytical expression for Π^{rep} (total repulsion force/area) between two charged particles using the parallel plate model of Figure 4.15. Express your answer in terms of Φ_0, κ, and $w/2$. Use the Maxwell stress tensor method and the surface marked "C" in Figure 4.15. Assume that the plates are perfect conductors (equipotential surfaces).

(**b**) Use a "lumped parameter" energy method approach to find Π^{rep} and show that your answer agrees with part (a). To do this, proceed as follows: (1) Define the constitutive relation between terminal voltage Φ_0 and terminal charge/area σ_d. This constitutive law can be viewed as the consequence of charging the plates up to potential Φ_0 with respect to $x \to \pm\infty$ in the surrounding electrolyte (Hint: σ_d can be related to Φ_0 using the interacting double layer potential solution). (2) Find the energy per area $W(\sigma_d, w/2)$ by appropriate assembly of the thermodynamic subsystem. (3) Find an analytical expression for Π^{rep} using $\Pi^{rep} = -\partial W/\partial(w/2)$, where the derivative is evaluated at constant charge.

4.11 REFERENCES

[1] Melcher JR (1981) *Continuum Electromechanics*. MIT Press, Cambridge, MA. (This book is also available online at http://ocw.mit.edu/resources/res-6-001-continuum-electromechanics-spring-2009/textbook-contents/).

[2] Pohl H (1978) *Dielectrophoresis*. Cambridge University Press, Cambridge.

[3] Dukhin AS, Ulberg R, Karamushka VI, and Gruzina TG (2010) Peculiarities of live cells' interaction with micro- and nanoparticles. *Adv. Coll. Interf. Sci.* **159**, 60–71.

[4] Vahey MD and Voldman J (2008) An equilibrium method for continuous-flow cell sorting using dielectrophoresis. *Anal. Chem.* **80**, 3135–3143.

[5] Gouy G (1910) Sur la constitution de la charge électrique à la surface d'un électrolyte. *J. Phys. Radium* **9**, 457–468.

[6] Debye P and Hückel E (1923) Zur Theorie der Elektrolyte. I. Gefrierpunktserniedrigung und verwandte Erscheinungen. *Phys. Z.* **24**, 185–206.

[7] Attard P (2002) *Thermodynamics and Statistical Mechanics: Equilibrium by Entropy Maximisation*. Academic Press, London.

[8] Haydon DA (1964) The electrical double layer and electrokinetic phenomena. In *Recent Progress in Surface Science*, Volume 1 (Danielli JF, Pankhurst KGA, and Riddiford AC, eds). Academic Press, New York, pp. 94–158.

[9] Chu VB, Bai Y, Lipfert J, et al. (2008) A repulsive field: advances in the electrostatics of the ion atmosphere. *Curr. Opin. Chem. Biol.* **12**, 619–625.

[10] Grahame DC (1941) Properties of the electrical double layer at a mercury surface. I. Methods of measurement and interpretation of results. *J. Am. Chem. Soc.* **63**, 1207–1214.

[11] Fuoss RM (1959) Polyelectrolytes. In *Molecular Science and Molecular Engineering* (Von Hippel SR, ed.). MIT Press, Cambridge, MA, pp. 376–389.

[12] Morawetz M (1965) *Macromolecules in Solution*. Wiley-Interscience, New York, p. 325.

[13] Oosawa F (1971) *Polyelectrolytes*. Marcel Dekker, New York, p. 13.

[14] Fuoss RM, Katchalsky A, and Lifson S (1951) The potential of an infinite rod-like molecule and the distribution of the counter ions. *Proc. Natl Acad. Sci. USA* **37**, 579–589.

[15] Gliozzi A, Vittoria V, and Ciferri A (1972) Transport properties of charged membranes. *J. Membrane Biol.* **8**, 149–162.

[16] Mysels KJ (1959) *Introduction to Colloid Chemistry*. Wiley-Interscience, New York, pp. 378–383.

[17] Katchalsky A (1964) Polyelectrolytes and their biological interactions. In *Connective Tissue: Intercellular Macromolecules. Proceedings of the Symposium of the New York Heart Association*. Little, Brown, Boston, p. 9.

[18] Katchalsky A and Zwick M (1955) Mechanochemistry and ion exchange. *J. Polymer Sci.* **16**, 221–223.

[19] Elliott GF (1968) Force-balances and stability in hexagonally-packed polyelectrolyte systems. *J. Theor. Biol.* **21**, 71–87.

[20] Adamson AW (1967) *Physical Chemistry of Surfaces*, 2nd ed. Wiley-Interscience, New York, Chapter 4.

[21] Anderson JC and Eriksson C (1970) Piezoelectric properties of dry and wet bone. *Nature* **227**, 491–492.

[22] Overbeek JThG (1952) The interaction between colloidal particles. In *Colloid Science*, Volume I: *Irreversible Systems* (Kruyt HR, ed.). Elsevier, Amsterdam, pp. 245–277.

[23] Overbeek JThG (1952) Stability of hydrophobic colloids and emulsions. In *Colloid Science*, Volume I: *Irreversible Systems* (Kruyt HR, ed.). Elsevier, Amsterdam, pp. 302–341.

[24] Shear DB (1970) Electrostatic forces in muscle contraction. *J. Theor. Biol.* **28**, 531–546.

[25] Palmer KJ and Schmitt FO (1941) X-ray diffraction studies of lipide emulsions. *J. Cell. Comp. Physiol.* **17**, 385–394.

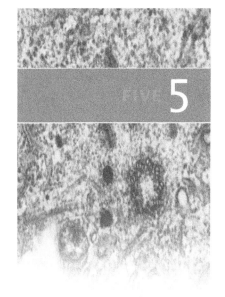

Newtonian Fluid Mechanics

5.1 INTRODUCTION

The fluxes and concentration profiles of neutral and charged solutes have been described thus far in the context of their response to gradients in chemical concentrations and electrical potentials. Examples include an incredibly broad array of applications involving diffusion and electrical migration of low-molecular-weight compounds and complex biological macromolecules within and across tissues, cytoskeletal networks, and cell membranes. However, the effects of fluid convection are often dominant in determining the dynamics of systems involving mass transport.

On a macrocontinuum scale, many important problems in biomedical engineering focus on the study of fluid mechanics relating to respiratory, cardiovascular, musculoskeletal, and other physiological systems. On the nanoscale of individual proteins, nucleic acids, and polysaccharides, convective flows are also essential in understanding transduction processes that link electrical, mechanical, and chemical subsystems. Examples include the electrophoresis of proteins, electroosmotic fluid flow through membranes and tissues, flow-induced electric fields in tissues containing charged macromolecular constituents, and electrokinetic interactions within micro- and nanofluidic systems (see Chapter 6). In these examples, convection, diffusion, and electrical migration processes within and near the electrical double layers that surround charged macromolecules are of equal importance.

In the following sections, the continuum laws for a classical fluid are developed, including conservation of mass and momentum. The effects of finite viscosity focus the discussion to the case of Newtonian fluids. The investment made in formulating this special case is a useful one for later development of non-Newtonian fluid mechanics. Low-Reynolds-number viscous-dominated creeping flows are of particular interest in our studies, and motivate the use of stream function representations for a variety of examples such as Stokes drag on a solid particle. Finally, convective diffusion and diffusion boundary layers are described in the context of specific biological examples.

5.2 CONSERVATION OF MASS

The classical theory of fluid mechanics deals with the average motion of many molecules. Thus, the mass density ρ $(\mathrm{kg\,m^{-3}})$ involves a sum over all of the species within a given volume. A fluid volume of fixed identity is defined by requiring that for such a volume V the total mass is conserved:

$$\frac{d}{dt}\int_V \rho\, dV = 0 \qquad (5.1)$$

The time derivative in (5.1) can be taken inside the time-varying integral by exploiting the generalized Leibnitz rule ((A.9) in Appendix A):

$$\int_V \frac{\partial \rho}{\partial t}\, dV + \oint_S \rho\, \boldsymbol{v}\cdot \boldsymbol{n}\, da = 0 \qquad (5.2)$$

The first term of (5.2) is the rate of accumulation of mass in the control volume; the second term is the net transport of mass out of the control volume. In this form, this *integral law of mass conservation* applies whether the volume V and enclosing surface S are moving or not. This is true because the integration is performed at a single instant in time after the time and space rates of change have been evaluated. The differential form of mass conservation then follows by using Gauss' theorem to convert the surface integral of (5.2) to a volume integral. Because the volume V is arbitrary, it follows that the integrand can be set equal to zero [1]:

$$\boxed{\frac{\partial \rho}{\partial t} + \nabla \cdot (\rho \, \boldsymbol{v}) = 0} \tag{5.3}$$

An *incompressible fluid* is defined as one in which the time rate of change of the mass density ρ for an observer moving with the fluid (at the velocity $\boldsymbol{v}$) is zero:

$$\frac{D\rho}{Dt} = 0 \tag{5.4}$$

It follows from (5.3) and (5.4) that the fluid velocity is solenoidal:

$$\nabla \cdot \boldsymbol{v} = 0 \tag{5.5}$$

In general, an incompressible fluid does not have to have a uniform mass density. Rather, the rate of change of the mass density in the neighborhood of a volume of fixed identity must vanish as required by (5.4). Finally, a general consequence of mass conservation, (5.3), is that any variable X (X can be a scalar or the ith component of a vector) will satisfy the following identity, which will be useful when considering other conservation laws:

$$\frac{\partial X}{\partial t} + \nabla \cdot (\boldsymbol{v}X) = \rho \left(\frac{\partial}{\partial t} + \boldsymbol{v} \cdot \nabla \right) \left(\frac{X}{\rho} \right) \tag{5.6}$$

5.3 CONSERVATION OF MOMENTUM

A statement of Newton's law for the material within a volume V of fixed identity is

$$\frac{d}{dt} \int_V \rho \, \boldsymbol{v} \, dV = \int_V \boldsymbol{F} \, dV \tag{5.7}$$

where $\rho \boldsymbol{v}$ is the momentum per unit volume and $\boldsymbol{F}$ is the sum of all force densities acting on the volume.

To convert (5.7) to a form for which the Leibnitz rule is appropriate, we first write the ith component. It then follows from (A.9) of Appendix A with $\zeta \rightarrow \rho v_i$ that

$$\int_V \frac{\partial (\rho \, v_i)}{\partial t} \, dV + \oint_S \rho \, v_i \, \boldsymbol{v} \cdot d\boldsymbol{a} = \int_V F_i \, dV \tag{5.8}$$

This *integral law of momentum conservation* states that the rate of accumulation of momentum plus the net momentum transport out of the control volume equals the sum of the forces acting on the control volume.

With the use of Gauss' theorem, the surface integral of (5.8) is converted to a volume integral. The volume is arbitrary, and so the integrands are equal:

$$\frac{\partial(\rho\,v_i)}{\partial t} + \nabla \cdot (\rho\,v_i\,\boldsymbol{v}) = F_i \tag{5.9}$$

Introducing the constraint of conservation of mass, (5.6) with $X \to \rho v_i$ can be combined with (5.9) to give

$$\boxed{\rho\left(\frac{\partial}{\partial t} + \boldsymbol{v}\cdot\nabla\right)\boldsymbol{v} = \boldsymbol{F}} \tag{5.10}$$

5.4 INVISCID, INCOMPRESSIBLE FLOW: BERNOULLI'S EQUATION

In this section, we treat that subset of fluid mechanical interactions that can be modeled by neglecting viscous forces on fluid elements due to neighboring fluid motions. In a sense, the fluid becomes a continuum of inertia. At the outset, we assume that the mass density ρ is constant. (This is essentially an equation of state for the fluid; for many compressible fluids, on the other hand, $p = \rho RT$ is the equation of state). Therefore, (5.4) ($D\rho/Dt = 0$) is automatically satisfied and the discussion thus concerns "incompressible" fluids. The equation of motion for fluid flow is based on conservation of momentum, (5.10). Our job is simply to identify the force densities $\boldsymbol{F}$ of interest (the right-hand side of (5.10)) for any given problem.

For inviscid fluid flow, a fluid element can only sustain normal stresses (i.e., there is no mechanism by which externally applied shear stresses can be balanced by internal or fluid shear stresses, since we have not included viscosity and related viscous shear stresses in our model). We know from hydrostatics that pressure acts normally to any surface elements with equal magnitude independent of surface orientation—pressure is isotropic and can thus be represented by the tensor and force density:

$$\sigma_{ij}^{p} = -\delta_{ij}\,p \tag{5.11}$$

$$F^{p} = -\nabla p \tag{5.12}$$

where $F_i^p = \partial\sigma_{ij}^p/\partial x_j$. We will see in the next section that the pressure force due to adjacent fluid elements is augmented by other normal and shear forces when viscosity is included.

In general, the gravitational force density $\boldsymbol{F}^g = \rho\boldsymbol{g}$ must be included on the right-hand side of the momentum equation (5.10). For the case where ρ = constant, $\boldsymbol{F}^g$ can be represented as

$$\boldsymbol{F}^g = \rho\,\boldsymbol{g} = \nabla(\rho\,\boldsymbol{g}\cdot\boldsymbol{r}) \tag{5.13}$$

where $\boldsymbol{r}$ is the position vector,

$$\boldsymbol{r} = \boldsymbol{i}_x r_x + \boldsymbol{i}_y r_y + \boldsymbol{i}_z r_z \tag{5.14}$$

If there are no other forces present, (5.10) becomes

$$\rho\left(\frac{\partial}{\partial t} + \boldsymbol{v}\cdot\nabla\right)\boldsymbol{v} = -\nabla p + \nabla(\rho\,\boldsymbol{g}\cdot\boldsymbol{r}) \tag{5.15}$$

A vector identity can be used to replace the "convective momentum" term $(v \cdot \nabla)v$ by terms involving the vorticity $\omega \equiv \nabla \times v$ and the gradient of $v \cdot v$:

$$(v \cdot \nabla)v \equiv (\nabla \times v) \times v + \tfrac{1}{2}\nabla(v \cdot v) \tag{5.16}$$

$$\rho\left(\frac{\partial v}{\partial t} + \omega \times v\right) = -\nabla\left(p - \rho\, g \cdot r + \tfrac{1}{2}\rho\, v \cdot v\right) \tag{5.17}$$

assuming $\rho \equiv$ constant.

If we take the scalar product of v with all the terms in (5.17), then, for the case of steady flow ($\partial/\partial t = 0$), we have

$$v \cdot \nabla(p - \rho\, g \cdot r + \tfrac{1}{2}\rho\, v \cdot v) = 0 \tag{5.18}$$

where the second term on the left-hand side of (5.17) vanishes, since $v \cdot (\omega \times v) \equiv 0$. Equation (5.18) says that the combination of terms in the parentheses must be constant as long as we move along the direction of v; thus, along a *streamline*,

$$p - \rho\, g \cdot r + \tfrac{1}{2}\rho\, v \cdot v = \text{const} \tag{5.19}$$

which is Bernoulli's equation for *steady, inviscid, incompressible* flow. Equation (5.19) is often interpreted in terms of an energy conservation equation—the pressure work done on the fluid as it moves an incremental length along a streamline is taken up as an increase of potential or kinetic energy.

A word of caution is necessary at this point. Equation (5.19) was derived by integrating the conservation-of-momentum equation, (5.17). We could have proceeded from another direction—writing a general statement of conservation of energy and taking the limit of negligible dissipation (i.e., no viscous losses or heat flow). We would find that such an energy conservation statement would lead to exactly (5.19). In general, however, the total energy of a fluid element moving along a streamline is *not* conserved, because of heat conduction into or out of the element, or viscous dissipation. Yet, a momentum conservation statement still has physical meaning. We would then have a situation where the integrated momentum conservation statement was not identical with a statement of energy conservation.

Another way of showing the results (5.18) or (5.19), is to integrate (5.17) along a streamline. This time, we will keep the $\partial v/\partial t$ term and thus include the case of nonsteady flow:

$$\int_a^b \rho\, \frac{\partial v}{\partial t} \cdot d\ell + \left[p - \rho\, g \cdot r + \frac{1}{2}\rho\,(v \cdot v)\right]_a^b = 0 \tag{5.20}$$

where a and b are points along the same streamline. The left-hand term of (5.20) is crucial to the study of pulsatile blood flow in the arterial system and other time-varying flows where inertial effects are important. (In this regard, it is often more useful to think of the left-hand integral in terms of inertia rather than energy.)

Example 5.4.1 Oscillations of a Fluid Pendulum A simple fluid mechanical problem which accentuates the role of fluid inertia is the "fluid pendulum" shown in Figure 5.1. An inviscid incompressible liquid with mass density ρ is confined in a U-tube and given an initial displacement about its equilibrium position at $t = 0$. We would like to describe the free motions of the liquid, $\xi(t)$, for $t > 0$. The equation of motion for the liquid is simply (5.17). When (5.17) is integrated between points a and b along the same streamline, (5.20) results. Since v and $d\ell$ are collinear, the integral in (5.20) is easily evaluated:

$$\left(\rho\frac{\partial v}{\partial t}\right)\ell + (p^b - p^a) + \rho g\xi^b - (-\rho g\xi^a) + \frac{1}{2}\rho\left[(v^2)^b - (v^2)^a\right] = 0 \tag{5.21}$$

From conservation of mass ($\nabla \cdot v = 0$), we know that $\xi^b = \xi^a = \xi$, and $v^b = \partial\xi^b/\partial t = v^a = \partial\xi^a/\partial t$. Further, force balances at the two fluid/air interfaces, neglecting the effect of surface tension, show that

$$p^a = p^d$$
$$p^b = p^c$$

where

$$p^c = p^d = p_0 \equiv \text{atmospheric pressure} \tag{5.22}$$

(surface tension γ would introduce another force normal to the interface having a magnitude $\gamma(2/R)$, where R is the radius of curvature of the interfacial meniscus). Thus, (5.21) becomes

$$\rho\ell\frac{\partial^2\xi}{\partial t^2} + 2g\rho\xi = 0 \tag{5.23}$$

which is the equation for oscillatory motion about the equilibrium position with an angular frequency $\omega_0 = \sqrt{2g/\ell}$. Equation (5.23) represents the interplay between fluid inertia and a gravity "spring" force. When the experiment of Figure 5.1 is performed, it is obvious that the oscillatory motion is damped. This is not accounted for by the inviscid Bernoulli model of (5.23). We conclude that frictional forces must be included in the momentum equation—a subject for the next section. In terms of a "lumped parameter model," viscous dissipation can be represented by a lumped damping term:

$$\rho\ell\frac{\partial^2\xi}{\partial t^2} + B\frac{\partial\xi}{\partial t} + 2g\rho\xi = p^d - p^c \tag{5.24}$$

In this equation, the right hand side has been generalized to the case of an arbitrary pressure excitation $p^d - p^c$ applied across the U-tube.

Another form of Bernoulli's equation can be derived from (5.17) corresponding to the very important case of *irrotational* flow, $\omega = \nabla \times v \equiv 0$. Here, the $\omega \times v$ term in (5.17), integrated between any two points a and b in the fluid (not necessarily along a streamline), drops out for an entirely different reason. The case of irrotational flow will be explored further in Problem 5.1 at the end of the chapter.

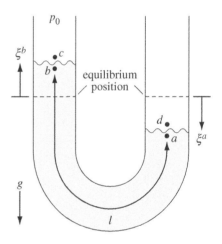

Figure 5.1 Fluid oscillations in a U-tube: a "fluid pendulum."

5.5 VISCOUS FORCES AND STRESS–STRAIN RATE RELATIONS; THE NAVIER–STOKES EQUATION

A fluid may be distinguished from a solid in that the former cannot support a shear stress in equilibrium—acceleration will result, no matter how small the shear stress. Intermolecular attraction or cohesion forces lead to frictional losses when adjacent fluid elements are forced to move at different velocities with respect to each other. Such effects were neglected in the "inertia continuum" model of a fluid discussed in Section 5.4. In this section, we use an empirical approach to find the relation between fluid rates of deformation and the resulting viscous stresses. The resulting viscous stress tensor is then used to derive a viscous force density that, included in the conservation-of-momentum equation, (5.10), will yield the Navier–Stokes equation of motion for a viscous, compressible fluid. The limiting case of incompressibility is immediately apparent.

Rate of Strain in a Fluid Continuum

To examine the types of fluid deformations that can give rise to viscous stresses, we focus on fluid elements located at $\mathbf{r}$ and $\mathbf{r} + \Delta\mathbf{r}$ in Figure 5.2, moving with respective velocities $\mathbf{v}(\mathbf{r}, t)$ and $\mathbf{v}(\mathbf{r} + \Delta\mathbf{r}, t)$. The ith component of the velocity of the element at $\mathbf{r} + \Delta\mathbf{r}$ can be expanded by means of a Taylor series to give

$$v_i(\mathbf{r} + \Delta\mathbf{r}, t) \simeq v_i(\mathbf{r}, t) + \left[\frac{\partial v_i}{\partial x_j}\bigg|_{(\mathbf{r},t)} \Delta x_j \right] + \cdots \qquad (5.25)$$

To linear terms, (5.25) can be written as

$$v_i(\mathbf{r} + \Delta\mathbf{r}, t) = v_i(\mathbf{r}, t) + \frac{1}{2}\left(\frac{\partial v_i}{\partial x_j} - \frac{\partial v_j}{\partial x_i} \right) \Delta x_j + \frac{1}{2}\left(\frac{\partial v_i}{\partial x_j} + \frac{\partial v_j}{\partial x_i} \right) \Delta x_j \quad (5.26)$$

The first term in (5.26) corresponds to pure translational motion, i.e., if all neighboring elements move at the same velocity $\mathbf{v}(\mathbf{r}, t)$, then all such elements are undergoing "rigid body" translation. The second term corresponds to rigid body rotation of the element at $\mathbf{r} + \Delta\mathbf{r}$ about the position $\mathbf{r}$, noting that $(\partial v_i/\partial x_j) - (\partial v_j/\partial x_i) \equiv \nabla \times \mathbf{v}$. The last term of (5.26) serves to define the strain rate $\dot{e}_{ij}$:

$$\boxed{\dot{e}_{ij} \equiv \frac{1}{2}\left(\frac{\partial v_i}{\partial x_j} + \frac{\partial v_j}{\partial x_i} \right)} \qquad (5.27)$$

Since rigid body translation and rotation do not involve relative motion of adjacent fluid elements, such motion does not lead to frictional stresses and dissipation. The combination of spatial derivatives in the strain rate does imply relative motion, however, and the viscous stresses on an incremental fluid volume must be related to $\dot{e}_{ij}$.

For example, Figure 5.3 pictures an experiment that, in circular geometry, is commonly used to measure fluid viscosity. The plane parallel plates are situated as shown: the bottom one fixed and the top moving in the x_1 direction at constant velocity U. The resulting fluid velocity profile is found to transmit a force to the fixed bottom plate that is directly proportional to U and inversely proportional to the plate separation d. Using the stress tensor notation of Section 4.2,

$$\sigma_{21}^{v} = \mu \frac{U}{d} \qquad (5.28)$$

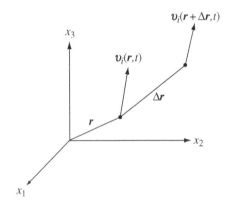

Figure 5.2 Velocities of two neighboring fluid elements.

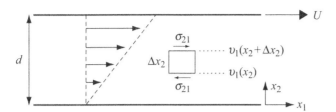

Figure 5.3 Parallel shear flow between plates.

The constant of proportionality μ is the dynamic viscosity (with units of $\text{kg s}^{-1}\,\text{m}^{-1}$), and is found to be a function of fluid mass density and temperature, $\mu(\rho, T)$.

We can also consider the relative motion between fluid layers at x_2 and $x_2 + \triangle x_2$ with the same empirical result:

$$\sigma_{21}^v = \mu \frac{v_1(x_2 + \triangle x_2) - v_1(x_2)}{\triangle x_2} \tag{5.29}$$

and, in the limit where $\triangle x_2 \to 0$,

$$\sigma_{21}^v = \mu \frac{\partial v_1}{\partial x_2} \tag{5.30}$$

Thus, σ_{21}^v can be considered as the viscous shear stress transmitted to the fluid at x_2 by the relative motion at $x_2 + \triangle x_2$, or vice versa.

Newtonian fluids are defined as that class of fluids for which stress and strain rate are *linearly* related; i.e., the viscosity is independent of stress and strain rate. For many important physiological fluid mechanical problems, the results of a Newtonian description are directly applicable (e.g., some aspects of flow in the circulatory system, many aspects of flow through porous membranes, and flow through hemodialysers). We should be aware, however, that some biofluids are non-Newtonian in general. Examples are blood (whose particulate nature due to erythrocytes and proteins makes the viscosity a function of shear rate) and synovial fluid (whose long-chain polysaccharide molecules not only result in non-linear stress–strain rate properties but viscoelastic effects) [2].

Stress–Strain Rate Relations

Empirical relations such as (5.30) corresponding to physical experiments such as that of Figure 5.3 guide us in finding the general relation between viscous stresses (normal and shear) and strain rates (normal and shear). In general, the tensors σ_{ij}^v and $\dot{e}_{kl}$ are related by a fourth-rank tensor c_{ijkl}:

$$\sigma_{ij}^v = c_{ijkl}\dot{e}_{kl} \tag{5.31}$$

For example, c_{1112} relates the normal stress σ_{11} to the shear strain rate $\dot{e}_{12}$. Our job is now to find the 81 components of c_{ijkl}. For example, we consider the relation between σ_{11} and $\dot{e}_{12}$ in Figure 5.4. If we suppose that a normal stress σ_{11} can lead to a shear strain $\dot{e}_{12}$, we should obtain the same relation between σ_{11} and $\dot{e}_{12}$ for arbitrary orientation of the cubic element on the basis of isotropy. However, a rotation of $180°$ about the x_1 axis leads to a different picture of the strain rate while leaving the stress σ_{11} unchanged. We conclude that $c_{1112} = 0$. A combination of such experiments leads to the generalized relation between shear stresses and strain rates:

$$\sigma_{ij}^v = 2\mu\dot{e}_{ij} \quad (i \neq j) \tag{5.32}$$

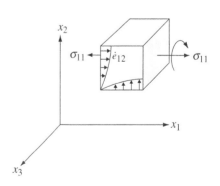

Figure 5.4 Evaluation of c_{1112} based on fluid isotropy.

Thus, shear stress can only lead to shear strain rates, and vice versa. (Think of an experiment to prove that a shear stress σ_{ij}^v can only produce a shear strain rate $\dot{e}_{kl}$ such that $i = k$ and $j = l$.)

To describe the relation between normal stresses and strain rates, we see, for example, that various combinations of σ_{11}^v, σ_{22}^v, and σ_{33}^v can lead to an $\dot{e}_{11}$, since "pushing" in the x_2 or x_3 direction will produce the same results as pulling in the x_1 direction (Figure 5.5). We can represent such an experiment analytically in the form

$$\dot{e}_{11} = k_1 \sigma_{11}^v - k_2 \sigma_{22}^v - k_2 \sigma_{33}^v \tag{5.33}$$

(note that k_2 multiplies both σ_{22}^v and σ_{33}^v—isotropy once again). Similar equations of the form (5.33) can be written for $\dot{e}_{22}$ and $\dot{e}_{33}$. Physical experiments can again be used to show that normal $\dot{e}_{ij}$ will not depend on any shear components σ_{ij}^v, which we assumed in writing (5.33). We can therefore write the general relation between $\dot{e}_{ij}$ and σ_{ij}^v by collecting (5.32) and (5.33):

$$\dot{e}_{ij} = \begin{cases} (k_1 + k_2)\sigma_{ij}^v - k_2 \sigma_{mm}^v \delta_{ij} & (i = j) \tag{5.34a} \\ \dfrac{1}{2\mu}\sigma_{ij}^v & (i \neq j) \tag{5.34b} \end{cases}$$

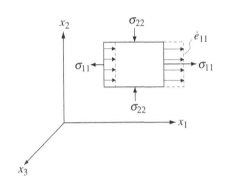

Figure 5.5 $\dot{e}_{11}$ is related to σ_{11} and σ_{22} (and σ_{33}).

We see from (5.34) that there are three coefficients (k_1, k_2, μ) left of the original 81 components c_{ijkl}. However, we will presently find that only two of the three are independent. This arises because of isotropy and the way in which the tensors transform (isotropy implies that $c_{ijkl} = c'_{ijkl}$). Both σ_{ij}^v and $\dot{e}_{ij}$ are symmetric tensors and transform from an unprimed to a primed coordinate system according to the rule

$$\sigma_{ij}^{v\prime} = a_{ik} a_{jl} \sigma_{kl}^v \tag{5.35}$$

$$\dot{e}'_{ij} = a_{ik} a_{jl} \dot{e}_{kl} \tag{5.36}$$

Further, σ_{ij}^v and $\dot{e}_{ij}$ are invariant physical quantities, since they describe the same phenomena independent of the reference frame in which they are written. We found in Problem 4.1 that there are cases in which a σ_{ij}^v (or $\dot{e}_{ij}$) that appears to be purely normal in one frame transforms to a σ'_{ij} ($\dot{e}'_{ij}$) that appears in the form of a shear in the primed frame. They must still refer to the same physical quantity. This fact introduces a relation between k_1, k_2, and μ such that only two of the three are independent. We proceed to find this relation by transforming (5.34) into a primed reference frame by two independent methods and comparing the results. First, we use (5.36) and write

$$\dot{e}'_{ij} = \begin{cases} a_{ik} a_{jl}(k_1 + k_2)\sigma_{kl}^v - k_2 a_{ik} a_{jl} \delta_{kl} \sigma_{mm}^v & (k = l) \tag{5.37a} \\ a_{ik} a_{jl} \dfrac{1}{2\mu}\sigma_{kl}^v & (k \neq l) \tag{5.37b} \end{cases}$$

In (5.37a), we note that since $a_{ik} a_{jl} \delta_{kl}$ is nonzero only when $k = l$, this combination can be replaced by $a_{ik} a_{jk} \equiv \delta_{ij}$ [3]. Therefore, (5.37) becomes

$$\dot{e}'_{ij} = \begin{cases} a_{ik} a_{jl}(k_1 + k_2)\sigma_{kl}^v - k_2 \sigma_{mm}^v \delta_{ij} & (k = l) \tag{5.38a} \\ a_{ik} a_{jl} \dfrac{1}{2\mu}\sigma_{kl}^v & (k \neq l) \tag{5.38b} \end{cases}$$

We can also accomplish this same transform by noting that (5.34) must be equivalent in all reference frames, and thus can be written in the primed frame as

$$
\dot{e}'_{ij} =
\begin{cases}
(k_1 + k_2)\sigma^{v'}_{ij} - k_2\sigma^{v'}_{mm}\delta_{ij} & (i = j) \tag{5.39a} \\
\dfrac{1}{2\mu}\sigma^{v'}_{ij} & (i \neq j) \tag{5.39b}
\end{cases}
$$

$\dot{e}'_{ij}$ can now be written in terms of the unprimed stresses using (5.35), in order to compare with (5.38):

$$
\dot{e}'_{ij} =
\begin{cases}
(k_1 + k_2)a_{ik}a_{jl}\sigma^{v}_{kl} - k_2\sigma^{v}_{mm}\delta_{ij} & (i = j) \tag{5.40a} \\
\dfrac{1}{2\mu}a_{ik}a_{jl}\sigma^{v}_{kl} & (i \neq j) \tag{5.40b}
\end{cases}
$$

In writing the second term of (5.40a), we have used the fact that $\sigma^{v'}_{mm} = a_{mk}a_{ml}\sigma^{v}_{kl} \equiv \delta_{kl}\sigma^{v}_{kl} = \sigma^{v}_{mm}$. Finally, we realize that (5.40) must represent the same $\dot{e}'_{ij}$ as (5.38)—only the order of accomplishing the transformation has differed. Thus, if we consider $\dot{e}'_{ij}$ for the case $i = j$, for example, then the coefficients multiplying both shear and normal stresses in (5.40a) must match those of (5.38). For shear stresses ($k \neq l$), a comparison of (5.38b) and (5.40a) shows that

$$
k_1 + k_2 = \frac{1}{2\mu} \tag{5.41}
$$

A similar conclusion is drawn by comparing the coefficients of normal σ^{v}_{kl} for the case $i \neq j$.

Equation (5.40) can now be inverted to write the viscous stress tensor in terms of the strain rate and two independent coefficients of viscosity. Such an inversion shows that

$$
\sigma^{v}_{ij} =
\begin{cases}
2\mu\dot{e}_{ij} & (i \neq j) \tag{5.42a} \\
2\mu\dot{e}_{ij} + K\delta_{ij}\dot{e}_{kk} & (i = j) \tag{5.42b}
\end{cases}
$$

where K is a constant related to μ and the second coefficient of viscosity λ. The total mechanical stress σ_{ij} on a fluid is the sum of the pressure $-\delta_{ij}p$ and the viscous stress σ^{v}_{ij} of (5.42):

$$
\boxed{\sigma_{ij} = (\sigma^{p}_{ij} + \sigma^{v}_{ij}) = -\delta_{ij}p + 2\mu\dot{e}_{ij} + (\lambda - \tfrac{2}{3}\mu)\delta_{ij}\dot{e}_{kk}} \tag{5.43}
$$

In (5.43), K has been taken as $\lambda - \tfrac{2}{3}\mu$ purely as a matter of definition, so that the pressure p in (5.43) has the same meaning for the moving liquid as for a liquid at rest. (For water, $\mu = 10^{-3}\,\text{kg s}^{-1}\,\text{m}^{-1}$.)

Navier–Stokes Equation

The force density consistent with (5.43) is found by evaluating $\nabla \cdot \underline{\underline{\sigma}}$:

$$
F_i = \frac{\partial \sigma_{ij}}{\partial x_j} = -\frac{\partial(\delta_{ij}p)}{\partial x_j} + \mu\frac{\partial}{\partial x_j}\left(\frac{\partial v_i}{\partial x_j} + \frac{\partial v_j}{\partial x_i}\right) + \left(\lambda - \tfrac{2}{3}\mu\right)\frac{\partial}{\partial x_j}\delta_{ij}\frac{\partial v_k}{\partial x_k} \tag{5.44}
$$

$$
F_i = -\frac{\partial p}{\partial x_i} + \mu\frac{\partial^2 v_i}{\partial x_j \partial x_j} + \left(\lambda + \tfrac{1}{3}\mu\right)\frac{\partial}{\partial x_i}\frac{\partial v_k}{\partial x_k} \tag{5.45}
$$

$$
\boldsymbol{F} = -\nabla p + \mu\nabla^2\boldsymbol{v} + (\lambda + \tfrac{1}{3}\mu)\nabla(\nabla \cdot \boldsymbol{v}) \tag{5.46}
$$

The conservation-of-momentum equation, (5.10), augmented by $\boldsymbol{F}^{v}$ is then

$$\rho\left(\frac{\partial}{\partial t} + \boldsymbol{v} \cdot \nabla\right)\boldsymbol{v} = -\nabla p + \mu\nabla^2\boldsymbol{v} + \left(\lambda + \tfrac{1}{3}\mu\right)\nabla(\nabla \cdot \boldsymbol{v}) + \boldsymbol{F}(\text{other}) \qquad (5.47)$$

Equation (5.47) is the Navier–Stokes equation of motion for a viscous, compressible liquid of density ρ (which is not assumed to be constant). For incompressible liquids, $\nabla \cdot \boldsymbol{v} = 0$ and the third term on the right of (5.47) vanishes. $\boldsymbol{F}(\text{other})$ may include electric, magnetic, or other relevant force densities acting on the fluid.

5.6 PLANE, FULLY DEVELOPED FLOW OF INCOMPRESSIBLE, VISCOUS FLUIDS; LOW-REYNOLDS-NUMBER FLOW

For a wide range of fluid flow problems, the fluid can be treated as incompressible. This approximation is valid provided that fluid velocities are everywhere much less than the speed of sound in the fluid. Under these conditions, the density of a fluid element, observed in a frame moving with the fluid, does not change:

$$\frac{\mathrm{D}\rho}{\mathrm{D}t} = 0 \qquad (5.48)$$

Conservation of mass therefore becomes

$$\nabla \cdot \boldsymbol{v} = 0 \qquad (5.49)$$

To complete the description governing fluid flow, the Navier–Stokes equation for an incompressible fluid is

$$\rho\left(\frac{\partial}{\partial t} + \boldsymbol{v} \cdot \nabla\right)\boldsymbol{v} = -\nabla p + \mu\nabla^2\boldsymbol{v} + \boldsymbol{F} \qquad (5.50)$$

where $\boldsymbol{F}$ is another force density acting on the fluid element in addition to the pressure and viscous force densities. (Note that it is sometimes more convenient to express one or more of the force densities on the right-hand side of (5.50) as the divergence of a stress tensor.) Owing to the presence of the convective momentum term, $\boldsymbol{v} \cdot \nabla\boldsymbol{v}$, (5.50) is in general nonlinear.

An important class of fluid flow problems deal with flow velocities that are unidirectional and do not vary in the direction of flow. Examples are the "plane Couette flow" discussed in Section 5.5 and shown in Figure 5.6, and the fully developed Poiseuille flow, or "pipe flow,"

Figure 5.6 Plane Couette flow.

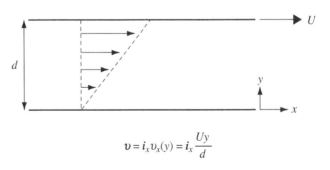

$$\boldsymbol{v} = \boldsymbol{i}_x v_x(y) = \boldsymbol{i}_x \frac{Uy}{d}$$

of Figure 5.7. The term $\boldsymbol{v} \cdot \nabla \boldsymbol{v}$ is identically zero for these unidirectional flows, as can be shown by direct evaluation; for example, from Table B.4 applied to the pipe flow of Figure 5.7,

$$(\boldsymbol{v} \cdot \nabla \boldsymbol{v})_r \text{ component} = v_r \frac{\partial v_r}{\partial r} + \frac{v_\phi}{r} \frac{\partial v_r}{\partial \phi} - \frac{v_\phi^2}{r} + v_z \frac{\partial v_r}{\partial z} = 0 \qquad (5.51)$$

$$(\boldsymbol{v} \cdot \nabla \boldsymbol{v})_z \text{ component} = v_r \frac{\partial v_z}{\partial r} + \frac{v_\phi}{r} \frac{\partial v_z}{\partial \phi} + v_z \frac{\partial v_z}{\partial z} = 0 \qquad (5.52)$$

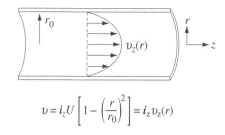

$$v = i_z U \left[1 - \left(\frac{r}{r_0} \right)^2 \right] = i_z v_z(r)$$

Figure 5.7 Poiseuille flow.

In addition, if the fluid motion is steady ($\partial/\partial t = 0$), then all inertia terms drop out of (5.50), leaving a linear differential equation. Even if the motion is unsteady, the Navier–Stokes equation is still linear, but then inertial terms can be important in the general force balance.

Finally, for the general case in which both inertial terms are nonzero in (5.50), we may still be justified in neglecting these terms compared with viscous forces, which again would lead to a linear limit of (5.50), but here for an entirely different reason. An estimate of the relative importance of the inertial and viscous terms is the magnitude of the dimensionless Reynolds number R_y, expressed in terms of a characteristic dimension L and speed U for a given problem. We can derive the form of R_y from scaling arguments, with $\partial/\partial t \to 1/\tau \to (L/U)^{-1}$ and $\nabla \to 1/L$:

$$\rho \frac{\partial \boldsymbol{v}}{\partial t} \to \frac{\rho U^2}{L}$$

$$\rho(\boldsymbol{v} \cdot \nabla \boldsymbol{v}) \to \frac{\rho U^2}{L} \qquad (5.53)$$

$$\mu \nabla^2 \boldsymbol{v} \to \frac{\mu U}{L^2}$$

The dimensionless ratio of inertial to viscous forces using (5.53) thus gives

$$\boxed{R_y = \frac{\rho U L}{\mu}} \qquad (5.54)$$

In the limit $R_y \ll 1$, the left-hand (inertia) terms of (5.50) are negligible, and the resulting fluid velocity profiles are solutions of the "Stokes equation"

$$0 = -\nabla p + \mu \nabla^2 \boldsymbol{v} + \boldsymbol{F}$$

The form of (5.54) corresponds to the case in which only one velocity and one characteristic length are needed to represent the boundary conditions. For $R_y \ll 1$, inertia can be neglected in (5.50), even though such terms are not identically zero. In this limit, viscous forces are balanced by the pressure distribution and any other forces present. For $R_y \gg 1$, inertial forces dominate viscosity, and the velocity profile results from a playoff between inertia and pressure. Quite often, the fluid flows in this limit will take the form of the "plug flow" problems found in the incompressible, inviscid classification (Section 5.4). However, the boundary conditions are very different from the inviscid limit. Whereas an inviscid fluid can slip by a rigid boundary, a fluid with finite viscosity (e.g., all real fluids) is found empirically to stick to rigid walls. Therefore, the fluid velocity goes to zero at the walls. High-Reynolds-number flows are therefore characterized by velocity profiles having a plug flow nature (below the limit of turbulence), with a "boundary layer" near the walls

Figure 5.8 Transition to fully developed flow in a channel. For channel length L much greater than L_B, flow can be approximated as fully developed throughout.

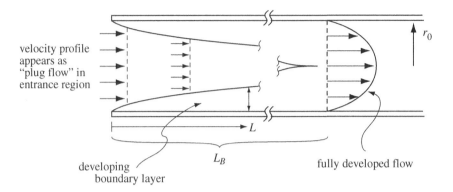

velocity profile appears as "plug flow" in entrance region

r_0

L

developing boundary layer

L_B

fully developed flow

where the velocity must change to zero. The effects of viscosity are thus mostly confined to the boundary layers.

In summary, the velocity profiles of Figures 5.6 and 5.7 apply to both (a) steady, unidirectional flow where the laminar profiles follow from identically zero inertial effects, and (b) finite inertia but $R_y \ll 1$. Both cases are classified as fully developed when the velocity does not change in the direction of flow. Typically, this would result if the channel were long enough so that entrance and exit effects were confined to relatively small distances (Figure 5.8).

In Figure 5.8, after fluid enters a channel, a boundary layer begins to develop because of the no-slip boundary condition at the walls. This boundary layer embodies the playoff between viscous and inertial forces. To emphasize their role in the time required for the development of the boundary layer, we rewrite (5.50), neglecting all other forces.

Neglecting for now the $v \cdot \nabla v$ term,

$$\frac{\partial v}{\partial t} = \frac{\mu}{\rho} \nabla^2 v \tag{5.55}$$

The "viscous diffusion time" based on a distance r_0 (the time for diffusion of vorticity in from the walls) can be found from the "diffusion equation" (5.55), which has an effective diffusion coefficient μ/ρ:

$$\tau_{\text{vd}} \sim \frac{r_0^2}{2\mu/\rho} \tag{5.56}$$

The transport time based on a characteristic length L_B and speed U is

$$\tau_{\text{tran}} \sim \frac{L_B}{U} \tag{5.57}$$

Therefore, the flow in the channel can be thought of as fully developed if L_B found from $\tau_{\text{tran}} \sim \tau_{\text{vd}}$ is much less than the total channel length.

Another general result relates the boundary layer thickness δ at a distance L from the channel opening to R_y. From (5.56),

$$\delta = \sqrt{2\tau_{\text{vd}} \frac{\mu}{\rho}} \tag{5.58}$$

But in a time τ_{vd}, fluid has been transported a length L:

$$L = \tau_{\text{tran}} U \sim \tau_{\text{vd}} U \tag{5.59}$$

$$\delta = \sqrt{\frac{2L^2 \mu}{\rho U L}} = L \sqrt{\frac{2}{R_y}} \tag{5.60}$$

5.7 STREAM FUNCTIONS

The creep flow approximation can be regarded as a quasistatic model. That is, it can be used to describe slowly time-varying as well as static physical systems. It is particularly useful because the pressure and viscous force densities remaining in the Stokes equation are linear:

$$\nabla(p - \rho \boldsymbol{g} \cdot \boldsymbol{r}) = \mu \nabla^2 \boldsymbol{v} + \boldsymbol{F} \tag{5.61}$$

Stream Function Representations

Because the flow is incompressible and hence $\nabla \cdot \boldsymbol{v} = 0$ is solenoidal, it is convenient to use a vector stream function $\boldsymbol{C}$ defined such that

$$\boldsymbol{v} = \nabla \times \boldsymbol{C}, \quad \nabla \cdot \boldsymbol{C} = 0 \tag{5.62}$$

Such a function is particularly convenient if the flow is two-dimensional or has symmetry such that a single component of $\boldsymbol{C}$ is all that is required. Examples are summarized in Table 5.1. In terms of the scalar quantities defined, the creep flow in the absence of external forces is represented by a biharmonic equation. In general, the curl of (5.61) is

$$\mu \nabla^2 \nabla^2 \boldsymbol{C} = \nabla \times \boldsymbol{F} \tag{5.63}$$

Table 5.1 Important configurations represented by a single component of the vector potential.

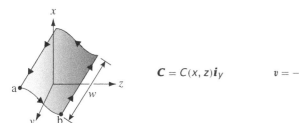

$$\boldsymbol{C} = C(x, z)\boldsymbol{i}_y \qquad \boldsymbol{v} = -\frac{\partial C}{\partial z}\boldsymbol{i}_x + \frac{\partial C}{\partial x}\boldsymbol{i}_z \qquad \text{(a)}$$

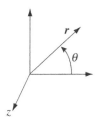

$$\boldsymbol{C} = \frac{\Psi(r, z)}{r}\boldsymbol{i}_\theta \qquad \boldsymbol{v} = -\frac{1}{r}\frac{\partial \Psi}{\partial z}\boldsymbol{i}_r + \frac{1}{r}\frac{\partial \Psi}{\partial r}\boldsymbol{i}_z \qquad \text{(b)}$$

$$\boldsymbol{C} = C(r, \theta)\boldsymbol{i}_z \qquad \boldsymbol{v} = -\frac{1}{r}\frac{\partial C}{\partial \theta}\boldsymbol{i}_r + \frac{\partial C}{\partial r}\boldsymbol{i}_\theta \qquad \text{(c)}$$

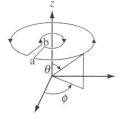

$$\boldsymbol{C} = \frac{\Psi(r, \theta)}{r \sin \theta}\boldsymbol{i}_\phi \qquad \boldsymbol{v} = -\frac{1}{r \sin \theta}\left(\frac{1}{r}\frac{\partial \Psi}{\partial \theta}\boldsymbol{i}_r - \frac{\partial \Psi}{\partial r}\boldsymbol{i}_\theta\right) \qquad \text{(d)}$$

where we have exploited the identity $\nabla^2 \boldsymbol{C} = -\nabla \times \nabla \times \boldsymbol{C}$, which is valid because $\nabla \cdot \boldsymbol{C} = 0$. In dealing with the vector Laplacian, it is perhaps best to regard it as a shorthand way of writing $-\nabla \times \nabla \times \boldsymbol{C}$, where the curl operators are as defined in Appendix B. The four curl operations implicit in (5.63) do lead to differential equations having a Laplacian character. In cartesian coordinates, the scalar function defined in equation (a) of Table 5.1 is introduced into (5.63) to obtain

$$\mu \nabla^2 \nabla^2 C = (\nabla \times \boldsymbol{F})_y \qquad (5.64)$$

In Section 5.8, an expression similar to (5.63) is obtained for the *Stokes stream function* Ψ for axisymmetric flow in spherical coordinates. The stream function Ψ is introduced as a matter of convenience and convention so that the volume rate of flow through appropriate surfaces can be represented by the stream function evaluated at two different positions. In general, we can write the volume rate of flow through a surface S as

$$Q_v = \oint_S \boldsymbol{v} \cdot \boldsymbol{n} \, da = \int_S \nabla \times \boldsymbol{C} \cdot \boldsymbol{n} \, da = \oint_C \boldsymbol{C} \cdot d\boldsymbol{\ell} \qquad (5.65)$$

and, in particular for the two-dimensional flow in Cartesian coordinates (Table 5.1),

$$Q_v = \int_C \boldsymbol{C} \cdot d\boldsymbol{\ell} = w \left[C(a) - C(b) \right] \qquad (5.66)$$

where w is the width in the y direction. In spherical coordinates, the volume rate of flow through the axisymmetric surface having inner and outer radii at b and a, respectively, becomes

$$Q_v = \oint \boldsymbol{C} \cdot d\boldsymbol{\ell} = 2\pi \left[\psi(a) - \psi(b) \right] \qquad (5.67)$$

5.8 CREEP FLOW TRANSFER RELATIONS: MOLECULAR, CELL, AND TISSUE SURFACES

Low-Reynolds-number flows constitute an extremely important class of problems in hydrodynamics that are encountered in physiology and state-of-the-art microelectromechanical and nanoelectromechanical device applications. The small pore size encountered in the extracellular matrix of tissues, the intracellular cytoskeleton, and cell membranes, and in microfluidic channels, typically results in fluid transport processes that are viscous-dominated. In this section, we focus on relations between the total mechanical stress and fluid velocity evaluated on surfaces of biological interest. For the case of spherical surfaces, these relations exemplify the nature of low-Reynolds-number flows and are used as a case study for subsequent sections. Our approach will closely follow that pioneered by James R Melcher in his text *Continuum Electromechanics* [4], in which transfer relations in a convenient matrix format enable treatment of a wide class of boundary value problems. In this manner, the classic example of Stokes drag on a rigid sphere (Section 5.9) can be easily expanded to include intra- and extracellular flows as well as electrophoresis of charged molecules and cells without having to re-solve a different fluid mechanics problem each time.

For the case of a spherical surface, in the absence of external rotational forces, equation (d) of Table 5.1 is introduced into (5.63) to obtain

$$\left[\frac{\partial^2}{\partial r^2} + \frac{\sin\theta}{r^2}\frac{\partial}{\partial\theta}\left(\frac{1}{\sin\theta}\right)\frac{\partial}{\partial\theta}\right]^2 \psi = 0 \tag{5.68}$$

This equation makes no restriction on the time dependence, which is therefore arbitrary. An important class of solutions have the θ dependence

$$\psi = (\sin^2\theta)\tilde{\psi}(r, t) \tag{5.69}$$

where substitution into (5.68) shows that the radial dependence is predicted by

$$\left(\frac{d^2}{dr^2} - \frac{2}{r^2}\right)^2 \tilde{\psi} = 0 \tag{5.70}$$

With the assumed dependence on θ, we limit consideration to one of a set of solutions. Others have a higher number of oscillations in the interval $0 < \theta < \pi$.

Two solutions to (5.70) take the form r^n, $n = 2, -1$. Because these are double roots, two additional solutions are $r^2 r^n$, as can be seen by substitution. Thus, the linear combination of the four solutions is

$$\tilde{\psi} = \tilde{\psi}_1\left(\frac{r}{R}\right)^2 + \tilde{\psi}_2\left(\frac{r}{R}\right)^{-1} + \tilde{\psi}_3\left(\frac{r}{R}\right)^1 + \tilde{\psi}_4\left(\frac{r}{R}\right)^4 \tag{5.71}$$

In terms of the stream function, the velocities in the radial and azimuthal directions follow from Table 5.1:

$$v_r = \tilde{v}_r 2\cos\theta, \quad \tilde{v}_r \equiv \frac{\tilde{\psi}}{r^2}$$
$$\tag{5.72}$$
$$v_\theta = \tilde{v}_\theta \sin\theta, \quad \tilde{v}_\theta = -\frac{1}{r}\frac{d\tilde{\psi}}{dr}$$

The pressure distribution implied by the stream function of (5.69) and (5.71) is found by integrating the force equation. The radial component of (5.61) is (neglecting $\rho \boldsymbol{g}\cdot\boldsymbol{r}$):

$$\frac{\partial p}{\partial r} = \frac{d\tilde{p}}{dr}2\cos\theta, \quad \frac{d\tilde{p}}{dr} = \frac{\mu}{r^2}\left[\frac{d}{dr}\left(r^2\frac{d\tilde{v}_r}{dr}\right) - 4\tilde{v}_r - 2\tilde{v}_\theta\right] \tag{5.73}$$

and, after substituting (5.71) and (5.72) into (5.73), integration gives

$$p = \tilde{p}2\cos\theta, \quad \tilde{p} = \mu\left(\tilde{\psi}_3\frac{r^{-2}}{R} + 10\tilde{\psi}_4\frac{r}{R^4}\right) \tag{5.74}$$

The stress functions $\sigma_r \equiv \sigma_{rr}$ and $\sigma_\theta \equiv \sigma_{\theta r}$ representing the net viscous and pressure stress ((5.44) with $\partial v_k/\partial x_k = 0$) are written in spherical coordinates (Appendix B, Table B.5) as

$$\sigma_r = \tilde{\sigma}_r 2\cos\theta, \quad \tilde{\sigma}_r = 2\mu\frac{d\tilde{v}_r}{dr} - \tilde{p}$$
$$\tag{5.75}$$
$$\sigma_\theta = \tilde{\sigma}_\theta \sin\theta, \quad \tilde{\sigma}_\theta = \mu\left(r\frac{d}{dr}\frac{\tilde{v}_\theta}{r} - \frac{2}{r}\tilde{v}_r\right)$$

In writing transfer relations, complex amplitudes are used, with the understanding that the θ dependence of the radial velocity, stress, and pressure distributions is given by multiplying by $2\cos\theta$, while the θ components of velocity and stress are multiplied by $\sin\theta$. These definitions are summarized by (5.73)–(5.75). We now consider flow both interior and exterior to the spherical surface at $r = R$.

Case Study: Interior Flow

Biological "particles" or cells are often not rigid, but viscous or viscoelastic in nature. In general, if such a spherical particle were to migrate through a liquid, or if liquid were to flow past the particle, there would be a velocity distribution associated wth the *interior* of the particle (picture a spherical amoeba, for which $R_y \ll 1$ is certainly valid). The exterior and interior flows would then be matched by appropriate boundary conditions at $r = R$. Here, we derive the stress–velocity relations for the *interior fluid flow problem*.

Velocity components at $r = R$ are defined as $(\tilde{v}_r^\alpha, \tilde{v}_\theta^\alpha)$. With the origin included in the volume, the velocity is finite there only if $\tilde{\psi}_2$ and $\tilde{\psi}_3 \rightarrow 0$. Then evaluation of (5.73) using (5.71) at $r = R$ gives two equations in the unknowns $(\tilde{\psi}_1, \tilde{\psi}_4)$. These can be inverted so that the stream function is specified in terms of velocity amplitudes:

$$\begin{bmatrix} \tilde{\psi}_1 \\ \tilde{\psi}_4 \end{bmatrix} = R^2 \begin{bmatrix} 2 & \dfrac{1}{2} \\ -1 & -\dfrac{1}{2} \end{bmatrix} \begin{bmatrix} \tilde{v}_r^\alpha \\ \tilde{v}_\theta^\alpha \end{bmatrix} \tag{5.76}$$

The desired stress components are written in terms of the stream function amplitudes by substituting (5.71), (5.73), and (5.74) into (5.75):

$$\begin{bmatrix} \tilde{\sigma}_r^\alpha \\ \tilde{\sigma}_\theta^\alpha \end{bmatrix} = \frac{\mu}{R^3} \begin{bmatrix} 0 & -6 \\ 0 & -6 \end{bmatrix} \begin{bmatrix} \tilde{\psi}_1 \\ \tilde{\psi}_4 \end{bmatrix} \tag{5.77}$$

and so we obtain the transfer relation by substituting (5.76) into (5.77):

$$\begin{bmatrix} \tilde{\sigma}_r^\alpha \\ \tilde{\sigma}_\theta^\alpha \end{bmatrix} = \frac{\mu}{R} \begin{bmatrix} 6 & 3 \\ 6 & 3 \end{bmatrix} \begin{bmatrix} \tilde{v}_r^\alpha \\ \tilde{v}_\theta^\alpha \end{bmatrix} \tag{5.78}$$

Case Study: Exterior Flow

For the case of cells or macromolecules migrating at constant velocity with the fluid, we wish to include the possibility that there is a uniform flow at infinity, $v = U\boldsymbol{i}_z = U(\cos\theta\,\boldsymbol{i}_r - \sin\theta\,\boldsymbol{i}_\theta)$. The stream function representing this flow is the first term in (5.71), with $\tilde{\psi}_1 = U/2R^2$. The last term in (5.71) represents an unbounded velocity at $r \rightarrow \infty$, and so we set $\tilde{\psi}_4 \rightarrow 0$ at the outset. The two remaining amplitudes are specified in terms of the velocity amplitudes $(\tilde{v}_r^\beta, \tilde{v}_\theta^\beta)$ at $r = R$ by evaluating (5.71) and (5.73). Inverted for the stream function amplitudes, these equations are

$$\begin{bmatrix} \tilde{\psi}_2 \\ \tilde{\psi}_3 \end{bmatrix} = \frac{R^2}{2} \begin{bmatrix} 1 & 1 \\ 1 & -1 \end{bmatrix} \begin{bmatrix} \tilde{v}_r^\beta - \frac{1}{2}U \\ \tilde{v}_\theta^\beta + U \end{bmatrix} \tag{5.79}$$

Equations (5.71) and (5.73)–(5.75) specify the complex stress amplitudes in terms of these stream function amplitudes:

$$\begin{bmatrix} \tilde{\sigma}_r^\beta \\ \tilde{\sigma}_\theta^\beta \end{bmatrix} = -\frac{\mu}{R^3} \begin{bmatrix} 6 & 3 \\ 6 & 0 \end{bmatrix} \begin{bmatrix} \tilde{\psi}_2 \\ \tilde{\psi}_3 \end{bmatrix} \tag{5.80}$$

Finally, substitution of (5.79) into (5.80) gives the desired stress–velocity relations:

$$\begin{bmatrix} \tilde{\sigma}_r^\beta \\ \tilde{\sigma}_\theta^\beta \end{bmatrix} = -\frac{\mu}{2R} \begin{bmatrix} 9 & 3 \\ 6 & 6 \end{bmatrix} \begin{bmatrix} \tilde{v}_r^\beta - \frac{1}{2}U \\ \tilde{v}_\theta^\beta + U \end{bmatrix} \tag{5.81}$$

Of course, in making use of (5.78) and (5.81), there is no restriction on the time dependence. These relations are limiting forms of more general transfer relations relating velocity and stress components on the spherical surfaces of a fluid shell having outer and inner surfaces associated with the α and β variables, respectively.

5.9 STOKES DRAG ON A RIGID SPHERE

Certainly the most celebrated low-Reynolds-number flow is that around a rigid sphere placed in what would otherwise be a uniform flow. Of particular interest is the total drag force on the sphere, found by integrating the z component of the traction, $\sigma_r \cos \theta - \sigma_\theta \sin \theta$, over its surface:

$$f_z^{\text{TOT}} = \int_0^\pi [\sigma_r \cos \theta - \sigma_\theta \sin \theta] 2\pi R^2 \sin \theta \, d\theta \tag{5.82}$$

We can identify the exterior flow of Section 5.8 with that around the sphere. U is the uniform z-directed velocity far from the sphere. Because the surface at $r = R$ is rigid, both velocity components vanish there. In (5.81),

$$\tilde{v}_r^\beta = 0, \quad \tilde{v}_\theta^\beta = 0 \tag{5.83}$$

and the stress components are

$$\begin{bmatrix} \tilde{\sigma}_r^\beta \\ \tilde{\sigma}_\theta^\beta \end{bmatrix} = -\frac{\mu}{2R} \begin{bmatrix} -\frac{3}{2}U \\ 3U \end{bmatrix} \tag{5.84}$$

Using these amplitudes, as well as the θ dependence given by (5.75), (5.82) is integrated to become

$$\boxed{f_z = 6\pi\mu RU} \tag{5.85}$$

This "Stokes drag" force is a good approximation provided the Reynolds number based on the particle radius is small compared with unity. The flow and pressure distributions can be evaluated by using (5.79) evaluated with the conditions of (5.83). For a frictionless rather than no-slip surface (5.85) becomes $f_z = 4\pi\mu RU$ [1].

5.10 CONVECTIVE DIFFUSION: THE ROLE OF CONVECTIVE MASS TRANSPORT

Thus far, the effect of fluid convection on the distribution of solute species $c(\mathbf{r}, t)$ has not been accounted for in a systematic manner. The general equation of convective diffusion, neglecting electrical effects, was derived in Chapter 1:

$$\frac{\partial c_i}{\partial t} + (\mathbf{v} \cdot \nabla)c_i = D_i \nabla^2 c_i \tag{5.86}$$

However, the electrochemical coupling examples of Chapter 3 were treated for limiting cases in which the convective mass transport term $v \cdot \nabla c$ was negligible (leaving (5.86) in the standard form of a molecular diffusion equation), or cases in which the fluid velocity was assumed to be a known constant for simplicity.

In the electrokinetic interactions of Chapter 6, we will see that the distribution of chemical species in the electrical double layer will be viewed as only slightly perturbed by fluid motion, and is still essentially that due to the Boltzmann equilibrium consistent with the intense double layer electric field (see Problem 2.3). (Of course, electrokinetic coupling may occur in inhomogeneous systems such as those found in the examples of Section 6.5, in which a detailed knowledge of $c(\mathbf{r}, t)$ may be of utmost importance.)

There are many situations in biology in which the effect of convective mass transport on the distribution of solutes plays a dominant role. It is often necessary to know and control the interplay between convective transport and molecular diffusion at boundaries, as typified in the examples of Figure 5.9. We will begin our discussion of convective diffusion by examining (5.86) for cases where the effects of electric fields on the distribution of solute molecules either are unimportant or are relevant only at surfaces surrounding the fluid. The problem of finding $c(\mathbf{r}, t)$ is then a boundary value problem defined by (5.86) (with v either known at the outset or found from the Navier–Stokes equation) subject to boundary conditions, which may include the effects of electric fields.

The relative importance of convective mass transport and molecular diffusion can be interpreted by a dimensional consideration of the ratio between $v \cdot \nabla c$ and $D\nabla^2 c$ (for steady state problems with $\partial/\partial t \to 0$). For

Figure 5.9 Examples involving convective diffusion. (a) Convective diffusion of reacting proteins in blood: adsorption and reaction on vessel walls in the face of convection is important in the study of thrombogenesis with native and synthetic biomaterials. (b) Convective mass transport in hemodialysers: diffusion of solutes across polymeric membranes. (c) Convective diffusion boundary layers and the Nernst film model.

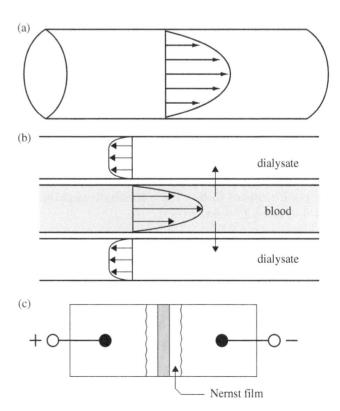

a characteristic length L and velocity U, this ratio is

$$\frac{Uc/L}{Dc/L^2} = \frac{UL}{D} \equiv P_e \tag{5.87}$$

For large values of the Péclet number P_e, corresponding to relatively large velocities and dimensions of interest, molecular diffusion can be neglected compared with convective transport of fluid-entrained solutes, and thus $v \cdot \nabla c \sim 0$. In the opposite limit of small P_e, convection may be neglected and therefore $c(r, t)$ is found from the steady molecular diffusion equation, $\nabla^2 c \sim 0$. The Péclet number is entirely analogous to the Reynolds number of fluid flow, repesenting the interplay between convective momentum transport and viscous diffusion:

$$R_y = \rho \frac{UL}{\mu} = \frac{UL}{\mu/\rho} \equiv \frac{UL}{\nu} \tag{5.88}$$

where ν is defined as the kinematic viscosity and constitutes a "diffusion coefficient" for vorticity.

Another characterization of the relative importance of convection and diffusional transport is the dimensionless Schmidt number, defined as the ratio of the Péclet and Reynolds numbers:

$$S_c \equiv \frac{P_e}{R_y} = \frac{\nu}{D} \tag{5.89}$$

Note that S_c is independent of dimension and velocity, and represents the relative diffusivities of fluid vorticity and solute. For typical solutes in water, $\nu = \mu/\rho = 10^{-3}/10^3 \text{ m}^2 \text{ s}^{-1}$, $D \sim 10^{-9} \text{ m}^2 \text{ s}^{-1}$, and thus $S_c \sim 10^3$. Since

$$P_e = S_c R_y = 10^3 R_y \tag{5.90}$$

we conclude that convective mass transport can be important even with very low-R_y flows; i.e., convective mass transport associated with viscous creep flow can still compete with or dominate molecular diffusion.

5.11 DIFFUSION BOUNDARY LAYERS

The competition between convective and diffusional transport found in many systems of interest is typified by the membrane diffusion potential experiment pictured in Figure 5.9(c). In the bulk on each side of the membrane, where vigorous stirring generally results in large Péclet numbers, (5.86) reduces to

$$v \cdot \nabla c_i \sim 0 \tag{5.91}$$

assuming that a steady state does, in fact, exist. Equation (5.91) is satisfied by the condition $c_i = \text{const}$. As the velocity approaches zero at the membrane surface, large concentration gradients may exist there (depending on the boundary condition on $c(v)$ at the surface), and the full equation (5.86) must be solved to find the distribution of c_i for all space. This distribution can, therefore, be thought of in terms of a two-region problem: a bulk region of uniform concentration and a thin fluid region adjacent to the membrane where molecular diffusion becomes important. This latter region is embodied in the concept of a diffusion boundary layer, analagous to the fluid mechanical boundary layer associated with viscous diffusion in high-Reynolds-number flows.

To fully examine the diffusion boundary layer concept, we focus on the two-dimensional problem shown in Figure 5.10, where a source of solute molecules (gas, liquid, or solid) constrains the concentration at the surface of a layer of water flowing with *uniform* bulk velocity U in the z direction:

$$c = c_0 \quad \text{at} \quad x = 0, \ z > 0 \tag{5.92}$$

We wish to find the steady state concentration $c(x, z)$ in the water for $z > 0$, and dependence of the molecular diffusion boundary layer thickness δ_m on z and other physical parameters of interest.

Although the simplified geometry of Figure 5.10 does not correspond to the membrane problem of Figure 5.9(c), the two-dimensional problem is far less formidable mathematically and will allow us to examine more easily the competing processes responsible for the developing diffusion boundary layer.

Before we begin the details of the solution, we note that dimensional arguments can give us important information concerning δ_m. The thickness of the diffusion boundary layer δ_m in the x direction at any point $z > 0$ will depend on the amount of time particles have had to *diffuse* in the x direction while being *convected* in the z direction away from the "inlet" $z = 0$, where the molecular diffusion and convection transport

Figure 5.10 Molecular diffusion and viscous boundary layers.

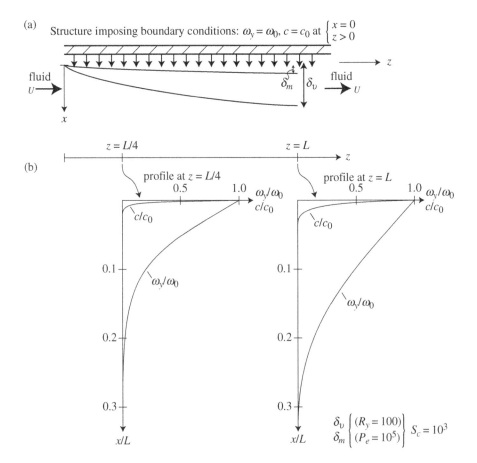

times are

$$\tau_{md} \sim \frac{\delta_m^2}{D} \tag{5.93}$$

$$\tau_t \sim \frac{z}{U} \tag{5.94}$$

By setting $\tau_{md} = \tau_t$, we find the order-of-magnitude estimate for δ_m:

$$\delta_m \sim \sqrt{\frac{Dz}{U}} \tag{5.95}$$

The boundary layer thickness increases with z for a given U, and decreases with increased fluid speed U at any point $z = L$, as we would expect from physical reasoning. The exact value of δ_m will depend on the exact $c(\mathbf{r})$ for any given problem (δ_m may be defined as the thickness over which the concentration changes by a certain percentage of its known surface or bulk value).

For the two-dimensional problem at hand, (5.86) becomes

$$v_x \frac{\partial c}{\partial x} + v_z \frac{\partial c}{\partial z} = D\left(\frac{\partial^2 c}{\partial x^2} + \frac{\partial^2 c}{\partial z^2}\right) \tag{5.96}$$

The fluid velocity is known at the outset to be $\mathbf{v} = U\mathbf{i}_z$ and is assumed to be independent of solute concentration, thus eliminating the leftmost term of (5.96). Further, for characteristic lengths L of interest in the z direction, the diffusion time is so much longer than the convective transport time that we will neglect molecular diffusion in the z direction, i.e., the rightmost term in (5.96). (Alternatively, we may say that derivatives in the z direction are much less than in the x direction for the thin, developing boundary layer region.) Using these two assumptions, (5.96) becomes

$$U \frac{\partial c}{\partial z} = D \frac{\partial^2 c}{\partial x^2} \tag{5.97}$$

One method of approaching the solution of (5.97) is to use a similarity transformation in order to transform the partial differential equation into an ordinary differential equation. This may be accomplished by normalizing the x and z dimensions to the same variable; one choice is to normalize the x dimension to the thickness δ_m:

$$\underline{x} = \frac{x}{2\delta_m} = \frac{x}{2\sqrt{Dz/U}} \tag{5.98}$$

Then the derivatives in (5.97) can be expressed in terms of $\underline{x}$ by recognizing that

$$\frac{\partial \underline{x}}{\partial x} = \frac{1}{2\sqrt{Dz/U}}, \quad \frac{\partial \underline{x}}{\partial z} = -\frac{xz^{-3/2}}{4\sqrt{D/U}} \tag{5.99}$$

Substitution of (5.99) into (5.97) gives

$$\frac{\partial^2 c}{\partial \underline{x}^2} + 2\underline{x}\frac{\partial c}{\partial \underline{x}} = 0 \tag{5.100}$$

It can be seen by direct substitution that (5.100) has the solution

$$c(\underline{x}) = c_0 \left[1 - \mathrm{erf}(\underline{x})\right] \equiv c_0 \, \mathrm{erfc}(\underline{x}) \tag{5.101}$$

which satisfies the boundary condition (5.92) at $x = 0, z > 0$, and the condition

$$c(x \to \infty, z) \to 0$$

or (5.102)

$$c(\underline{x} \to \infty, z) \to 0$$

The error function has the properties

$$\mathrm{erf}(\underline{x}) \equiv \frac{2}{\sqrt{\pi}} \int_0^{\underline{x}} e^{\underline{x}^2} \, d\underline{x} \begin{cases} \mathrm{erf}(\underline{x} = 0) = 0 \\ \mathrm{erf}(\underline{x} = \infty) = 1 \end{cases} \quad (5.103)$$

The concentration distribution (5.101) is plotted in Figure 5.10(b) as c/c_0 for two different positions along the z axis. The plot has been normalized to the Reynolds and Péclet numbers shown. (A typical flow condition of interest was chosen as $U = 1\,\mathrm{cm\,s}^{-1}$, $L = 1$ cm, $\nu = 10^{-6} \to R_y = 100$; therefore, for a $S_c = 10^3$ typical of diffusivities $D \sim 10^{-5}\,\mathrm{cm^2\,s^{-1}}$, $P_e = 10^5$.) From (5.98), we may rewrite (5.101) for a characteristic dimension of interest $z = L$:

$$c(x, z) = c_0 \left[1 - \mathrm{erf}\left(\frac{x}{2} \sqrt{\frac{U}{DL}} \right) \right] = c_0 \left[1 - \mathrm{erf}\left(\frac{1}{2} \frac{x}{L} \sqrt{P_e} \right) \right] \quad (5.104)$$

Figure 5.10(b) shows $c(x)$ for $z = 0.25L$ and $z = L$ with $P_e = 10^5$.

If δ_m is defined as the distance over which $c(x, z)$ decreases to 10% of its value at $x = 0$ (i.e., c_0), then $\delta_m \approx 0.01L$ from the graph of Figure 5.10(b). By comparison, the order-of-magnitude estimate for δ_m based on (5.95) gives $\delta_m \sim \sqrt{DL/U} = 0.003L$.

The analysis used to find the diffusion boundary layer thickness and the concentration profile has its analog in other fields where one is interested in the competing effects of convection and some other physical process. For example, the distribution of a magnetic field inside a moving conducting material (see Section 2.6.4) can be described by the "magnetic diffusion" (Bullard's) equation

$$\frac{\partial \boldsymbol{B}}{\partial t} + \nabla \times (\boldsymbol{B} \times \boldsymbol{v}) = \frac{1}{\mu_m \sigma} \nabla^2 \boldsymbol{B} \quad (5.105)$$

where μ_m is the magnetic permeability of the material and σ is the conductivity.

For steady state interactions, the competition between convection and diffusion is embodied in the "magnetic Reynolds number" [4]

$$R = \frac{LU}{1/\mu_m \sigma} = \mu_m \sigma LU \quad (5.106)$$

In the area of fluid mechanics, a relation for the diffusion of vorticity in a viscous liquid can be found by taking the curl of both sides of the Navier–Stokes equation, giving

$$\frac{\partial \boldsymbol{\omega}}{\partial t} + \nabla \times (\boldsymbol{\omega} \times \boldsymbol{v}) = \frac{\mu}{\rho} \nabla^2 \boldsymbol{\omega} \quad (5.107)$$

In general, finding the viscous boundary layer requires the solution of a nonlinear problem. However, we can construct a "linearized" example analogous to the configuration of Figure 5.10(a). With the same liquid moving at a uniform velocity U in the z direction, the vorticity is zero everywhere in the bulk at the outset. At $x = 0, z > 0$, a stress σ_0 is applied to the free surface of the water (at $x = 0$), resulting in a small perturbation in the fluid velocity about U, and a concomitant perturbation vorticity that tends to diffuse into the bulk ($x > 0$) with a viscous "diffusivity" $\nu = \mu/\rho$.

Since v_y and $\partial/\partial y$ are zero (two-dimensional), $\boldsymbol{\omega} = \boldsymbol{i}_y \omega_y$. The convective transport term $\nabla \times (\boldsymbol{\omega} \times \boldsymbol{v}) = \boldsymbol{i}_y(\omega_y U)$ to lowest order (neglecting the perturbation velocities v_x and v_z). Thus, the y component of (5.107) becomes

$$U\frac{\partial \omega_y}{\partial z} = \nu \frac{\partial^2 \omega_y}{\partial x^2} \tag{5.108}$$

which is analogous to (5.97). The solution to (5.108) matching the boundary conditions

$$\begin{aligned} \omega_y &= 0 & (x \to \infty) \\ \omega_y &= \omega_0 = \sigma_0/\mu & (x = 0, z > 0) \end{aligned} \tag{5.109}$$

is then

$$\omega_y = \omega_0[1 - \text{erf}(\underline{x}')] \tag{5.110}$$

where

$$\underline{x}' = \frac{x}{2\sqrt{\nu z'/U}} \tag{5.111}$$

The solution ω_y/ω_{y0}, (5.110), is plotted on the same graphs as c/c_0 in Figure 5.10(b), again for the case $R_y = 100$, $\nu = 10^{-6}$. We find that the viscous boundary layer δ_v develops much sooner (is much thicker for any given z) than the molecular diffusion boundary layer δ_m. From the graph of ω_y/ω_0, $\delta_v = 0.23L$ for the given conditions.

For many physiological situations of interest, and in microfluidic device applications, both molecular and viscous diffusion layers develop simultaneously in the fluid. For the case of flow in cylindrical vessels, we may consider three regimes of interest. In the first regime, *neither* the fluid flow *nor* the species concentration is fully developed— both profiles may be considered in terms of boundary layers, with the viscous boundary layer developing more rapidly from the inlet as shown in Figure 5.11(a). For channels many times longer than their diameter, there is often a second regime in which the flow has become fully developed, but the diffusional boundary layer still occupies a thin section near the vessel wall (Figure 5.11(b)). In general, a third regime may be defined in which both flow and concentration are fully developed for a vessel of sufficient length. (A quick calculation will show that for $r_0 = 0.5$ cm and $U = 1$ cm s^{-1}, the concentration profile will become fully developed in about 100 m!)

Finally, we note that the no-slip boundary condition at $r = r_0$ results in a nonlinear problem for the solution of the viscous boundary layer. A

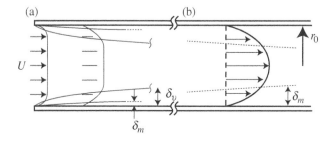

(a) (b)

U

δ_v

δ_m

r_0

δ_m

Figure 5.11 Simultaneous development of viscous and diffusional boundary layers, δ_v and δ_m. ($v_r = v_z = 0$ at $r = r_0$; $c = c_0$ at $r = r_0$, $z > 0$).

linearized equation of the form (5.108) is not valid, since the velocity at $r = r_0$ is by no means a small perturbation about U. Similarly, the solution of the molecular diffusion boundary layer pictured in Figure 5.11 no longer corresponds to the development in this section. Equation (5.96) now becomes a nonlinear partial differential equation with fluid velocities entering as space-varying coefficients. Physically, the smaller fluid velocities in the neighborhood of the wall allow for greater diffusion in the radial direction for a given transport time. A numerical solution for the relation between δ_v and δ_m with respect to two-dimensional flow above a rigid plane [5] shows that $\delta_m \sim 0.1\delta_v$ for this nonlinear problem, as compared with $\delta_m \sim 0.04\delta_v$ for the linearized boundary layer seen in this section.

5.12 PROBLEMS

Problem 5.1 In Section 5.4, we examined incompressible, inviscid flow and a form of Bernoulli's equation relevant to points along a given streamline. Here, we extend the discussion to include *irrotational* flow, $\nabla \times v = 0$, and another form of Bernoulli's equation consistent with such flow.

(a) With $\nabla \times v = 0$, v can be represented as $-\nabla \psi$, where ψ is a scalar potential. Use this relation in (5.17) and integrate between *any* points a and b in the fluid to obtain a corresponding form of Bernoulli's equation analogous to (5.20). Note that a and b do *not* have to be along the same streamline.

(b) Use conservation of mass to find a differential equation that can be used to find $\psi(\mathbf{r}, t)$. Note that a solution of this equation automatically satisfies mass conservation, and that v and p can be found from ψ.

(c) There are many important applications involving inviscid, irrotational flows in homogeneous fluids. The *theorem of vorticity* of Helmholtz and Kelvin shows that a flow that is known to be irrotational at some instant in time is irrotational for all time. Prove this theorem as follows:

  (i) Write a differential equation for the vorticity $\boldsymbol{\omega} = \nabla \times v$ by taking the curl of (5.17).

  (ii) Use the integral theorem (A.8) of Appendix A to convert the differential equation to an integral relation in terms of

$$\frac{d}{dt} \int_S \boldsymbol{\omega} \cdot \mathbf{n} \, da \qquad (5.112)$$

  Intepret the results.

(d) With viscosity included in an incompressible model (no other forces are present),

$$\rho \left(\frac{\partial}{\partial t} + v \cdot \nabla \right) v = -\nabla p + \mu \nabla^2 v \qquad (5.113)$$

Take the curl of this equation and find a differential equation for ω. What is the "effective diffusion coefficient" in your diffusion equation?

Conclusion: Viscosity provides a mechanism for the diffusion of vorticity if such vorticity is produced in a given situation (by a boundary condition or the like).

Problem 5.2 (Adapted from a problem in Lightfoot [2])
Bernoulli's equation is often used in a "pseudo-quantitative" approach to explain pressure changes occurring at aneurisms (local enlargement of blood vessel diameter) and stenoses (local constrictions resulting, for example, from atherosclerotic deposits). Find the changes in pressure and velocity at the center of the aneurism and stenosis (refer to Figure 5.12). Is this vessel stable or unstable for the two situations shown in Figure 5.12?

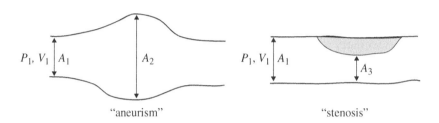

Figure 5.12 Aneurism and stenosis.

Problem 5.3

(a) Find the upward z dependence of the pressure p of the atmosphere as a function of the ground level pressure $p(z = 0)$. Assume that the equation of state for the atmosphere is

$$p = \rho RT$$

where T is constant.
(b) Repeat part (a) for an atmosphere of constant density ρ_0.
(c) Why is the Zefal tire pump better than "Brand X" if your tire needs 70–100 psi of air pressure? (Figure 5.13 is drawn aproximately to scale, with the exception of the tire, which is flat.)

Figure 5.13 Zefal Pump Co. advertisement.

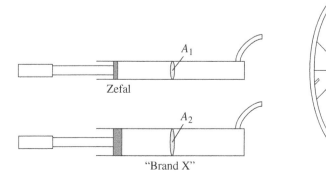

Problem 5.4 The Couette Viscometer

(a) Find the velocity and shear stress distributions $v_\phi(r)$ and $\sigma_{\phi r}$ for the tangential laminar flow of an incompressible viscous fluid (viscosity μ) between two vertical coaxial cylinders, as shown in Figure 5.14. The inner cylinder is fixed and the outer cylinder rotates at angular velocity Ω_O. (Neglect end effects.)

(b) Calculate the torque per unit length required to turn the outer cylinder at an angular speed $V_\phi = R\Omega_O = 5\,\mathrm{cm\,s^{-1}}$ for a liquid having $\mu = 3.5\mu_{H_2O}$ ($\mu_{H_2O} = 10^{-3}\,\mathrm{kg\,m^{-1}\,s^{-1}}$), the approximate viscosity of blood under certain limiting conditions. ($R = 5$ cm, $a = 4$ cm.)

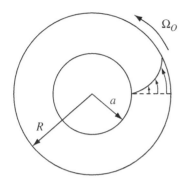

Figure 5.14 Couette viscometer.

Problem 5.5 Laminar Flow in a Square Microchannel Find the velocity profile $v_z(x, y)$ for fully developed laminar flow of a fluid (viscosity μ) in a square channel (Figure 5.15). Note that since $\nabla p = \text{constant} \equiv K$, the z component of the Navier–Stokes equation is equivalent to Poisson's equation. Hence, the problem is analogous to finding the two-dimensional electrical potential $\Phi(x, y)$ inside a box whose walls are perfectly conducting and grounded ($\Phi = 0$ at the walls), and whose cross-section is filled with a uniform charge density ($-\rho_u/\epsilon \equiv K/\mu$). To find $v_z(x, y)$, first assume a particular solution v_p (hint: suppose the channel were infinitely wide in either the x or y directions). Then, choose a homogeneous (Laplacian) solution v_h so that the total velocity satisfies the appropriate boundary conditions.

Application: In flow cytometers, cells and cell volume are typically measured by means of light scattering or fluorescence techniques. Flow visualization and flow cytometry can also be implemented using holographic video microscopy (Figure 5.16) [6]. It is thereby possible to use particle-image velocimetry to visualize the Poiseuille flow profile with a rectangular microchannel (Figure 5.17).

Problem 5.6 The stress–velocity relations characteristic of the "creeping flow" of a spherical particle moving relative to a liquid with speed U are given by (5.81). For a rigid sphere, both tangential and normal velocity components vanish at $r = R$:

$$\left. \tilde{v}_r^\beta \right|_{r=R} = \left. \tilde{v}_\theta^\beta \right|_{r=R} = 0$$

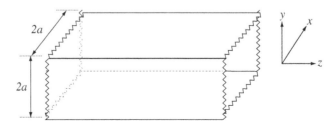

Figure 5.15 Square microchannel.

(**a**) *Find* the velocity distribution associated with the flow around a rigid sphere by using the results of Sections 5.7 and 5.8, and the derived relation between stream function and velocity amplitudes:

$$\begin{bmatrix} \tilde{\psi}_2 \\ \tilde{\psi}_3 \end{bmatrix} = \frac{R^2}{2} \begin{bmatrix} 1 & 1 \\ 1 & -1 \end{bmatrix} \begin{bmatrix} \tilde{v}_r^{\beta} - \frac{1}{2}U \\ \tilde{v}_{\theta}^{\beta} + U \end{bmatrix}$$

This flow is important for a host of problems concerned with velocities small enough so that $R_y = 2R\rho U/\mu \ll 1$.

(**b**) Find the pressure distribution for $r \geq R$. Show that the r and θ components of the Navier–Stokes equation are identically satisfied using the velocity and pressure distributions you have derived.

Problem 5.7 Dispersion of cells flowing through a rectangular chamber

A steady stream of cells is introduced from a "point source" located at the origin $x = y = z = 0$, in the center of a rectangular chamber shown in Figure 5.18. The cells are spherical (osmotically swollen) and their concentration is dilute. Physiological saline flows in the rectangular cell between two glass plates spaced $2h$ apart. A fully developed laminar profile exists throughout the region of interest, including the point $x = y = z = 0$. Electrodes on the two side walls spaced W apart impose a uniform electric field $\boldsymbol{E} = E_0 \boldsymbol{i}_x$.

Make reasonable assumptions to answer the following questions: (you should not have to solve the "Taylor diffusion" [7] problem in detail). (The electric field terms will be derived in detail in Chapter 6.)

(**a**) Find an approximate analytical expression for the downstream concentration profile of cells, c, in a convenient reference frame (set of coordinates). Express your answer in terms of fluid properties (μ, ϵ, σ), cell properties

PROBLEM

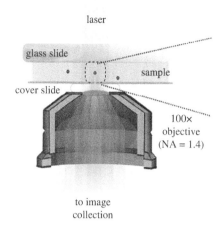

Figure 5.16 In-line holographic video microscope [6]. A collimated laser beam illuminates the sample. Light scattered by the sample interferes with the unscattered portion of the beam in the focal plane of the objective lens.

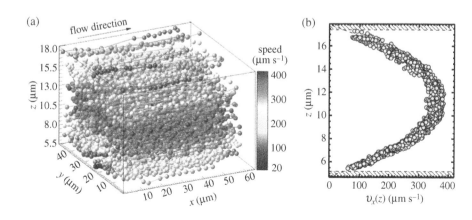

Figure 5.17 Holographic particle-image velocimetry [6]. (a) Measured three-dimensional trajectories of 500 colloidal spheres travelling down a microfluidic channel in a pressure-driven flow. Each sphere represents the position of a particle in one field of a holographic snapshot. Features from a sequence of fields are linked into trajectories that are shaded by the particle's measured speed. (b) Poiseuille flow profile along the vertical direction obtained from the data in (a). Particles are excluded from the cross-hatched regions by their interactions with the upper and lower glass walls of the channel. The dashed curve is a fit to the anticipated parabolic flow profile.

Figure 5.18 Dispersion of cells flowing through a rectangular chamber.

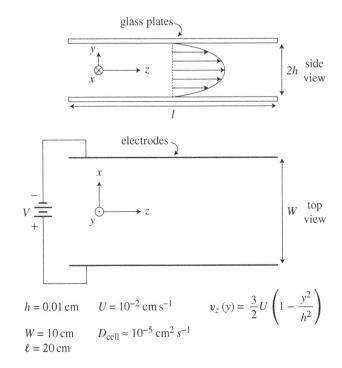

$h = 0.01\,\text{cm}$ $U = 10^{-2}\,\text{cm s}^{-1}$ $v_z(y) = \dfrac{3}{2}U\left(1 - \dfrac{y^2}{h^2}\right)$

$W = 10\,\text{cm}$ $D_{\text{cell}} \approx 10^{-5}\,\text{cm}^2\,s^{-1}$
$\ell = 20\,\text{cm}$

(diffusivity D, surface charge σ_d, and/or ζ potential, radius R_0), and other necessary parameters.

Assume *for now* that in the measurement time of interest, transport along the particle direction of flow is characterized by a high Péclet number $P_e \gg 1$ (show by calculation that this is reasonable). That is, Taylor diffusion results in a uniform concentration in the y direction. Secondly, the Taylor spreading along the net longitudinal direction of particle flow is negligible at first.

(**b**) Show that the analytical expression for the concentration profile can be used to calculate:

> (i) the mean speed of the electrolyte (if it were not given);
>
> (ii) cell diffusivity D (if it were unknown);
>
> (iii) σ_d or ζ of the cell.

(**c**) How would you measure the concentration profile of cells in space and time? Your answer depends, of course, on the spatial sensitivity required, i.e., on the characteristic length L over which the greater part of the concentration change takes place.

Assume that the longitudinal (Taylor) dispersion parameter K is known to be

$$K = \frac{U^2 h^2}{20D} \qquad (5.114)$$

and estimate L for this experimental chamber. (Note that an exact calculation might give a different factor in (5.114). The problem does *not* ask you to do this calculation.)

(**d**) From the value of K in part (c), discuss the accuracy of the approximation in part (a) that Taylor spreading colinear with the net velocity can be neglected.

5.13 REFERENCES

[1] Batchelor GK (1967) *An Introduction to Fluid Dynamics.* Cambridge University Press, Cambridge.

[2] Lightfoot EN (1974) *Transport Phenomena and Living Systems: Biomedical Aspects of Momentum and Mass Transport.* Wiley, New York, Chapter 1.

[3] Long RR (1961) *Mechanics of Solids and Fluids.* Prentice Hall, Englewood Cliffs, NJ, pp. 9–10.

[4] Melcher JR (1981) *Continuum Electromechanics.* MIT Press, Cambridge, MA (this book is also available online at http://ocw.mit.edu/resources/res-6-001-continuum-electromechanics-spring-2009/textbook-contents/).

[5] Levich V (1962) *Physicochemical Hydrodynamics.* Prentice Hall, Englewood Cliffs, NJ, Chapter 2.

[6] Cheong FC, Sun B, Dreyfus R, et al. (2009) Flow visualization and flow cytometry with holographic video microscopy. *Optics Express* **17**, 13071–13079.

[7] Taylor GI (1954) Conditions under which dispersion of a solute in a stream of solvent can be used to measure molecular diffusion. *Proc. R. Soc. Lond.* **A225**, 473–477.

Electrokinetics: MEMS, NEMS, and Nanoporous Biological Tissues

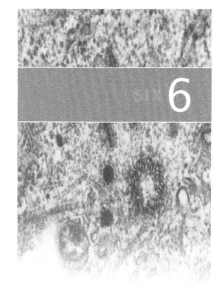

6.1 INTRODUCTION

Charge separation naturally occurs at phase boundaries in electrochemical systems and at biological interfaces. This separation leads to the formation of an electrical "double layer" of charge, most often with one sign of charge in each adjacent phase (see Problem 2.3 in Chapter 2 and Chapter 4). The resulting charge density and electrical field strength can be of such magnitude as to play a crucial role at electrochemical interfaces (e.g., electrode/electrolyte), as well as nanoscale interfaces such as those at the surfaces of cells and biomacromolecules. The electrical double layer is thus a common and important site for the coupling of electrical, mechanical, and chemical events in biological tissues.

In this chapter, a brief survey is first given of several classic electromechanical interactions associated with the double layer. These interactions involve chemically homogeneous environments in which there is no transfer of chemical energy except that needed to maintain an equilibrium electrochemical double layer at relevant charged interfaces. We then discuss a cylindrical pore membrane model that can be used to describe a class of biological or synthetic polyelectrolyte materials. Electroosmosis and streaming potential, two classical electrokinetic phenomena, are developed in the context of this model. This is followed by an analysis of the electrophoretic motion of charged particles (e.g., proteins and cells) caused by an applied electric field. Electrocapillary motions of conducting particles in an electric field introduce several phenomena associated with metal/electrolyte interfaces that are in distinct contrast to interfaces in biomacromolecular, cell, and tissue systems. All of the examples treated here make use of the fluid mechanical developments of Chapter 5.

6.2 ELECTROCAPILLARY AND ELECTROKINETIC PHENOMENA

Charge separation at a boundary between phases can occur for several reasons. Discontinuities in chemical and physical properties may lead to different affinities for the various mobile charged species present. Thus, a potential difference across a metal/electrolyte interface is accounted for by the separation of mobile electrons and electrolyte ions which may not be free to cross the boundary (complete insulation to flow of carriers is the case of "ideal polarization").

In other cases, charged species may be adsorbed from the solution phase onto a particle or surface, charging the surface, and causing attraction of oppositely charged counter-ions in the solution. Classic examples of this phenomenon are the adsorption of surface-active ions on an air/water interface and crystal ions on certain salt crystal/electrolyte systems. With biological polyelectrolytes, it is the *dissociation* of ionizable groups that can lead to the presence of net charge bound within the molecular structure. For example, the amino acid residues of proteins include acidic and basic groups (carboxyl and amino groups being the most common), which dissociate in a manner consistent with the pH and ionic strength of the electrolytic environment. In all such cases, there is an interfacial potential difference consistent with the separation of positive and negative charge layers.

Helmholtz proposed the first and simplest model of this double layer (1853)—a parallel-plate capacitor representing the two equal and opposite charge layers, with plate separation equal to double layer thickness (Figure 6.1). Many qualitative and quantitative electrochemical studies are still based on this model. A more detailed picture proposed by Gouy (1910) and then Chapman (1913) included the "diffuse" spatial distribution of ions in the electrolyte phase. Given that the nonelectrolyte phase is somehow charged, electrostatic forces result in attraction of the oppositely charged electrolyte counter-ions. However, thermal motion (diffusion) opposes the formation of a rigid plane of counter-ions as exists in the Helmholtz model. Rather, a diffuse layer forms, extending into the bulk of the solution. Mathematically, the Gouy–Chapman diffuse double layer looks much like the Debye–Hückel ionic atmosphere theory for electrolytes and the electron/hole distribution at a semiconductor junction. We will discuss the relevance of this model in greater detail in the next section.

Stern's approach adds two further modifications to the general theory. First, the charge distribution is modeled as a series combination of two layers (capacitances). The compact, or Helmholtz, layer includes counter-ions at a "plane of closest approach," postulated because of finite ion diameter. Next a diffuse (Gouy) layer begins at approximately one ionic radius from the interface. Thus, the potential drop across the entire double layer consists of two components.

A second modification concerns specific (nonelectrostatic) adsorption of ions (Figure 6.1(c)) and dipoles (Figure 6.1(d)) at the interface owing to van der Waals forces and chemisorption: the former is an effect classified as physical adsorption—a reversible process; the latter involves irreversible chemical association or reaction, and exhibits greater free energy changes. Both of these effects may radically change the shape of the potential close to the interface, as well as the total potential

Figure 6.1 Charge and potential distribution in the electrical double layer: (a) according to Helmholtz (compact); (b) according to Gouy and Chapman (diffuse); (c) with inclusion of ion adsorption (Stern) and ζ potential; (d) with inclusion of dipole adsorption and ζ potential. (Adapted from Vetter KJ, *Electrochemical Kinetics*, Academic Press, New York, 1967, Chapter 1.)

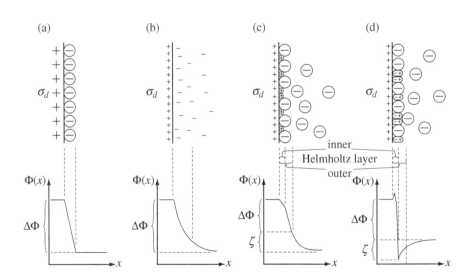

difference and/or the general shape of the diffuse layer. Since adsorbed ions and dipoles are of different sizes, more than one plane of closest approach is postulated.

Electrocapillarity

The study of surface tension at a metal/electrolyte phase boundary and the analysis of ideally polarizable electrodes is generally referred to as *electrocapillarity.* This is distinguished from *electrokinetic phenomena* (discussed in the next subsection) which are defined as the electromechanical transduction effects consistent with the relative motion between an insulating solid and adjacent electrolyte. Metal/electrolyte interfaces offer many advantages for quantitative physicochemical studies of the double layer. First, the potential difference across the interface can often be varied without changing the composition of the two phases, simply by using a voltage source and appropriate electrodes. Second, in some cases, the interface can be modeled quite accurately as a capacitor or, more broadly, a diode—i.e., no current will cross the interface for a certain range of applied potential difference. This is the case of an "ideally polarized" liquid mercury/electrolyte interface, for which the current–voltage characteristics are exemplified in Figure 6.2. In the potential range where no current crosses the interface, it is as if a perfectly insulating membrane separates the conducting electrolyte from the perfectly conducting mercury. Because the normal component of the current density, J_n, must be zero at the surface, the normal electric field E_n must also be zero there. Therefore, any electric field present at the interface must be *tangential* and must decrease to *zero* over the thickness of the double layer, since E_{tan} is essentially zero in mercury.

This presence of a tangential electric field within the double layer can cause electromechanical shear interactions of an electrocapillary nature. An applied tangential field will "see" only one sign of double layer charge—that in the solution phase—since the electric field in the highly conducting metal phase is zero. The electric shear stress will result in fluid and mercury motion, consistent with viscous shear equilibrium, which can be incorporated into the Stokes equation (Chapter 5).

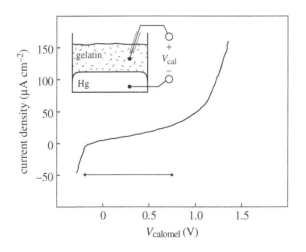

Figure 6.2 Static current–voltage curve for a mercury/gelatin electrolyte interface ($0.02\,M\,KNO_3$, 5% gelatin); stationary mercury pool of area $= 10\,cm^2$. The arrow indicates the approximate range of ideal polarization. The voltage drop was measured using a calomel electrode (see Chapter 3).

Electrokinetics

Electrokinetic phenomena also involve electromechanical coupling associated with double layers, but are common to *insulating solid/electrolyte* boundaries, as opposed to the previous *metal/electrolyte* case. Electrokinetic phenomena manifest themselves in *relative motion between the insulating solid and electrolyte phases*, where the solid may be deformable (e.g., macromolecules) or rigid (e.g., the channel walls of microelectromechanical systems (MEMS) devices). To appreciate what is unique to a *deformable* electrokinetic system such as a charged hydrogel scaffold or membrane, we first summarize the four classical electrokinetic effects [1] found at interfaces with nondeformable solids: electroosmosis, streaming potential, electrophoresis, and sedimentation potential (Figure 6.3).

These electrokinetic phenomena can be interpreted in terms of a balance between viscous and electrical shear forces. For example, consider *electroosmotic* motion of an electrolyte in a fixed glass or MEMS channel (Figure 6.3(a))—an example of electrical-to-mechanical energy conversion. A double layer forms, with one charge layer at the wall surface and the other extending into solution. (For instance, charges at a glass surface are due to the presence of silicic acid in the glass, which will dissociate to form silicate (SiO_3^{2-}) ions at neutral or higher pH.) Part of the outer diffuse double layer is mobile, while the inner compact layer, roughly two to three molecular diameters thick, is immobile owing to the no-slip boundary condition at the wall.

Figure 6.3 Electrokinetic phenomena: (a) electroosmosis; (b) streaming potential; (c) electrophoresis; (d) sedimentation potential.

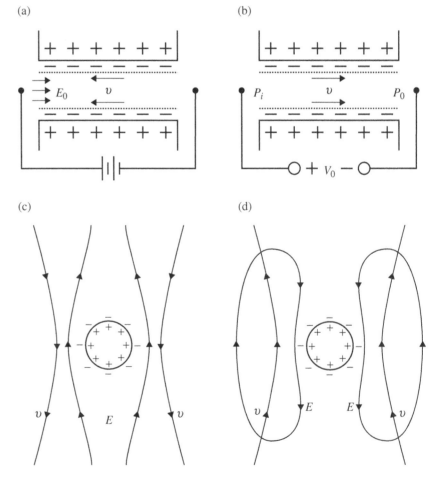

An electric field applied tangential to the interface as shown gives rise to a $\rho_e \boldsymbol{E}$ force density on the mobile portion of the double layer. In contrast to the case of electrocapillarity, E_{tan} is continuous at the interface, and there is also a force of electrical origin on the glass wall surface charge. The channel, however, is constrained to be immobile, while the liquid electrolyte can move. From the Stokes equation (Section 5.6), the balance of electrical and viscous forces in the diffuse double layer region, neglecting inertia, reduces to:

$$\rho_e \boldsymbol{E} + \mu \nabla^2 \boldsymbol{v} = \nabla p \tag{6.1}$$

The resulting velocity of the bulk solution in the channel can easily be derived via two integrations of (6.1) (see the next section). This velocity can be expressed in terms of the applied axial electric field E_0 and the interfacial potential drop, or ζ potential, across the *mobile* portion of the double layer. In the limit where the pressure rise in (6.1) can be neglected compared with viscous stresses, the bulk electroosmotic fluid velocity is directly proportional to E_0 and ζ:

$$U_{\text{eo}} = -\frac{\epsilon \zeta}{\mu} E_0 \tag{6.2}$$

where ζ takes the sign of the surface (wall) charge. We will see that the form of (6.2) is valid when the channel (pore) width is much greater than a Debye length. If the diffuse layer is modeled as a simple Helmholtz capacitor of thickness d, such that

$$\sigma_d = \frac{\epsilon}{d} \zeta \tag{6.3}$$

then the bulk velocity can be rewritten in terms of the surface charge density σ_d:

$$v = -\left(\frac{\sigma_d \, d}{\mu} \right) E_0 \tag{6.4}$$

The converse electrokinetic effect, commonly referred to as the *streaming potential*, also involves a mobile electrolyte moving with respect to a fixed solid, as shown in Figure 6.3(b). In this case, the fluid velocity is imposed, inducing a potential drop across the length of the channel. The same theoretical model given by (6.2)–(6.4) can be used to find the magnitude of the potential drop.

The remaining two electrokinetic effects involve motion of solid particles with respect to the fluid electrolyte. Electrical-to-mechanical energy conversion is typified by *electrophoresis* of charged particles in a liquid subjected to an electric field (Figure 6.3(c)). Mechanical-to-electrical conversion is evidenced by the *sedimentation potential* generated by the settling of charged particles through a liquid (Figure 6.3(d)). Thus, in Figures 6.3(c) and (d), a spherical insulating particle is shown with positive charge bound to the particle, and a surrounding cloud of negative charge entrained in the liquid. In actuality, the particle may be more complex in shape, like that of a single polyelectrolyte molecule. But the model of a spherical particle is useful in establishing the coupling of electrical and viscous stresses in the diffuse double layer surrounding the particle. Levich's analytical model (see [2], p. 475) gives substance to qualitative pictures of this electromechanical coupling. Electrophoresis of a particle results because of the electrical force *on the diffuse part of the double layer in the liquid*, as well as the electrical force on the particle. From a frame of reference fixed to the particle, the application of an electrical field E causes a pumping of the liquid along lines indicated by v, relative to the particle (Figure 6.3(c)). From a frame of reference with respect to the liquid, the result is an electrically driven motion, or electrophoresis, of the particle. The inverse coupling is illustrated by Figure 6.3(d), where there is no applied field. Rather, the particle is one of many settling under the influence of gravity. The diffuse layer of charge in the liquid is convected along the particle boundary from one pole to the other, and bled away as a conduction current in the surrounding liquid. Hence, a sedimentation potential is generated by the aggregate of particles.

6.3 ELECTROOSMOSIS AND STREAMING POTENTIALS IN CHARGED, POROUS MEMBRANES AND TISSUES

Biological tissues, hydrogels, and synthetic polyelectrolyte membranes display a wide variety of internal structures. Porous membranes are thought of as being three-dimensional networks bathed in electrolyte. Somewhat more useful for our purpose is a model based on a multitude of channels or pores containing the electrolyte. Pore models are widely used for many well-characterized polyelectrolyte membranes such as ion exchange, sintered-particle, and fibrous gel membranes (Figure 6.4). Such models relate nano- and microscopic flow parameters to averaged, observable, macroscopic parameters, and have been used with success to characterize many physiological membranes [3]. In addition, they can accurately describe electroosmotic flow critically important in MEMS devices and applications (Figure 6.5).

Internal Transduction at the Pore Level

The total volume flux Q_p and total current i_p through a pore (Figure 6.4(c)) are now related to the pressure and potential drops across the pore, denoted by Δp and ΔV, respectively. The channel walls have an average surface charge (positive in Figure 6.4(c)) consistent with the given polyelectrolyte and bathing solution. Thus, a diffuse space charge layer having a corresponding electrical potential $\Phi(r)$ resides inside each pore in the aqueous phase. The fluid contains only neutral salt and pH-determining ions—no other solvents or macromolecules. The pores have a radius r_0, length ℓ_p, and average interpore spacing s.

If the Debye length is small compared with r_0, then the electrical double layer potential $\Phi(r)$ quickly decays away from the pore wall and is at the zero reference potential throughout most of the channel cross section. On the other hand, with $1/\kappa \sim r_0$, $\Phi(r)$ may never reach zero inside the pore. In this case, electrical repulsion forces will tend to expand the pore in the radial direction until a balance is achieved with structural-mechanical forces. This is an interaction between "internal" electric fields that can cause swelling if the material is deformable. The radius r_0, is here taken to be the equilibrium pore radius after such an internal balance has occurred.

Fluid mechanical motions of the pore liquid are treated as Newtonian (viscosity μ) and incompressible (mass density ρ). For pore diameters on the order of micrometers or larger, the mean free path for collisions of water molecules is much less than the pore radius, and the use of a continuum model is appropriate. (Even for nanometer-size pore diameters, a continuum model can often successfully predict

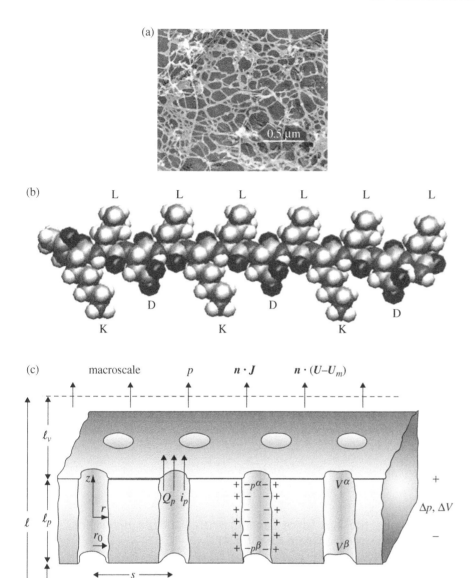

(a)

0.5 µm

(b)

L   L   L   L   L   L

D   D   D

K   K   K

(c)

Figure 6.4 (a) (KLDL)$_3$ self-assembling peptide hydrogel scaffold based on repeats of lysine (K), leucine (L), and aspartic acid (D). (b) (KLDL)$_3$ sequence (from Kisiday J, Jin M, Kurz B, et al. Self-assembling peptide hydrogel fosters chondrocyte extracellular matrix production and cell division: Implications for cartilage tissue repair. *Proc. Natl Acad. Sci. USA* **99**, 9996–10000 (2002)). (c) Cylindrical pore model of micro- or nanoporous membranes with fixed positive charges on the pore wall and negative counter-ions shown with the pore (the parameters are defined in the text).

the trends of electrokinetic behavior, although molecular level effects may need to be considered.) The Reynolds number for flow through a pore, $R_y = \rho r_0 v_p / \mu$, based on a typical pore fluid velocity v_p, is small enough to neglect inertia in the Navier–Stokes equation, leading to a fully developed "creep flow" model for fluid flow:

$$-\nabla p + \mu \nabla^2 v + \rho_e \boldsymbol{E} = 0 \qquad (6.5)$$

Inherent to a fully developed flow are fluid velocity and electric field distributions that vary only with r and not in the z direction. Therefore, we look for solutions of the form $v = v_z(r)$, $F_z^e = \rho_e(r)E_z = F_z^e(r)$, and $p = p(r,z)$. (Transverse force equilibrium is consistent with T_{rr}^e and the radial dependence of the hydrostatic pressure within the pore; we assume that radial swelling equilibrium has been achieved and is not altered by z-dependent flows.)

Figure 6.5 Rectangular MEMS μ-channel with (a) hyperbolic constriction that results in (b) greatly enhanced electric field strength in the x direction with the constriction. An applied x-directed potential drop causes electroosmotic fluid flow within the channel as well as an electrophoretic migration of any charged macromolecules within the fluid stream. (From Kim JM and Doyle PS. Design and numerical simulation of a DNA electrophoretic stretching device. *Lab Chip* **7**, 213–225 (2007).)

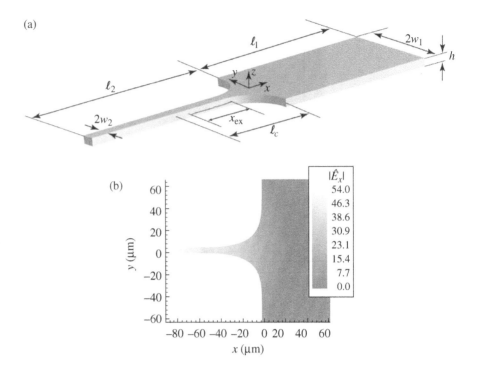

The $\rho_e \mathbf{E}$ electrical force density in the z direction, to be included in the Stokes equation (6.5), can be obtained using Gauss' law (2.59) to substitute for ρ_e:

$$(\rho_e \mathbf{E})_z = \underbrace{\left\{ \frac{1}{r} \frac{\partial}{\partial r} [r \epsilon E_r(r)] \right\}}_{\nabla \cdot \epsilon \mathbf{E}} E_z \tag{6.6}$$

Alternatively, the $\rho_e \mathbf{E}$ force can be written as the divergence of the Maxwell stress tensor (from Chapter 4):

$$F_z^e = \frac{1}{r} \frac{\partial}{\partial r}(r \sigma_{zr}^e), \quad \sigma_{zr}^e = \epsilon E_z E_r \tag{6.7}$$

where E_r is the radial component of the electrical field and E_z the component collinear with the pore axis (normal to the plane of the membrane).

For typical double layer field strengths of interest, E_r is much greater than the magnitude of any field applied or generated in the z direction. We therefore treat E_r as due to the ionic atmosphere alone; the radial dependence of the space charge is essentially unperturbed by, and decoupled from, fluid flow in the channel, consistent with the assumption $v = v_z(r)$. Thus, with E_r written in terms of $\Phi(r)$, and E_z in terms of the potential drop $\Delta V = V^\alpha - V^\beta$ across the pore, the z component of the Stokes equation is finally written as

$$\frac{\partial p}{\partial z} = \frac{1}{r} \frac{\partial}{\partial r} \left\{ r \left[\mu \frac{\partial v_z(r)}{\partial r} + \epsilon \frac{\Delta V}{\ell_p} \frac{\partial \Phi(r)}{\partial r} \right] \right\} \tag{6.8}$$

For a fully developed flow, $\partial p/\partial z$ is independent of z, and here equal to $\Delta p/\ell_p$.

To obtain the desired pore transduction relations, the velocity profile in a single pore is first found by integrating (6.8) twice using the boundary conditions $\partial v_z(r)/\partial r = \partial \Phi(r)/\partial r \equiv 0$ at $r = 0$ by symmetry, and $v_z = 0$, $\Phi \equiv \zeta$ at the mechanical "slip plane" $r = r_0 - \delta$ (i.e., the position at which the "no-slip" condition is applicable):

$$v_z(r) = \frac{[r^2 - (r_0 - \delta)^2]}{4\ell_p \mu} \Delta p + \frac{\epsilon[\zeta - \Phi(r)]}{\ell_p \mu} \Delta V \qquad (6.9)$$

The second boundary condition recognizes that electromechanical coupling occurs in the mobile region of the space charge layer, thought to extend to within a few molecular diameters of the pore wall. (The "no-slip" condition occurs not precisely at the wall, but several water molecule diameters away from the wall at $r = r_0 - \delta$. The potential at the position of this "slip plane" is called the zeta (ζ) potential, which can differ greatly from the potential at the pore wall.)

Electroosmotic fluid velocity profile

Before continuing, we note that (6.9) immediately reduces to the well-known electroosmotic fluid velocity of (6.2), where U is the macroscale (superficial) velocity just outside the porous membrane (Figure 6.4):

$$U = -\left(\frac{\epsilon\zeta}{\mu}\right) E_0 \qquad (6.10)$$

when $\Delta p \to 0$ in (6.9), $\Delta V/\ell_p \to -E_0$, and the pore radius $r_0 \gg 1/\kappa$, so that $\Phi(r)$ is essentially zero throughout the pore except near the wall.

The current through a pore can now be defined in terms of a sum of convection and conduction terms, respectively:

$$i_p = \int_0^{r_0} 2\pi r \rho_e(r) v_z(r)\, dr - \int_0^{r_0} 2\pi r \sigma(r) \frac{\Delta V}{\ell_p}\, dr \qquad (6.11)$$

The electrolyte conductivity $\sigma(r)$ will vary with r in general, since ion concentrations vary in the space charge regions. Similarly, the volume flux Q_p can be defined as

$$Q_p = \int_0^{r_0} 2\pi r v_z(r)\, dr \qquad (6.12)$$

Thus, the incorporation of (6.9) into (6.11) and (6.12) yields transfer relations between Q_p, i_p and Δp, ΔV, provided one knows the radial dependence of $\Phi(r)$, $\rho_e(r)$, and $\sigma(r)$. This is equivalent to knowing the exact radial potential and charge distribution within the pore, which would lead to exact solutions of (6.11) and (6.12).

The simplest double layer model for the purpose of obtaining a closed-form analytical solution is a "Helmholtz double layer" (Figure 6.1(a)), for which the space charge is approximated by an impulse, or surface charge density $-\sigma_d$ located a distance from the pore wall equivalent to the double layer thickness $d = 1/\kappa$:

$$\rho_e \simeq -\sigma_d\, u_0(r_0 - d) \qquad (6.13)$$

where u_0 is the notation for a unit impulse located at $r = r_0 - d$. For the case $d \ll r_0$, $\Phi(r)$ can be approximated by

$$\Phi(r) \simeq \begin{cases} 0 & r \leq (r_0 - d) \qquad (6.14) \\[2ex] \dfrac{\zeta[r - (r_0 - d)]}{d} & (r_0 - d) \leq r \leq r_0 \qquad (6.15) \end{cases}$$

and the surface charge density σ_d is linearly related to the potential drop ζ across the mobile portion of the double layer:

$$\sigma_d \approx \frac{\epsilon}{d}\zeta \tag{6.16}$$

Using (6.13)–(6.15) along with (6.9), the integrals in (6.11) and (6.12) can be carried out simply (see Problem 6.1 at the end of the chapter), giving the internal pore transduction relations:

$$\begin{bmatrix} Q_p \\ i_p \end{bmatrix} = \begin{bmatrix} -\dfrac{\pi r_0^4}{8\mu\ell_p} & \dfrac{\pi r_0^2 \epsilon \zeta}{\mu\ell_p} \\ \dfrac{\pi r_0^2 \epsilon \zeta}{\mu\ell_p} & -\left(\dfrac{\pi r_0^2 \sigma}{\ell_p} + \dfrac{2\pi r_0 \epsilon^2 \zeta^2}{\mu d\ell_p}\right) \end{bmatrix} \begin{bmatrix} \Delta p \\ \Delta V \end{bmatrix} \tag{6.17}$$

The following approximations have been made in order to arrive at (6.17): $d \ll r_0$ and $\delta \ll d$ in (6.9); the conductivity $\sigma(r)$ is taken to be uniform ($\equiv \sigma$) over the cross section in (6.11), consistent with $d \ll r_0$ and neglecting variations of conductivity in the space charge region.

Identifying the matrix of entries in (6.17) as coefficients g_{ij}, the electrokinetic coupling manifests itself in the cross coefficients g_{12} and g_{21}. The term $1/g_{11}$ is the hydrodynamic resistance of a single pore, while $1/g_{22}$ is the effective electrical resistance of the pore liquid. Although the relations (6.17) correspond to the restriction $d \ll r_0$, the g_{ij} will have the same basic form for larger d, since (6.14)–(6.16) would merely have a more complicated spatial dependence.

Note that the velocity profile (6.9) is composed of a term due to the pressure drop Δp alone and another term due to ΔV alone. These terms are sketched individually in Figure 6.6. With $\Delta V = 0$, the pressure-induced velocity is the familiar parabolic pipe flow, or "Poiseuille flow," as seen in Section 5.6. The only additional complication included in the first term of (6.9) is that of the slip plane, i.e., the plane at which $v_z = 0$, thought to occur at a distance δ equal to a few molecular diameters from the wall.

Figure 6.6 Single-channel velocity profile: (a) contribution of Δp term in (6.9), sketched for the case $\Delta p < 0$; (b) contribution of ΔV term in (6.9), sketched for the case $\Delta V > 0$.

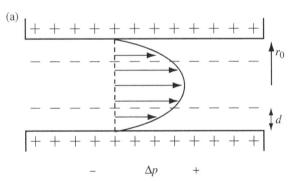

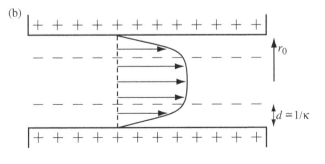

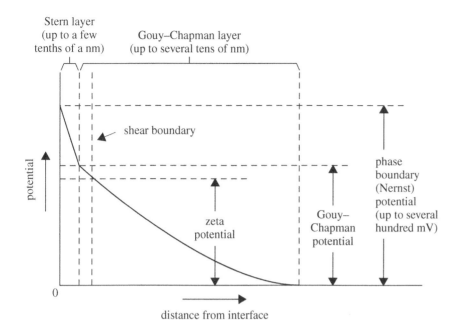

Figure 6.7 The zeta, Guoy–Chapman, and phase boundary (Nernst) potentials. The Stern layer is the plane of "closest approach" for finite-size ions. The Guoy–Chapman layer is the "diffuse double layer zone." (Adapted from Jain MK. *The Biomolecular Lipid Membrane.* Van Nostrand, New York, 1972. p. 38.)

The second term in (6.9) is the electroosmotic flow term, as stated above. The critical double layer parameter is now the potential at the slip plane, $\Phi(r_0 - \delta) = \zeta$, the so-called "zeta potential" as pictured in Figure 6.7. Thus, (6.16) relates the relevant surface charge to the potential drop across the mobile portion of the double layer. Figure 6.6(b) is a sketch of this electroosmotic velocity profile, for a double layer thickness $1/\kappa$ slightly less than r_0. Figure 6.8 pictures the Helmholtz model used in the integrals of (6.11) and (6.12) to arrive at the transduction matrix (6.17).

6.4 MEMBRANE ELECTROKINETIC COUPLING: PHENOMENOLOGICAL AND PHYSICAL MODELS

The electroosmotic flow discussed in Section 6.3 is one example of a class of electrokinetic coupling phenomena involving relative motion between a charged solid and liquid containing some ionic species. In this section, we also focus on the "inverse" effect, called the "streaming potential." This potential drop can be induced across a (pore) membrane by the flow of liquid and entrained counter-ions through the pores. We also generalize the approach to account for membranes that are moving and deformable.

Of the various approaches to membrane modeling, those based on a nonequilibrium thermodynamic approach [3] result in phenomenological models relating measurable mass and current flows through a membrane to pressure and potential drops Δp and ΔV across the membrane, which can be cast in the form

$$
\begin{bmatrix} \boldsymbol{n} \cdot (\boldsymbol{U}^a - \boldsymbol{U}_m) \\[2mm] \boldsymbol{n} \cdot \boldsymbol{J} \end{bmatrix} = \begin{bmatrix} L_{11} & L_{12} \\[2mm] L_{21} & L_{22} \end{bmatrix} \begin{bmatrix} \Delta p \\[2mm] \Delta V \end{bmatrix} \tag{6.18}
$$

where $\boldsymbol{U}^a$ is the liquid velocity on the a side of the membrane, $\boldsymbol{U}_m$ the membrane velocity, and $\boldsymbol{n}$ the unit vector normal to the membrane, so that $\boldsymbol{n} \cdot (\boldsymbol{U}^a - \boldsymbol{U}_m)$ is the mass flow relative to the membrane. Similarly, $\boldsymbol{n} \cdot \boldsymbol{J}$ is the current flow normal to the surface, which may be movable

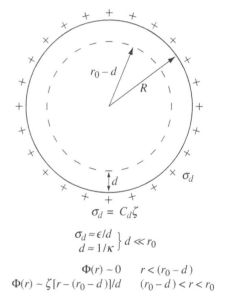

$$\sigma_d = C_d \zeta$$

$$\left. \begin{array}{l} \sigma_d \approx \epsilon/d \\ d \approx 1/\kappa \end{array} \right\} d \ll r_0$$

$$\Phi(r) \sim 0 \qquad r < (r_0 - d)$$
$$\Phi(r) \sim \zeta[r - (r_0 - d)]/d \qquad (r_0 - d) < r < r_0$$

Figure 6.8 Helmholtz double layer model for the cross section of a cylindrical pore.

or deformable in general. The jumps Δp and ΔV are defined as in Section 6.3: $\Delta p = p^a - p^b$, with the normal pointing from the b side to the a side.

The phenomenological coupling coefficients L_{ij} can in principle be measured by static experiments. Table 6.1 summarizes the various electrical, mechanical, and electromechanical (electrokinetic) relations between the L_{ij} and the various "flows and forces." The experiments in Table 6.1 in which $\mathbf{J} = 0$ and $\Delta p = 0$ are the most commonly used measurements, since these constraints are easy to achieve in practice. Figures 6.9 and 6.10 picture the experimental configurations used to measure the streaming potential $(-L_{21}/L_{22})$, the hydrodynamic filtration coefficient $L_P = L_{11} - L_{12}L_{21}/L_{22}$ and the conductance L_{22}. Table 6.2 compares data for these quantities from various physiological membranes.

A word of caution should be included here. The experiments presented above are all static (dc) in nature, and their interpretation is inherently complicated owing to the presence of unwanted side effects—primarily rate processes at the electrodes (e.g., electrolysis) and electrolyte baths having nothing to do with the membrane. The use of ac measurements, where one is interested only in the response occurring at the frequency of the stimulus, can greatly increase the accuracy of the measurement of L_{ij}.

While the L_{ij} in (6.18) represent macroscopic, experimentally measurable coupling coefficients, it would be extremely useful to know the explicit functional dependence of the L_{ij} on membrane microstructure, electrical and viscous properties of the aqueous electrolyte bath, and membrane charge characteristics. This requires more than a phenomenological model; we are now asking for some physical interpretation of the transductive coupling on a macromolecular level. One commonly used approach employs a cylindrical pore model as begun in Section 6.3. To go from the pore transduction matrix of (6.17) to the full membrane matrix of (6.18), a model must be made for the way in which pores "stack" together in the membrane. This stacking is often pictured in terms of series, parallel, or combined series—parallel arrays of pores. Thus, the L_{ij} will finally be expressible in terms of averaged microstructural membrane parameters such as pore radius r_0 and interpore spacing s, along with membrane electrical properties as embodied in the zeta potential. Models for electrokinetic coupling in a fibril matrix have been derived, based on low-Reynolds-number flow through a bed of cylindrical rods [4, 5]. This approach will be explored in the next section in the context of a spherical particle analog.

Electrokinetic coupling is extremely important in the characterization and functioning of biomaterials under physiological conditions. *In vitro* experiments can be used to measure the charge properties of physiological specimens and *in vivo* electrokinetic transductive coupling is thought to play a major role in many instances. As most biological fluids are in a state of motion relative to fixed-charge media, electrokinetic effects must always be considered as mediators of coupling responses. There are also electrokinetic pore models being proposed to explain the voltage-variable conductances of nerve/muscle membrane, as well as to describe certain aspects of electromechanical coupling associated with specialized transducer cells.

While pore models have been widely used for "thick" physiological membranes and films, much controversy has ensued in the literature concerning 10 nm-thick plasma membranes, where facilitated and active transport across these membranes may occur by other means. The red blood cell membrane has been studied using various techniques (osmotic swelling, tracer diffusion), and there appears to be agreement

Table 6.1 Membrane electrokinetic transduction phenomena.

$$U \equiv (U^a - U_m)$$

	U	J	Δp	ΔV
U	—	"Streaming current" $\dfrac{n\cdot J}{n\cdot U}\bigg|_{\Delta V=0} = \dfrac{L_{21}}{L_{11}}$	Hydrodynamic filtration coefficient $\dfrac{\Delta p}{n\cdot U}\bigg|_{J=0} = \left(L_{11} - \dfrac{L_{12}L_{21}}{L_{22}}\right)^{-1}$	"2nd streaming potential" $\dfrac{\Delta V}{n\cdot U}\bigg|_{J=0} = \left(L_{12} - \dfrac{L_{11}L_{22}}{L_{21}}\right)^{-1}$
J	Electroosmotic flow $\dfrac{n\cdot U}{n\cdot J}\bigg|_{\Delta p=0} = \dfrac{L_{12}}{L_{22}}$	—	"2nd electroosmotic pressure" $\dfrac{\Delta p}{n\cdot J}\bigg|_{U=0} = \left(L_{21} - \dfrac{L_{11}L_{22}}{L_{12}}\right)^{-1}$	Electrical resistance $\dfrac{\Delta V}{n\cdot J}\bigg|_{U=0} = \left(L_{22} - \dfrac{L_{12}L_{21}}{L_{11}}\right)^{-1}$
Δp	Mechanical conductance $\dfrac{n\cdot U}{\Delta p}\bigg|_{\Delta V=0} = L_{11}$	"2nd streaming current" $\dfrac{n\cdot J}{\Delta p}\bigg|_{\Delta V=0} = L_{21}$	—	Streaming potential $\dfrac{\Delta V}{\Delta p}\bigg|_{J=0} = -\dfrac{L_{21}}{L_{22}}$
ΔV	"2nd electroosmotic flow" $\dfrac{n\cdot U}{\Delta V}\bigg|_{\Delta p=0} = L_{22}$	Electrical conductance $\dfrac{n\cdot J}{\Delta V}\bigg|_{\Delta p=0} = L_{22}$	"Electroosmotic pressure" $\dfrac{\Delta p}{\Delta V}\bigg|_{U=0} = -\dfrac{L_{12}}{L_{11}}$	—

Figure 6.9 Experimental set-up for measuring hydrodynamic filtration and streaming potential.

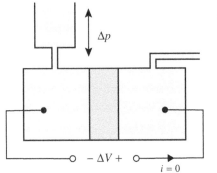

(1) hydrodynamic filtration

$$L_p = L_{11} - \frac{L_{12}L_{21}}{L_{22}}$$

$$= \left. \frac{\boldsymbol{n} \cdot (\boldsymbol{U}^a - \boldsymbol{U}_m)}{\Delta p} \right|_{\boldsymbol{J}=0}$$

(2) streaming potential

$$-\frac{L_{21}}{L_{22}} = \left. \frac{\Delta V}{\Delta p} \right|_{\boldsymbol{J}=0}$$

Figure 6.10 Set-up for measuring electrical conductance.

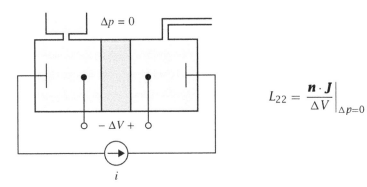

$$L_{22} = \left. \frac{\boldsymbol{n} \cdot \boldsymbol{J}}{\Delta V} \right|_{\Delta p=0}$$

between much of the data and a pore membrane model. This has led investigators to employ similar techniques to predict equivalent pore radii for various membranes. Table 6.3 gives predicted pore radii based on the difference in experimentally measured water permeability coefficients using osmotic and tracer diffusion techniques—each depending

Table 6.2 Phenomenological coefficients of physiological membranes.

Parameter	Measurement: rat tail tendon collagen membrane		Some comparative data from available literature		Units
Mechanical filtration coefficient	0.001 M HCl	5.0×10^{-12}	Human red blood cell	0.92×10^{-12}	
			Nitella	1.1×10^{-12}	
$L_{11} - \frac{L_{12}L_{21}}{L_{22}} \equiv L_p$	Distilled water	3.7×10^{-12}	Toad skin	0.4×10^{-12}	m^3 N^{-1} s^{-1}
			Visking dialysis tube	3.2×10^{-12}	
Conductance/area L_{22}	0.001M HCl:	1.25	Nerve, muscle membrane	1–10	S m^{-2}
L_{21}	0.001 M HCl:	1.36×10^{-6}	Comparison with Y_{21} predicted by dynamic collagen experiments in 0.001 M HCl: $L_{21} = 1.22 \times 10^{-6}$ $L_{12} = 1.30 \times 10^{-6}$		A N^{-1}

Table 6.3 The permeability coefficient for diffusion of water across various membranes.

Membrane	Permeability $\times 10^4$ (cm s^{-1})		Approximate "effective" pore radius (nm)
	Osmotic	Diffusional	
Self-diffusion	$(= D)^a$	2.41×10^{-5}	
Through hexadecane	$(= D)^a$	4.16×10^{-5}	
Bilayer lipid membrane (BLM)^b	10–12 at 25°C	9–11	
Liposomes	44 at 25°C		
Valonia	2.4	2.4	
Amoeba	0.37	0.23	0.21
Frog skin	3.3	1.1	0.43
Frog ovarian egg	89.1	1.28	3.0
Frog body cavity egg	1.30	0.75	0.28
Dog erythrocyte	400	63	0.75
Ox erythrocyte	152	51	0.43
Human erythrocyte:^{c}			
Adult	115–130	45–50	0.35–0.40
Fetal	61	23	0.39
Squid axon	250	30–35	0.85
Toad bladder	4.1	0.95	0.85
Toad bladder + vesopressin	230	10.8	1.8
Goat ventricular walls	270	2.8	3.6
Dialysis tubing	380	10.9	2.3

^{a}This value is in cm^2 s^{-1}. Considering a thickness of the water slab of 10 nm, the permeability coefficient can be obtained as 24.1 cm s^{-1}.
^{b}The permeability of a BLM is dependent upon its composition.
^{c}It may be noted that thiol reagents cause a substantial decrease in osmotic permeability and the ratio of osmotic to diffusional permeabilities. This may imply that the pores are modified in the treated cells.
From Jain MK. *The Bimolecular Lipid Membrane.* Van Nostrand Reinhold, New York, 1972.

in a different manner on the value of r_0. (In Table 6.3, permeability corresponds to the flux measurement N(mol m^{-2} s^{-1}) $= -D\Delta c/\Delta x \equiv P\,\Delta c$, so the units of P are m s^{-1}).

6.5 ELECTROPHORESIS

Another example of double-layer-mediated electrokinetic transduction is that of electrophoresis. The movement of charged particles in electrolyte solutions induced by applied electric fields has been known for two hundred years [6]. In the last 40 years, electrophoresis has seen wide application in biology and biochemistry for the separation of proteins and other macromolecules, an invaluable tool for preparative (i.e., purification) and analytical procedures. Nevertheless, standard publications on the theoretical principles underlying electrophoresis differ greatly, and there is still controversy regarding the representation of the electrical and mechanical force balances that are involved. Here, we are interested in the fundamental concepts underlying the transduction mechanisms as they may be related to physiological processes *in vivo*, as well as biochemical measurements *in vitro*.

Arne Tiselius [7] was first to use "free" electrophoresis in protein separation, with an apparatus similar to that shown in Figure 6.11. Proteins having different mobilities $b = U/E_0$ (where E_0 is the applied field and U is the electrophoretic particle velocity) could be detected

Figure 6.11 (a) Schematic view of the original Tiselius moving-boundary electrophoresis apparatus. (b) Electrophoretic pattern of human blood plasma proteins (pH 8.6): A, serum albumin; ϕ, fibrinogen; $\alpha_1, \alpha_2, \beta, \gamma$, various globulins. From Lehninger AL. *Biochemistry*. Worth Publishers, New York, 1970, pp. 131–132.

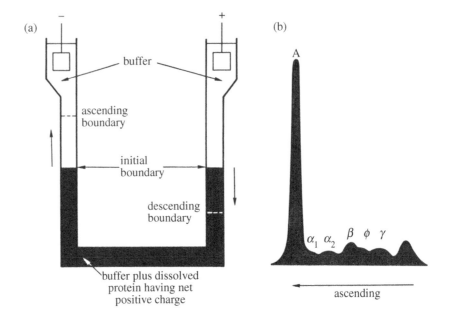

optically at the moving boundary between the protein/buffer solution (dark) and the pure buffer (light). Because of experimental problems concerning liquid convection and vibration, which would disturb the boundary, many improvements on the original technique have been made. These involve electrophoresis in "stabilizing" media such as the original gel columns (Figure 6.12), thick paper, or the now commonly used slab gel. At the same time, free electrophoresis saw a reemergence in the form of "capillary electrophoresis" instruments when engineering advances eliminated problems of thermally induced convection. Solutes are separated within a fluid stream moving (via electroosmosis) through 10–200 μm diameter charged, plastic capillaries. This electrophoretic separation of charged solutes within an electroosmotically convecting fluid stream has now become the paradigm for MEMS lab-on-a-chip devices. In addition, this configuration is often used to measure the charge on biological molecules, such as intracellular microtubules (Figure 6.13) [8].

Our present aim is to model the electrophoretic process for a spherical particle of radius R in a liquid of viscosity μ, using this simplified

Figure 6.12 Polyacrylamide gel (disk) electrophoresis columns (350 mm × 50 mm), in which a pH gradient along the column aids in separating proteins into narrow bands or disks. This result is from the original Tiselius experiments. From Lehninger AL. *Biochemistry*. Worth Publisher, New York, 1970, p. 132.

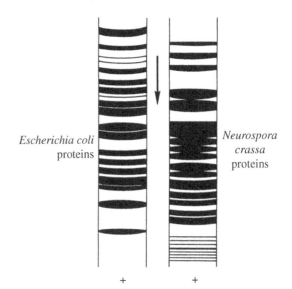

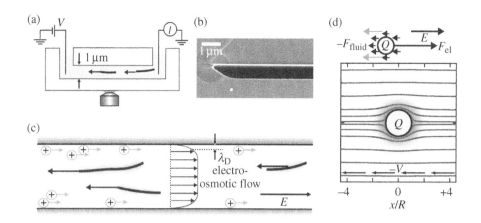

Figure 6.13 (a) Electrophoretic motion of microtubules in channels is observed using fluorescence microscopy. (b) Scanning electron microscope image of a part of the cross section of a channel. (c) The velocity of microtubules is a superposition of their electrophoretic velocity and an electroosmotic flow velocity. (d) During electrophoresis ($R/(1/\kappa) = 10$), the fluid is only perturbed within a much smaller distance around the particle. From van den Heuvel, MGL, de Graaff MP, Lemay SG, and Dekker C. Electrophoresis of individual microtubules in microchannels. *Proc. Natl Acad. Sci. USA* **104**, 7770–7775 (2007).

geometry to gain physical insight. One goal will be to realize that the particle motion is *not* simply the result of a "qE" force applied to the particle balanced by the viscous Stokes drag force $6\pi R\mu U$. We know, in fact, that the particle and surrounding double layer taken together contain no net charge. Rather, the coupling takes place in the *mobile* part of the double layer, where viscous and electrical shear stresses balance in a manner similar to that of electroosmotic flow (Section 6.4), resulting in a *relative motion between fluid and particle*. We will find that the fluid flow problem differs from that of simple flow around a rigid sphere because of the electromechanical coupling in the interfacial region. This will be apparent in terms of the fluid velocity boundary conditions at $r = R$. We use the approach detailed by Levich [1] to obtain a simplified analytical expression for the electrophoretic mobility in the limit $R \gg 1/\kappa$, based in part on the prior studies of Henry [9] and Booth [10].

The analysis begins in the particle frame of reference, in which the fluid has an apparent uniform flow at $z \to \pm\infty$, as shown in Figure 6.14. The electric field applied by means of external electrodes is also uniform in the z direction at $z \to \pm\infty$

$$\mathbf{E} = \mathbf{i}_z E_0 = E_0(\mathbf{i}_r \cos\theta - \mathbf{i}_\theta \sin\theta), \quad \Phi = -E_0 r \cos\theta$$

$$\mathbf{u} = -\mathbf{i}_z = -U(\mathbf{i}_r \cos\theta - \mathbf{i}_\theta \sin\theta)$$

(6.19)

With typical applied field strengths of E_0 in the range 1–50 V cm^{-1} (in bulk solution), it is found that typical values of measured mobilities for globular proteins are $b \simeq 0.1 \times 10^{-8}$ m^2 V^{-1} s^{-1}, consistent with speeds U on the order of 10^{-7} m s^{-1}. The very small Reynolds number $2RU\rho/\mu$ for such speeds justifies the use of a "creep flow" analysis for the fluid mechanics, as outlined in Sections 5.7–5.9. We will use the results of this analysis in conjunction with boundary conditions derived to account for the additional electrokinetic coupling. In this section, the double layer is assumed to be much thinner than the particle radius ($R \gg d \equiv 1/\kappa$). Therefore, we treat the double layer in terms of boundary conditions at $r = R + d \simeq R$. The electric field boundary value problem is that of an insulating particle in a conducting medium with an applied uniform field. Thus, we expect that the electrical potential solution should look like the superposition of uniform and dipole potentials, but modified by an electrical boundary condition at $R + d$, which will be seen to involve fluid motion.

In terms of a control volume that cuts through the double layer as shown on the right side of Figure 6.15, we arrive at a conservation-of-charge equation that requires that the normal conduction current in the

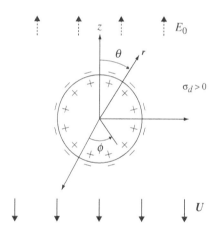

Figure 6.14 Schematic of the electrophoresis of a charged spherical particle subjected to a uniform electric field.

Figure 6.15 Schematic showing the double layer electrokinetic boundary condition as well as a control volume for conservation of charge at the surface of the sphere. Conservation of charge thereby assumes the role of an additional boundary condition.

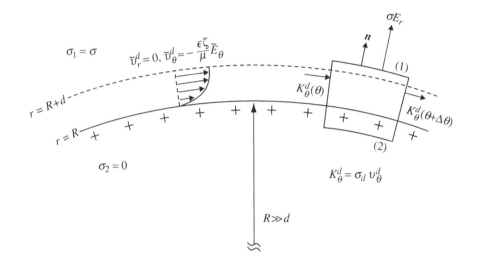

liquid phase be balanced by the divergence of the convective surface current that flows along the interface:

$$\boldsymbol{n} \cdot (\boldsymbol{J}_1 - \boldsymbol{J}_2) + \nabla_\Sigma \cdot \boldsymbol{K} = 0 \qquad (6.20)$$

where ∇_Σ is the surface divergence operator and $\boldsymbol{J}_2 = 0$ inside the insulating sphere. The additional effect of an Ohmic surface current is treated in Problem 6.11. In terms of a diffuse double layer model, the surface current $\boldsymbol{K} = \boldsymbol{i}_\theta K$ is

$$K = \int_R^\infty \rho_e(r) v_\theta(r, \theta) \, dr \qquad (6.21)$$

However, in keeping with the approximation $d \ll R$, we use a Helmholtz double layer model in which

$$\rho_e(r) \simeq -\sigma_d u_0[r - (R + d)] \qquad (6.22)$$

$$\sigma_d = \frac{\epsilon}{d} \zeta \qquad (6.23)$$

and $-\sigma_d$ represents all the *mobile* charge in the double layer. Thus, with (6.22) inserted into (6.21),

$$K \simeq -\sigma_d v_\theta^d \qquad (6.24)$$

Conservation of charge then becomes (using the divergence operator of Table B.5)

$$-\sigma \frac{\partial \Phi^d}{\partial r} + \frac{1}{r \sin \theta} \frac{\partial}{\partial \theta} (-\sigma_d v_\theta^d \sin \theta) = 0 \qquad (6.25)$$

The fluid velocity is zero at the electrokinetic slip plane $r = R + \delta$, where $\delta \ll d$; thus, the position $R + \delta$ really corresponds to a "no-slip" plane. Using Levich's approach, however, we propose that the velocity at the edge of the double layer at $r = R + d$ is *finite* and has the r and θ dependence (see Section 5.8)

$$v_r = \tilde{v}_r^d 2 \cos \theta \equiv 0, \quad v_\theta = \tilde{v}_\theta^d \sin \theta \qquad (6.26)$$

The tangential component of $\boldsymbol{E}$ causes the finite tangential fluid flow at the edge of the double layer, $R + d$, but the radial velocity at $r = R + d$ is zero. Therefore, (6.25) becomes

$$\frac{\partial \Phi^d}{\partial r} = -\frac{2\sigma_d \tilde{v}_\theta^d}{\sigma R} \cos \theta \qquad (6.27)$$

Equation (6.26) represents the r and θ dependence of the fluid mechanical boundary conditions to be found at $r = R + d$. If we know $\tilde{v}_\theta^d$, then (6.27) and (6.19) constitute sufficient information to determine Φ in the bulk fluid region. Since there is zero space charge in the bulk fluid, Φ satisfies Laplace's equation there. We use superposition to guess a solution to Laplace's equation of the form (Table B.7)

$$\Phi = \frac{A \cos \theta}{r^2} - E_0 r \cos \theta \qquad (6.28)$$

and use the boundary condition (6.27) to find A in terms of $\tilde{v}_\theta^d$:

$$A = \frac{R^3}{2} \left(-E_0 + \frac{2\sigma_d}{\sigma R} \tilde{v}_\theta^d \right) \qquad (6.29)$$

To complete (6.29), let us review the electrical and mechanical boundary conditions at the spherical surface (Figure 6.15). We recognize that because the sphere is rigid, $\tilde{v}_r^d = 0$. Of course, we really mean $\tilde{v}_r = 0$ at $r = R$, and the double layer is extremely thin. This approximation must be made with caution because we know that the tangential velocity $\tilde{v}_\theta = 0$ at $r = R$, but $\tilde{v}_\theta^d \neq 0$ because of the electrical shear force in the thin double layer region. To find $\tilde{v}_\theta^d$, which is the condition needed to find the fluid flow as well as the constant A in (6.29), we realize that, for a sphere with $d \ll R$, every section of the surface looks approximately like that of a plane surface. By analogy with the planar electroosmosis problem (see Figure 6.16),

$$\tilde{v}_\theta^d \simeq -\frac{\epsilon \zeta}{\mu} \tilde{E}_\theta^d \qquad (6.30)$$

If we now use the tangential field evaluated from (6.28) and (6.29) in (6.30), we can solve for $\tilde{v}_\theta^d$ in terms of the applied field E_0:

$$\tilde{v}_\theta^d = \frac{+\frac{3}{2}(\epsilon/\mu)\zeta E_0}{1 + \epsilon \zeta \sigma_d / \mu \sigma R} \qquad (6.31)$$

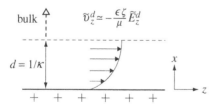

Figure 6.16 Planar electroosmosis.

$$\mu \frac{\partial^2 v_z}{\partial x^2} = -\frac{\partial \sigma_{zx}^e}{\partial x} = -\frac{\partial}{\partial x}(\epsilon E_z)\left[-\frac{d\Phi(x)}{dx} \right] \quad \text{Assuming } P = P_0 \text{ everywhere.}$$

$$\mu \frac{\partial v_z}{\partial x} = -\epsilon E_z \left[-\frac{d\Phi(x)}{dx} \right] + C_1 \quad \text{In bulk, } \frac{\partial v}{\partial x} = \frac{\partial \Phi}{\partial x} = 0. \text{ Therefore, } C_1 = 0.$$

$$\mu v_z(x) = +\epsilon E_z \Phi(x) + C_2 \quad \text{At slip plane, } \Phi = \zeta, v_z = 0.$$

$$v_z(x) = -\frac{\epsilon[\zeta - \Phi(x)]E_z}{\mu} \quad \text{Helmholtz model: } \Phi(x) \to 0 \text{ at } x = d.$$

We have assumed that the particle is fixed, and that the fluid is moving relative to the particle. Of course, if we consider the laboratory frame as one moving with the flow at infinity, the particle appears to move in the opposite direction, so we can interpret U as the particle velocity resulting from the given applied electric field intensity. In fact, if there are no other forces acting on the particle and the particle is to be in force equilibrium, we can compute the velocity U by recognizing that the net force in the z direction must vanish. This net force is given in terms of the viscous stresses by (5.82), with the stress components evaluated by using stress–velocity relations (5.81), with $v_r^d = 0$ and v_θ^d from (6.31).

Evaluation of the integral gives

$$f_z = \pi R \mu(-6U + 4\tilde{v}_\theta^d) = 0 \tag{6.32}$$

and hence we are left with the conclusion that the slip velocity adjacent to the surface is related to the uniform velocity at infinity by

$$U = +\tfrac{2}{3}\tilde{v}_\theta^d \tag{6.33}$$

We now have the final relation between the electrophoretic velocity U and the applied field E_0:

$$\boxed{U = \frac{\epsilon\zeta/\mu}{1 + \epsilon\zeta\sigma_d/\mu\sigma R}E_0} \tag{6.34}$$

6.6 PROBLEMS

Problem 6.1 Cylindrical Pore Model: Electrokinetic Transduction

(a) Carry out the integrations in (6.11) and (6.12), using (6.9) and the Helmholtz double layer model defined by (6.13)–(6.16). Assume in (6.9) that $d \ll r_0$ and $\delta \ll d$ and in (6.11) that the conductivity $\sigma(r)$ is uniform ($\equiv \sigma$) over the cross section, and show that these assumptions lead to the transductive coupling relation (6.17):

$$\begin{bmatrix} Q_p \\ i_p \end{bmatrix} = \begin{bmatrix} -\dfrac{\pi r_0^4}{8\mu\ell_p} & \dfrac{\pi r_0^2\epsilon\zeta}{\mu\ell_p} \\[2ex] \dfrac{\pi r_0^2\epsilon\zeta}{\mu\ell_p} & -\left(\dfrac{\pi r_0^2\sigma}{\ell_p} + \dfrac{2\pi r_0\epsilon^2\zeta^2}{\mu d\ell_p}\right) \end{bmatrix} \begin{bmatrix} \Delta p \\ \Delta V \end{bmatrix}$$

(b) Find

$$\left.\frac{Q_p}{i}\right|_{\Delta p=0} = \frac{g_{12}}{g_{22}} \tag{6.35}$$

Normalize both Q_p and i_p to the pore area πr_0^2, and the current i_p to the conductivity σ, to find

$$\left.\frac{U}{J/\sigma}\right|_{\Delta p=0} \tag{6.36}$$

PROBLEM

where U and J are the fluid velocity and current density over most of the pore, consistent with the Helmholtz model and the assumption $1/\kappa \ll r_0$ used to derive (6.17). (This is now analogous to the membrane coupling equation in Table 6.1; here we are treating a single pore rather than a complete series or parallel array of them.)

(c) Plot the ratio (6.36) as a function of ζ, showing the relevant dependence for large and small ζ; replace ζ with σ_d using the Helmholtz model (6.16). Discuss your results in terms of viscous-limited and relaxation-limited flow regimes.

(d) Show that in the limit of high conductivity σ, the expression (6.36) reduces to

$$\left.\frac{Q_p/\pi r_0^2}{\Delta V/\ell_p}\right|_{\Delta p=0} = \frac{\ell_p}{\pi r_0^2}g_{12} \qquad (6.37)$$

Evidently, $E_z = -\Delta V/\ell_p = J/\sigma$ in the high-conductivity limit. Why?

Show that (6.37) can be used to derive the complete form of (6.36) that you found in part (b), by replacing the denominator of (6.37) with an E_z that is consistent with both convection and conduction currents.

Problem 6.2

(a) A polyelectrolyte-based membrane is to be modeled as a hexagonal matrix of charged rods, as shown in Figure 6.17. The rods have uniform surface charge density σ_d and surface potential Φ_0. Show that the radially symmetric electrical potential ϕ between the rods has the form ($\phi = zF\Phi/RT$)

$$\phi(r) = \frac{\sigma_d}{\epsilon\kappa K_1(\kappa a)\left[\dfrac{I_1(\kappa R)}{K_1(\kappa R)} - \dfrac{I_1(\kappa a)}{K_1(\kappa a)}\right]}\left[I_0(\kappa r) + \frac{I_1(\kappa R)}{K_1(\kappa R)}K_0(\kappa r)\right]$$

$$(6.38)$$

where $1/\kappa$ is the Debye length and ϵ is the dielectric constant of the fluid (assumed to be homogeneous). In (6.38), I_0, I_1 and K_0, K_1 are modified Bessel functions of the first and second kinds, respectively.

(b) Assume that all the rods are oriented along the direction of fluid flow. Show that the fully developed velocity profile has the form

$$v_z(r) = \frac{(r^2 - a^2 - 2R^2\ln r/a)}{4\mu\ell}\Delta P + \frac{\epsilon}{\mu\ell}[\zeta - \Phi(r)]\Delta V \quad (6.39)$$

where ΔP and ΔV are the pressure and potential drops across the membrane in the direction of flow.

(c) Find the average fluid velocity U and current density J flowing in the axial direction in a unit cell (one charged rod and its surrounding fluid) as a function of ΔP and ΔV. For this calculation, assume that $1/\kappa \ll a \ll R$; assume that the double layer can be modeled by the Helmholtz approximation: $\Phi(r) \simeq 0$ for $r > a + 1/\kappa$, $\Phi(a + \delta) = \zeta$ at the slip plane.

Figure 6.17 Cylindrical rod unit cell model for electrokinetic transduction. (a) The membrane consists of a hexagonally packed matrix of charged rods of radius a, each within a unit cell of radius R. (b) The fluid flow profile (shown by arrows) is oriented along the direction co-axial with the rods. (c) Each rod has a uniform surface charge density, a surface potential Φ_0, and an associated zeta potential at the position of the slip plane. (d) The electrical potential between neighboring rods within the matrix has a minimum at the edge of the unit cell located at the radius R.

(a) Unit cell (b)

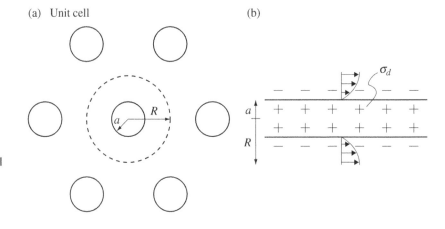

(c) (d)

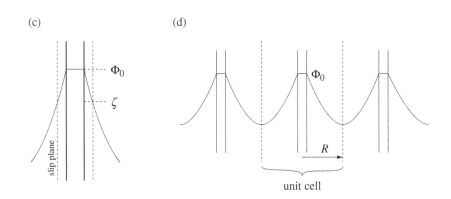

Write your solution in the form of an "electromechanical transduction matrix" of the form

$$
\begin{bmatrix} U \\ J \end{bmatrix} = \begin{bmatrix} L_{11} & L_{12} \\ L_{21} & L_{22} \end{bmatrix} \begin{bmatrix} \Delta P \\ \Delta V \end{bmatrix}
\tag{6.40}
$$

with L_{ij} written in terms of a, R, ϵ, κ, μ, ℓ, and ζ and/or σ_d. You should find that $L_{12} = L_{21}$. Give a physical motivation for this result.

(d) Now solve for the L_{ij} using the fully diffuse double layer potential derived in part (a). Assume that the slip plane coincides with $r = a$; hence, express your answer in terms of Φ_0 (not ζ) and/or σ_d.

Problem 6.3 Macrocontinuum Model for L_{ij} When $\Delta P = 0$ and $\Delta V \neq 0$, conservation of momentum for an interstitial fluid element can be written, in general, as

$$
\rho_{\text{mass}} \frac{d^2 \boldsymbol{u}}{dt^2} + b \frac{d\boldsymbol{u}}{dt} = \rho_e \boldsymbol{E}
\tag{6.41}
$$

where $\boldsymbol{E}$ is the field associated with ΔV, $\boldsymbol{u}$ is the displacement vector associated with fluid motion, b is a frictional coefficient, and ρ_e is the net space charge density of the fluid phase. For this

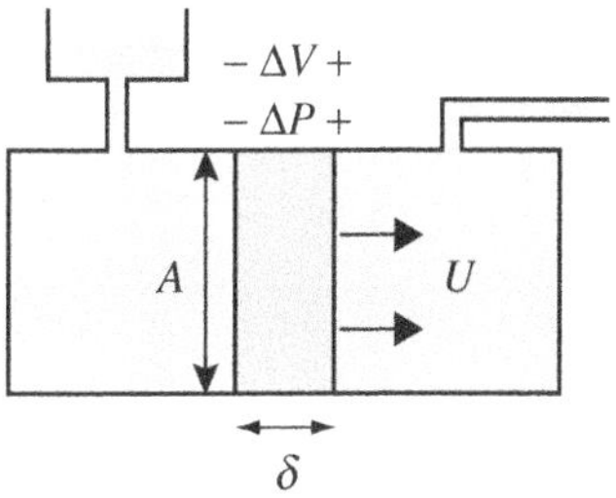

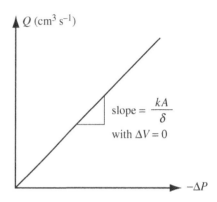

$$\begin{bmatrix} U \\ J \end{bmatrix} = \begin{bmatrix} L_{11} & L_{12} \\ L_{21} & L_{22} \end{bmatrix} \begin{bmatrix} \Delta P \\ \Delta V \end{bmatrix}$$

Figure 6.18 Homogeneous, isotropic charged tissue having cross-sectional area A and fixed charge density $\rho_m/F = 0.1$ M associated with the solid phase.

experiment, the solid matrix can be modeled as rigid, incompressible, and fixed in space even when fluid moves through it. The matrix has fixed charge density ρ_m.

Data from the experiment pictured in Figure 6.18 is shown in Figure 6.19 for the case $\Delta V = 0$, $\Delta P \neq 0$. For the range of ΔP that is applied across the tissue, the volume flux of fluid Q (cm^3 s^{-1}) measured by the capillary tube is found to be linearly proportional to ΔP.

(a) Assume that there are no concentration gradients of interest and that the fractional water content of the membrane can be approximated as unity. Find analytical expressions for the L_{ij} in terms of ρ_m, k, intramembrane ion concentrations $\bar{c}_i$, and any other appropriate parameters. (What is the relationship between b and k? What is the relationship between ρ_m and ρ_e?) State and justify all assumptions.

(b) Find an expression for the streaming potential based on your macro-model, and compare

$$-\frac{L_{21}}{L_{22}} \triangleq \left.\frac{\Delta V}{\Delta P}\right|_{J=0} \qquad (6.42)$$

with the streaming potential derived from the capillary pore micro-model of Section 6.3.

Figure 6.19 Experimentally observed hydraulic fluid flow through the membrane.

Problem 6.4 *Static* streaming potential measurements are often used in an attempt to estimate the charge and electromechanical transduction characteristics of biomembranes and synthetic biomaterials. However, rate processes having *nothing* to do with the electrokinetic coupling (i.e., the coupling modeled by (6.1)) often cloud the interpretation of experimental results—measured streaming potentials are often found to drift. One solution is to perform an ac rather than dc experiment. The purpose of this problem is to develop a linear transduction model for the ac experiment pictured in Figure 6.20. The model is valid for small membrane displacements ξ_m.

A plunger is moved vertically in Figure 6.20(a) in such a fashion as to produce a sinusoidal steady state pressure excitation:

$$P_0 = P^d - P^f = \text{Re}(\hat{P}_0 e^{jwt}) \qquad (6.43)$$

The linearized model to be developed must relate the measured sinusoidal transmembrane potential $\Delta V = \text{Re}(\Delta V e^{j\omega t})$ to the excitation P_0, where $\Delta V = V^a - V^b$ as shown.

PROBLEM

Figure 6.20 Pressure excitation applied to a membrane.

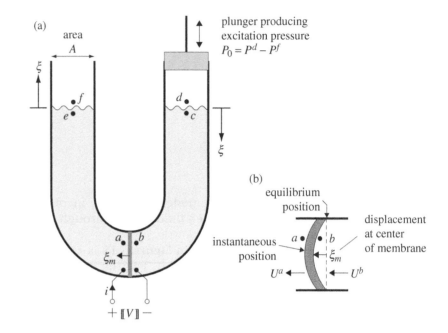

(a) Find a differential equation of the form of (5.24) relating the fluid column displacements ξ to the excitation P_0 and the unknown pressure jump across the membrane, $p^a - p^b = \Delta p$. Integrate Bernoulli's' equation separately between points c to b and a to e; neglect surface effects at the meniscus and assume $U^e = U^c(= j\omega\xi)$ and $U^a = U^b$. Note that, in general, the excitation P_0 results not only in membrane deformation, but also in fluid flow through the *permeable* membrane. Thus, U^a and U^b are defined as the *fluid* velocity just to the left and right of the membrane, respectively; the *relative* fluid velocity with respect to the membrane is then defined as $\boldsymbol{n} \cdot (\boldsymbol{U}^a - \boldsymbol{U}_m)$, where $\boldsymbol{U}_m = j\omega\xi_m$.

(b) Relate the unknown pressure drop Δp to the membrane displacement ξ_m (Figure 6.20(b)) by means of a lumped spring constant S, a linear model valid for small ξ_m. Neglect membrane inertia (justifiable for frequencies of interest).

(c) Find

$$\left(\frac{\Delta \hat{V}}{\hat{P}_0}\right)_{i=0} \tag{6.44}$$

by combining the results of parts (a) and (b) with an electrokinetic transduction model valid at *finite frequency*. Do this by writing your results in terms of four algebraic equations in the four unknown sinusoidal steady state *amplitudes* ΔV, Δp, ξ_m, and ξ, with excitation variables P_0 and $i/A = \boldsymbol{n} \cdot \boldsymbol{J}$, expressible in matrix form as

$$
\begin{bmatrix} \text{Fill in the} \\ \text{entries} \end{bmatrix}
\begin{bmatrix} \Delta \hat{V} \\ \Delta \hat{P} \\ \hat{\xi}_m \\ \hat{\xi} \end{bmatrix}
=
\begin{bmatrix} \hat{P}_0 \\ 0 \\ 0 \\ \hat{i}/A \end{bmatrix}
\tag{6.45}
$$

Discuss the frequency limitations of all constituents of the model as expressed in (6.45) above. What physical

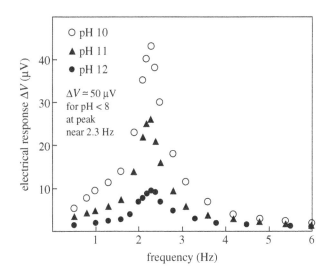

Figure 6.21 Mechanical-to-electrical transduction for a polylysine membrane with $L_{11} \simeq 10^{-12}\,\mathrm{m^3\,N^{-1}\,s^{-1}}$, $L_{22} \simeq 1\,\mathrm{S\,m^{-1}}$, and $L_{21} \simeq 8 \times 10^{-7}\,\mathrm{A\,N^{-1}}$ (at the peak in the response).

processes may lead to deviations from (6.45) for high enough frequency?

(**d**) Assume that the data shown in Figure 6.21 correspond to a membrane cast from polylysine (lysine is an amino acid). Interpret the data in terms of your model and sketch a rough titration curve for polylysine based on the data. Compare with the charge titration behavior shown in Figure 1.13.

Problem 6.5 Chemically Nonhomogeneous Systems In Sections 6.3 and 6.4, we derived transport relations such as (6.17) and (6.18) that describe electromechanical transduction with charged membranes in *homogeneous* electrolyte baths, i.e., $c' = c''$. In this problem, we relax the latter restriction and consider the additional effect of an applied concentration gradient across the membrane, $c' \neq c''$ as shown in Figure 6.22. The objective is to derive a set of linearized transport relations of the form

$$Q_p = g_{11}\frac{\partial p_\beta}{\partial z} + g_{12}\frac{\partial \psi}{\partial z} + g_{13}\frac{\partial c}{\partial z} \tag{6.46}$$

$$i_p = g_{21}\frac{\partial p_\beta}{\partial z} + g_{22}\frac{\partial \psi}{\partial z} + g_{23}\frac{\partial c}{\partial z} \tag{6.47}$$

$$j_s(\mathrm{mol\,s^{-1}}) = g_{31}\frac{\partial p_\beta}{\partial z} + g_{32}\frac{\partial \psi}{\partial z} + g_{33}\frac{\partial c}{\partial z} \tag{6.48}$$

where the g_{ij} are functions of z in the most general case and c in $\partial c/\partial z$ is the total concentration $c = c_+ + c_- = 2c_0$. Note that (6.17) and (6.18) are "phenomenological" relations in terms of Δp, ΔV, etc. Here, the transport equations are differential continuum equations applicable at any point (z). Assuming that $r_0 \ll \ell_p$ (pore radius $\ll$ pore length),

(**a**) Write the appropriate r and z components of the Stokes equation, following the method of Section 6.3. Note that in Section 6.3, the hydrostatic pressure could be considered

PROBLEM

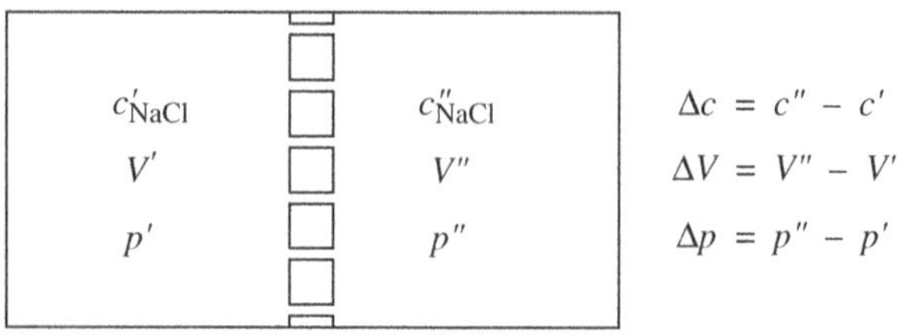

Figure 6.22 Applied concentration gradient across a membrane.

$$\Delta c = c'' - c'$$
$$\Delta V = V'' - V'$$
$$\Delta p = p'' - p'$$

as the sum of two terms: $p = p_\alpha(r) + p_\beta(z)$. p_α accounted for the osmotic contribution, which varied in the r direction owing to the radial double layer distribution; p_α did not depend on z. p_β represented the z dependence of hydrostatic pressure due to an applied pressure drop across the pore. Hence, $\partial p/\partial z = \partial p_\beta/\partial z$ in Section 6.3.

However, with $c' \neq c''$, the osmotic pressure varies with z as well as r:

$$p_\alpha(r, z) = RTc(r, z)$$

Therefore, the z component of the Stokes equation in this problem must contain an additional contribution to the ∇p term!

(**b**) Integrate the z component of the Stokes equation twice in order to obtain an expression for the fluid velocity profile $v = i_z v_z(r)$. Note that because $c' \neq c''$, the internal membrane concentration varies with z in general. Therefore, $\partial p/\partial z \neq$ constant in this problem (Why?), unlike Section 6.3, in which $\partial p/\partial z \equiv \Delta P/\ell_p =$ constant across the membrane.

Similarly, $\partial \psi/\partial z \neq$ constant inside the membrane (Why?), unlike Section 6.3, in which $E_z = -\partial V/\partial z \equiv \Delta V/\ell_p =$ constant.

(**c**) Assume $1/\kappa \ll r_0$ and $\delta \ll 1/\kappa$, where δ is the thickness of the "no-slip" plane. Using the Helmholtz double layer model (Section 6.3) and other appropriate approximations, find the $g_{ij}(z)$.

Problem 6.6

(**a**) Describe the effect of changes in ionic strength and pH on the electrophoretic mobility of charged, osmotically swollen (spherical) cells. Consider the effects on ζ for the cases when Φ_0 or σ_d are constrained (refer to Figure 6.7).

(**b**) Sketch the mobility as a function of charge σ_d and consider the effects of changes in ionic strength and pH in the limits,

$$\frac{d\sigma_d^2}{\sigma\mu R} \gg 1 \text{ and } \ll 1.$$

(**c**) Historically, some of the earliest measurements of cell electrophoretic mobilities were made using red blood cells. Experiments were performed to identify the cell-surface charge groups responsible for electrophoretic migration (Table 6.4 and Figure 6.23). Which region of the mobility-versus-σ_d curve of part (b) corresponds to these experiments?

PROBLEM

Erythrocyte donor	No. of sialic acid molecules ($\times 10^4\,\mu\text{m}^{-2}$)	EPM (TU)[b] at $I = 0.072\,\text{M}$, pH 6.4
Pig	5.3	11.8
Chicken	5.7	12.8
Calf	10.5	14.4
Lamb	12.3	16.9
Human	14.7	16.4
Horse	28.2	19.8

Table 6.4 Identification of sialic acid groups as a major contributor to erythrocyte cell surface charge: Relationship between electrophoretic mobility (EPM) and number of sialic acid molecules on erythrocyte surface[a]

[a]Data from Eylar et al. [11]. A statistical analysis of correlation between the sets of variables gave a correlation coefficient $r = 0.94$ ($p < 0.01$), i.e., the variables were completely interdependent.
[b]1 Tiselius unit (TU) $= 10^{-5}\,\text{cm}^2\,\text{V}^{-1}\,\text{s}^{-1}$.

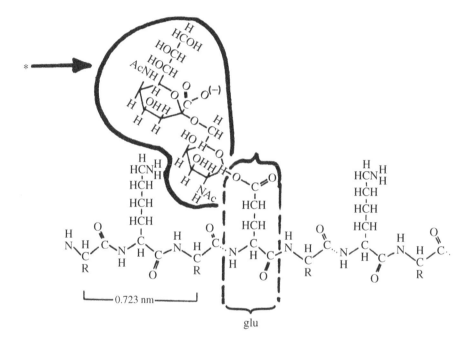

Figure 6.23 Sites containing surface charge groups in sheep erythrocyte membranes. *Disaccharide containing an ionizable carboxyl group: sialyl-*N*-acetylgalactosamine (sialic acid bonded to a galactosamine group). From Gottschalt A. *The Chemistry and Biology of Sialic Acids and Related Substances.* Cambridge University Press, Cambridge, p. 98.

Problem 6.7 Mobility of a Liquid Metal Drop in an Electrolyte Electrophoretic mobilities of liquid drops can be much greater than those of solid particles. For example, suppose that a liquid mercury drop having radius of 0.1 mm is placed in an electrolyte and subjected to an externally imposed electric field of $1\,\text{V}\,\text{cm}^{-1}$ (Figure 6.24). The drop will typically move with or against the field at an easily observed velocity. The mobility is much greater than that of solid particles, because the retarding viscous shear stresses are determined by the dimensions of the drop rather than the thickness of the double layer. The double layer in this example of electrocapillary transduction is no longer constrained on one side by a rigid boundary (the solid particle). Rather, the double layer is part of a deformable interface subjected to an electrical shear stress.

The new ingredient in this problem is the nature of the double layer. Although we are discussing a mercury drop, the development is relevant to *any highly conducting entity whose*

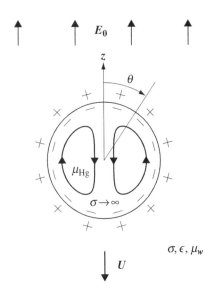

Figure 6.24 Mercury drop in a uniform electric field $E_0 i_z$ ($\sigma_d < 0$ here).

surface "polarizes ideally," i.e., *does not allow passage of current through the interface.* Because the current due to the applied $\boldsymbol{E}$ skirts around the drop, *there are tangential fields in the mobile region of the double layer, but* $E_{tan} \to 0$ *at* $r = R$ *since* the drop is modeled as a perfect conductor (Figure 6.25).

The component E_{tan} results in a potential profile which varies along the surface, which can be seen in the planar analog of Figure 6.25. Since $\nabla \times \boldsymbol{E} = 0$ or $\oint E \cdot dl = 0$ along the dotted line contour of Figure 6.25,

$$E_{tan} = -\frac{d\Phi}{dx} \qquad (6.49)$$

Generalizing to any curvilinear surface,

$$E_{tan} = -\nabla_\Sigma \Phi \qquad (6.50)$$

Thus, there is an electrical shear stress felt by the counter-ions,

$$F_{tan} = -\sigma_d E_{tan} \qquad (6.51)$$

but none felt by the particle charge, since $E_{in} \equiv 0$.

(a) Find the potential $\Phi(r,\theta)$ for $r > R$ by solving Laplace's equation subject to the boundary condition at $z = \pm\infty$ and the *conservation-of-charge* boundary condition at $r = R + d$, where $d \ll R$. Your answer should be expressed in terms of E_0 and $\tilde{v}_\theta$ (see (6.53) below).

(b) Write a shear stress equilibrium boundary condition at the interface, including electrical and viscous shears (Figure 6.26).

Note that $\tilde{\sigma}_\theta^a$ relates to the *external flow*, (5.81), while $\tilde{\sigma}_\theta^b$ relates to the *internal flow*, (5.78). In order to use these equations, we need to know $\tilde{v}_r$ and $\tilde{v}_\theta$ on both sides of the interface. It so happens that a small drop retains its spherical shape as it "swims" (to a good approximation). We will assume this (it can be checked afterward). Therefore,

$$\tilde{v}_r^a = \tilde{v}_r^b \approx 0 \qquad (6.52)$$

$$\tilde{v}^a = \tilde{v}^b \equiv \tilde{v}_\theta \qquad (6.53)$$

(Remember that $\sigma_\theta = \tilde{\sigma}_{\theta r} \sin\theta$ and $v_\theta = \tilde{v}_\theta \sin\theta$.)

(c) Write a normal stress equilibrium boundary condition at the interface, again by using (5.78) and (5.81) (remember:

Figure 6.25 Tangential electric field boundary condition.

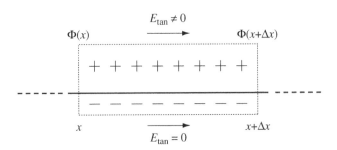

$\sigma_r = \tilde{\sigma}_{rr} 2 \cos \theta)$. In addition to the normal viscous stresses, there is another normal force that must be considered, namely, that due to surface tension as shown in Figure 6.26. Qualitatively, surface tension "pulls in" on a curved surface, tending to flatten it out:

$$\sigma_r(\text{surface tension}) = -\frac{2\gamma}{r} \qquad (6.54)$$

where γ is the surface tension with no double layer. Equation (6.54) is a well-known equation of Young and Laplace for a spherical surface [12]. Intuitively, we might guess that the presence of a double layer will change the surface tension, since both mercury and counter-ion charge repulsion in the plane of the surface will tend to "spread out" the surface (Figure 6.27):

$$\gamma^* = \gamma - \sigma_d[\Phi(R+d) - \Phi(R)] \qquad (6.55)$$

where γ^* is the surface tension of a (double layer) charged surface [13]. Although we have not derived (6.55), we reason physically that surface tension (which is a force/length $\equiv$ energy/area) is decreased by the energy associated with the electric double layer. Therefore, normal stress balance should take the form

$$\sigma_r^a - \sigma_r^b + \frac{2\sigma_d\Phi(r+d)}{R} + \underbrace{\left(-\frac{2\gamma}{R} - P^a + P^b\right)}_{=0 \text{ in equilibrium}} = 0 \qquad (6.56)$$

In (6.56), the pressures associated with the normal fluid stresses adjust so as to balance the surface tension (in the absence of the double layer). With the double layer, the third term of (6.56) represents the perturbation of the surface tension due to a Φ which varies around the surface. (Note: we have set $\Phi(R) = 0$ as a reference, since the drop is an equipotential; there is no loss of generality here.)

(**d**) The results of parts (b) and (c) constitute two equations in the *unknowns* U and $\tilde{v}_\theta$ since Φ was expressed in terms of $\tilde{v}_\theta$ and E_0 in part (a). Use these results directly to find v_θ in terms of U.

(**e**) Show that

$$U = \frac{\sigma_d E_0 R}{2\mu_w + 3\mu_{Hg} + \sigma_d^2/\sigma} \qquad (6.57)$$

and graph U versus σ_d.

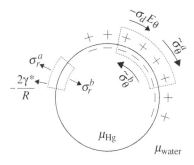

Figure 6.26 Liquid drop shear stress boundary condition.

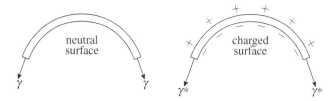

Figure 6.27 The presence of electrical charge at a surface or interface alters the surface tension according to (6.55).

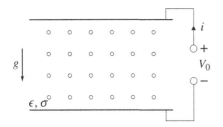

Figure 6.28 Electrical potential induced by sedimentation of falling (insulating) charged particles (e.g., globular protein macromolecules).

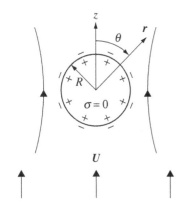

Figure 6.29 Insulating solid sphere falling at speed U.

Problem 6.8 The motion of *charged macromolecules* and *insulating particles* can give rise to macroscopic electric fields via a double-layer-mediated transduction process. In Figure 6.28, particles of volume density N are shown falling at constant velocity under the influence of gravity. This motion can give rise to a measurable "sedimentation potential," through a transduction process that is essentially the inverse of electrophoresis (Figure 6.29). The presence of such a potential can be used as yet another diagnostic tool for dealing with charged biomaterials. In addition, interaction between bioparticles and the physiological milieu may be influenced by such induced fields.

To find the measured V_0 of Figure 6.28, we focus on *one insulating particle* of radius R, whose motion can be assumed to be dominated by viscous forces (i.e., creep flow). In the frame of the particle, there is a uniform flow at $z = \pm\infty$, as shown in Figure 6.29. Fluid flow immediately adjacent to the particle causes convection of double layer counter-ions as pictured. This net displacement of counter-ions leads to macroscopic fields in the adjacent fluid. (Note that to be consistent, define σ_d as in (6.22) and (6.23).)

(a) Since there is no applied field but rather an applied flow (in the particle frame), one may proceed by first generalizing the boundary condition of (6.30). With Debye length $1/\kappa \equiv d$, show that

$$\tilde{\sigma}_\theta^d = \frac{\mu \tilde{v}_\theta^d}{d} + \frac{\epsilon \zeta}{d} \tilde{E}_\theta \qquad (6.58)$$

where $\tilde{\sigma}_\theta^d$ can be interpreted as an applied stress, or a stress due to the applied flow, even if ζ or E_θ is zero. (Refer to Figure 6.15 and extend the development.)

(b) Find the electrical potential $\Phi(r > R)$ outside the particle that results from the fluid flow. Express your answer in terms of the fluid velocity U as well as other geometrical and physical constants of interest. Assume $d \ll R$ and neglect terms of order d/R whenever possible. From your answer, define the effective *current dipole moment* of the particle, $I\Delta$ (refer to Problem 2.4).

(c) The aggregate field generated by the fall of N particles per unit volume is measured in Figure 6.28. Assume that the spacing between drops is large compared with R, so that their associated $\boldsymbol{E}$ fields are simply additive. Each particle is acted on by the force $-Mg\boldsymbol{i}_z$, where M is the mass corrected for buoyancy. We can write conservation of macroscopic current in the liquid as

$$\nabla \cdot \boldsymbol{J} = s \qquad (6.59)$$

where the average electrolyte current $\boldsymbol{J} = \sigma \boldsymbol{E}$ and s is the source density of current due to the dipole sources represented by the particles. Thus, we can define a dipole current density (analogous to the polarization density $\boldsymbol{P}$, where

$\rho_p = -\nabla \cdot \mathbf{P}$) such that

$$s = -\nabla \cdot \mathbf{J}, \quad \mathbf{J} = NI\mathbf{\Delta} \tag{6.60}$$

where N is the number of particles per unit volume. Assuming that the electrodes draw no net current, so that the net z-directed current is zero, find the macroscopic electric field between the plates in terms of the mass per particle, M.

(**d**) Find the potential difference V_0 generated by the fall of the particles between the electrodes.

(**e**) Suppose the "particles" have high enough conductivity that they can be modeled as highly conducting fluid drops with ideally polarized interfaces (i.e., modeled just as the mercury drops of Problem 6.7).

What are the normal and shear stress equilibrium relations? By comparing with the boundary condition in part (a), use physical reasoning to estimate whether the sedimentation potential should be higher or lower with ideally polarized fluid-like particles.

(**f**) In the limit for which the viscosity of the highly conducting particle (of part (e)) goes to infinity, $\mu_M \to \infty$, what is the sedimentation potential due to the fall of these "solid," conducting, polarizable particles?

Problem 6.9 Evidence has shown that ion pumps *inside* a developing cell could have an electrokinetic effect on distribution of cytoplasmic constituents of the cell [14]. Metabolic processes provide the energy for steady state current sources and current sinks within the cell [15].

Figure 6.30 is a schematic of a spherical cell whose inner cell membrane surface is charged ($+\sigma_d$), leading to the presence of a spherical double layer (counter-ion surface charge density $= -\sigma_d$). A *current dipole* source is located at the center of the spherical coordinate system. The current dipole has a moment Id, where

$$\text{moment} \stackrel{\Delta}{=} \lim_{\substack{I \to \infty \\ d \to 0}} (Id) \tag{6.61}$$

by analogy to a charge dipole source. (Assume that ion pumps generate Id.) This problem aims at modeling the extent of electroosmotic motion of cytoplasm that can exist *within* the cell, due to the current distribution *within* the cell produced by Id.

Assume the following:

(1) The cytoplasm can be modeled as a fluid of viscosity μ.

(2) The cell membrane is so much stiffer than the cytoplasm that the membrane may be modeled as a rigid, nondeformable sphere for the purposes of this problem. (This is obviously not true for many cells.)

PROBLEM

Figure 6.30 Spherical cell.

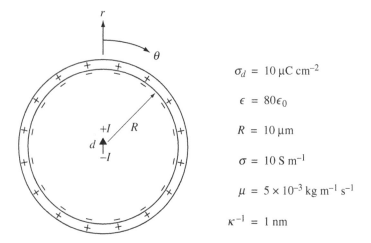

$$\sigma_d = 10\,\mu C\,cm^{-2}$$

$$\epsilon = 80\epsilon_0$$

$$R = 10\,\mu m$$

$$\sigma = 10\,S\,m^{-1}$$

$$\mu = 5 \times 10^{-3}\,kg\,m^{-1}\,s^{-1}$$

$$\kappa^{-1} = 1\,nm$$

(3) The cell membrane material is itself electrically insulating, and it supports an inner double layer. (The conditions outside the cell are *not* of interest to this problem.)

(a) Find an analytical expression for the "fluid" velocity profile $v(r,\theta) = \mathbf{i}_r v_r(r,\theta) + \mathbf{i}_\theta v_\theta(r,\theta)$ everywhere inside the cell, $r \leq (R - 1/\kappa)$, and the self-consistent electrical potential and current density everywhere inside the cell in terms of Id, σ_d, or ζ, and other appropriate parameters. State and justify any other assumptions you may wish to make. (Hint: The problem statement of Problem 2.4 should act as a guide to finding the form of the electrical potential attributable to a dipole current source located at the origin of a conducting sphere. Note, however, that Problem 2.4 does *not* deal with the additional complexity of the effect of fluid motion on the solution of the electrical potential.)

(b) Sketch the velocity profile and the current density inside the cell.

(c) Calculate the magnitude of J_θ at $(r = R,\ \theta = 90°)$ that will produce a θ-directed velocity at $(r = R,\ \theta = 90°)$ having the value $R/100\,s = 10^{-7}\,m\,s^{-1}$.

Problem 6.10 Impedance measurements have often been used to probe the structure and composition of cell surfaces. This problem deals with a suspension of osmotically swollen cells. You are to derive the phenomenological coupling representation (refer to Sections 5.8, 6.3, and 6.5 for definitions and nomenclature)

$$\begin{bmatrix} U \\ J \end{bmatrix} \begin{bmatrix} k_{11} & k_{12} \\ k_{21} & k_{22} \end{bmatrix} \begin{bmatrix} dP/dz \\ dV/dz \end{bmatrix} \tag{6.62}$$

for a *matrix* of cells placed between electrodes. In addition to the impedance seen by the electrodes (related to k_{22}), you are to derive the electrokinetic coefficients k_{12} and k_{21} as well as the hydraulic permeability of the suspension, k_{11}.

Your model should constitute a "microcontinuum" representation of electrokinetics. That is, the k_{ij} should be derived in

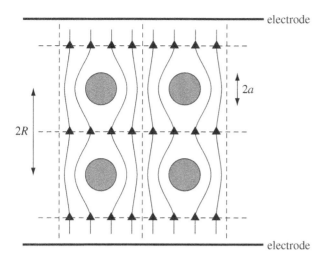

electrode

electrode

$2a$

$2R$

Figure 6.31 Depiction of unit cells, each defined as containing a spherically swollen cell in hypotonic saline.

terms of structural, compositional, and material parameters associated with the cell and its surrounding fluid.

Assume the following (see Figure 6.31):

(1) Identical cells have radius a and center-to-center separation distance $2R$. (therefore, the unit cell has sides of length $= 2R$).

(2) The cell surface charge has value σ_d.

(3) The hypotonic saline has viscosity μ, bulk conductivity σ, and dielectric constant ϵ.

(4) $1/\kappa \equiv d \ll a$; the slip plane thickness $\delta \ll 1/\kappa$; and all electromechanical coupling is assumed to occur at $r = a + d$ for each sphere.

(5) The cell separation distance $2R$ is large enough that cells can be considered noninteracting with regard to electrical potential profile as well as fluid flow. Therefore, each unit cell of Figure 6.31 can be thought of as a cell in an "infinite medium." Fluid velocity and electric field can be assumed *uniform* at the boundaries of each unit cell.

(6) The pressure drop within the double layer is negligible when using the Stokes equation within the double layer.

(7) Neglect the surface conductivity of the cell.

(a) Summary of the exterior flow problem around a single sphere:

(i) Show that the Stokes equation can be integrated across the double layer to give the following relation between the shear stress σ_θ^d at the outer edge of the Helmholtz double layer, the change in velocity across the double layer $(v_\theta^d - v_\theta^a = v_\theta^d)$, and the tangential field E_θ at the double layer that results from an applied field *or* applied fluid flow:

$$\tilde{\sigma}_\theta^d = \frac{\mu}{d}v_\theta^d + \frac{\epsilon\zeta}{d}E_\theta \qquad (6.63)$$

Assume $d \ll a$.

(ii) Write an expression for $\tilde{\sigma}_\theta^d$ in terms of U and $\tilde{v}_\theta^d$ associated with the problem of bulk flow exterior to a sphere. This must be continuous with the shear stress at the edge of the double layer.

(iii) Write a conservation-of-charge equation at $r = a + d$ in terms of $\tilde{\Phi}$ and $\tilde{v}_\theta^d$.

(b) Find k_{11} by assuming that the pressure gradient dP/dz across a unit cell is the force density required to keep the fluid moving at speed U:

$$k_{11} = \left.\frac{U}{dP/dz}\right|_{dV/dz=0}$$

This force density can be related to the drag force exerted by the cell on the fluid flowing within the unit cell.

To find the drag force, note that even though $dV/dz = 0$ (i.e., there is no net uniform field at infinity), there can still be a local dipole potential induced by fluid flow around each sphere. *Include* the effect of this E_0 on the drag force in your derivation. Does it turn out to be a significant effect?

(c) Find k_{12}, the ratio of flow velocity to applied field such that $dP/dz = 0$:

$$k_{12} = \left.\frac{U}{dV/dz}\right|_{dP/dz=0}$$

By examining one cell and the condition $dP/dz = 0$, require that the net drag force on each cell is equal to zero.

(d) Find k_{21}, the ratio of the short-circuit current J $(dV/dz = 0)$ to the pressure gradient dP/dz across the unit cell:

$$k_{21} = \left.\frac{J}{dP/dz}\right|_{dV/dz=0}$$

The approach to part (d) is as follows. First find the short-circuit current J due to an imposed U (flow) through a matrix of cells having number density $N = (2R)^{-3}$. Second, find the pressure gradient due to an imposed U with $dV/dz = 0$ (as in part (b)). The expression for k_{21} then follows. To find J, assume that the local dipole potential around a cell can be modeled as a current dipole consisting of a point current source I and a sink of equal magnitude spaced a distance Δ apart (see Problem 2.4, parts (a) and (b)). Conservation of macroscopic current in the liquid can be written as

$$\nabla \cdot \boldsymbol{J}_u = s = -\nabla \cdot \mathcal{J} \tag{6.64}$$

where $\mathcal{J}$ is the dipole moment density, with magnitude $NI\Delta$, and $\boldsymbol{J}_u$ is the free current density, with

$$\nabla \cdot \boldsymbol{J} = \nabla \cdot (\boldsymbol{J}_u + \mathcal{J}) = 0 \tag{6.65}$$

in steady state (note the analogy to polarization, where $-\nabla \cdot \mathcal{J} = s$ is analogous to $-\nabla \cdot \boldsymbol{P} = \rho_p$). Therefore, since $\boldsymbol{J}_u = 0$ in the liquid ($dV/dz = 0$), you should find an expression for J in terms of geometrical parameters R and a, and other relevant parameters.

Does $k_{21} = k_{12}$?

(**e**) Find the effective conductivity k_{22}:

$$k_{22} = \left.\frac{J}{dV/dz}\right|_{dP/dz=0}$$

The medium between the plates can be represented by an effective conductivity σ'. Hence, $J = \sigma' E_0 = -\sigma\, dV/dz$, and therefore $k_{22} = -\sigma'$. To find σ', use the *method* of Maxwell [16] (see Problem 2.1). The difference is that Maxwell considered little conducting spheres; we have insulating spheres (cells) with an electrokinetic boundary condition at the surface.

(**f**) The k_{ij} of parts (b)–(e) are valid only in the limit $a \ll R$. Why? Suppose the number density of cells was such that $R \sim 3a$. The fluid mechanics problem must now be resolved, since boundary conditions are no longer applicable at $r \to \infty$, but rather at the edge of a finite size unit cell. An approximate solution can be formulated based on a new unit cell composed of the cell inside a concentric spherical shell of radius R. Describe an appropriate set of fluid mechanical boundary conditions at $r = a$ and $r = R$ (this set may not be unique). Carefully outline a method of solution to the axisymmetric fluid flow problem (do not solve).

Problem 6.11 Electrophoresis of a Spherical Particle Including Surface Conductivity A spherical particle of radius R and surface charge density σ_d is immersed in a solution of NaCl having bulk viscosity μ, dielectric constant ϵ, and bulk conductivity σ (with ionic mobilities u_+ and u_-). Assume that the double layer surrounding the particle can be approximated by a Helmholtz model (ζ, σ_d, ϵ, $d = 1/\kappa$), where $d \ll R$. An applied field E_0 results in particle migration in the direction of the field at speed U.

(**a**) Find an expression for the electrophoretic mobility U/E_o including the effect of an *Ohmic* surface conductivity corresponding to the enhanced ion concentration in the double layer. (Use a model of ohmic surface conductivity appropriate to the Helmholtz model of the double layer).

(**b**) Sketch a graph of U/E_0 versus $\epsilon\zeta/\mu$ for the case $u > \epsilon\zeta/\mu$, where u is an appropriate ionic mobility. Identify and describe the physical processes associated with each region of the graph.

6.7 REFERENCES

[1] Wall S (2010) The history of electrokinetic phenomena. *Curr. Opin. Colloid Interf. Sci.* **15**, 119–124.

[2] Levich V (1962) *Physicochemical Hydrodynamics.* Prentice Hall, Englewood Cliffs, NJ.

[3] Katchalsky A and Curran PF (1965) *Nonequilibrium Thermodynamics in Biophysics.* Harvard University Press, Cambridge, MA.

[4] Eisenberg SR and Grodzinsky AJ (1988) Electrokinetic micromodel of extracellular matrix and other polyelectrolyte networks. *Physicochem. Hydrodyn.* **10**, 517–539.

[5] Chammas P, Federspiel WJ, and Eisenberg SR (1994) A microcontinuum model of electrokinetic coupling in the extracellular matrix: perturbation formulation and solution. *J. Colloid Interf. Sci.* **168**, 526–538.

[6] Reuss FF (1809) *Mémoires de la Société Impériale des Naturalistes de Moscou* **2**, 327.

[7] Tiselius A (1937) A new apparatus for electrophoretic analysis of colloidal mixtures. *Trans. Faraday Soc.* **33**, 524–541.

[8] van den Heuvel MGL, de Graaff MP, Lemay SG, and Dekker C (2007) Electrophoresis of individual microtubules in microchannels. *Proc. Natl Acad. Sci. USA* **104**, 7770–7775.

[9] Henry DC (1931) The cataphoresis of suspended particles, Part I. The equation of cataphoresis. *Proc. R. Soc. Lond.* **A133**, 106–129.

[10] Boothe F (1950) The cataphoresis of spherical, solid non-conducting particles in a symmetrical electrolyte. *Proc. R. Soc. Lond.* **A203**, 514–533.

[11] Eylar EH, Madoff MA, Brody OV, and Oncley JL (1962) The contribution of sialic acid to the surface charge of the erythrocyte. *J. Biol. Chem.* **237**, 1992–2000.

[12] Adamson AW (1967) *Physical Chemistry of Surfaces*, 2nd ed. Wiley-Interscience, New York, Chapter 1, pp. 1–52.

[13] Bockris JO'M and Reddy AKN (1970) *Modern Electrochemistry*, Volume 2. Plenum Press, New York, Chapter 7, pp. 718–790.

[14] Woodruff R and Telfer WH (1980) Electrophoresis of proteins in intercellular bridges. *Nature* **286**, 84–86.

[15] Jaffe LE (1979) Control of development by ionic currents. In *Membrane Transduction Mechanisms* (Cone RA and Dowling JE, eds). Raven Press, New York, pp. 199–231.

[16] Maxwell JC (1954) *Treatise on Electricity and Magnetism*, 2 Volumes. Dover Publications, New York, (originally published 1873), p. 440.

Rheology of Biological Tissues and Polymeric Biomaterials

7.1 INTRODUCTION

The deformation and swelling behavior of biological tissues and biomaterials requires an interdisciplinary approach that has seen revolutionary changes during the past decades. While research 20–50 years ago was centered primarily at the tissue (macro) scale, the emergence of new technologies in cell and molecular biophysics, biology, and engineering has led to rapid progress in studies of transductive coupling at cellular (micro) and molecular (nano) scales. In this chapter, we highlight specific advances in the understanding of tissue rheology motivated by our focus on interactions between chemical, electrical, and mechanical forces and flows. These interactions occur at the molecular level, but profoundly influence tissue level behavior.

Experimental observations of equilibrium and transient swelling behavior of connective tissues and hydrogels (Section 7.2) provide case studies of the underlying physicochemical and electromechanical processes. Important theoretical modeling issues immediately arise. Can the observed time dependence of swelling and deformation be described adequately by the macroscopic continuum laws of linear elasticity (Section 7.3) combined with solid phase viscoelasticity (Section 7.4)? How are electrical forces and flows to be incorporated into the governing equations? Are the observed rates of deformation limited primarily by mechanical, chemical, or electrical processes, or by a combination of these processes? What are the rate-limiting phenomena at the molecular level that control macroscopic deformation and energy dissipation in hydrated porous biomaterials?

While theories of solid phase viscoelasticity may describe a range of temporal processes, theories of poroelasticity (Section 7.5) involving the effects of fluid flow within molecular networks may also be needed to fully account for time-dependent dissipative behavior of tissues, hydrogels, and even cytoskeletal networks. Finally, in Section 7.6, we focus on phenomena involving the combined poroelastic and electrokinetic behavior of charged tissues, gels, and molecular networks as a means of integrating the multidisciplinary concepts discussed in all the preceding chapters.

7.2 SWELLING AND DEFORMATIONAL BEHAVIOR OF TISSUES: ILLUSTRATIVE EXAMPLES

An understanding of the *kinetics* of swelling and deformation of biological tissues is key to delineating mechanisms and rate-limiting processes that govern rheological behavior. We briefly highlight the swelling properties of selected connective tissues since their extracellular matrix constituents are also found in epithelial, nerve, and muscle tissue, i.e., all four anatomically defined tissue types. We then focus on nonequilibrium swelling processes in which diffusion, diffusion–reaction, electrostatic interactions, and intra-tissue fluid flow, as well as the intrinsic elastic and viscoelastic properties of the tissue matrix, combine to govern swelling kinetics. Natural and synthetic hydrogels provide additional applications of experimental and theoretical techniques to characterize swelling properties relevant to a wide range of biological tissues and macromolecular networks.

7.2.1 Swelling Behavior of Connective Tissues

Proteoglycans and their ionized glycosaminoglycan (GAG) constituents (Section 1.4), along with other proteins and glycoproteins within the extracellular matrix (ECM), are primarily responsible for the swelling pressures of connective tissues under physiological conditions. This swelling pressure maintains the hydration of the ECM and is crucial to the tissue's ability to withstand deformation *in vivo*. The total swelling stress is composed of several components:

(1) The stretching of macromolecular chains of the interconnected solid matrix opposes swelling, and its magnitude increases as swelling increases.

(2) Repulsive electrostatic interactions result from nanoscale double layer repulsive forces between charged molecules of the tissue (Chapter 4) or, alternatively, as a macro-scale thermodynamic (Donnan) osmotic pressure resulting from the increased concentration of counter-ions that must be present in the tissue to preserve electroneutrality (Chapter 3). These two interpretations of electrostatic interactions are closely related but are not identical in magnitude, and have different spatial distributions when viewed at the molecular versus macro-scale [1]. The magnitude of the electrostatic swelling stress decreases with increased swelling, increased external salt concentration, and decreased fixed charge density.

(3) Interactions between the solvent and matrix associated with the affinity between the two can also induce swelling. For "good" solvents, the matrix tends to imbibe fluid and increase swelling, while "poor" solvents decrease swelling.

(4) Thermal motion of the matrix macromolecular segments, often referred to as polymer-excluded volume effects, tends to swell the network; these effects decrease as swelling increases.

In many tissues, there are distinct constituents that are separately responsible for these different components of the total swelling stress. For example, in articular cartilage at physiological pH, it is accepted that the repulsive interactions between the charge groups of aggrecan glycosaminoglycans (GAGs) (Chapter 4) provide the positive swelling stress, while the stretched collagen fibrillar network constrains this swelling and maintains tissue integrity. It is not surprising, therefore, that changes in chemical environment such as pH and ionic strength can significantly alter the swelling state of connective tissues such as corneal stroma, cartilage, tendon, and the intervertebral disk (Figure 7.1). The collagen network architecture is strikingly different in these tissues as well, with the highly oriented fibrils of the individual lamellae of corneal stroma and axially oriented collagen of tendon, the hoop-like orientation in the annulus of the disk, and the more random orientation of fibrils within cartilage middle zone. Proteoglycan concentrations are higher in cartilaginous tissues than most other tissues: aggrecan accounts for about 5% of the net wet

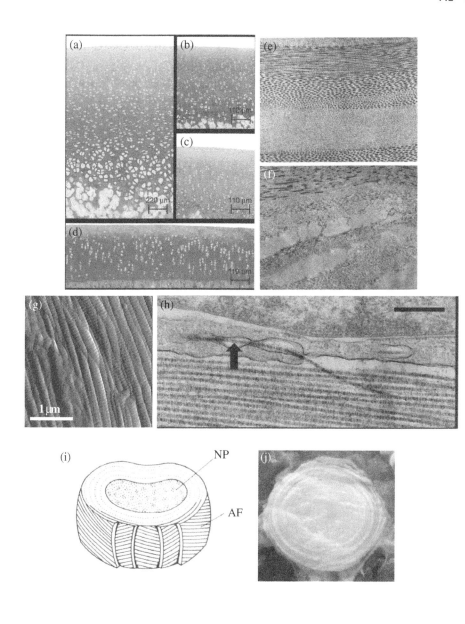

Figure 7.1 (a–d) Full-thickness images of rabbit articular cartilage during development from 1 month (a), 2 months (b), 3 months (c), and 8 months (d) after birth, showing decreased thickness with age and development of depth-dependent cell and tissue morphology. (e) Normal and (f) pathologically swollen corneal stroma showing layered architecture of approximately 2 μm-thick lamellae within which collagen fibrils are highly oriented. (g) Tapping-mode AFM image of rat tail tendon collagen fibrils after enzymatic removal of proteoglycans in ambient conditions. (h) 28 nm-diameter mouse tendon fibrils deposited by fibripositors (arrow) parallel to the axis of a newly forming tendon during development (bar = 500 nm). (i) Schematic of the collagen architecture of the intervertebral disk showing the nucleus pulposus (NP) and annulus fibrosis (AF). (j) Image of the cross section of a disk. ((a–d) from Hunziker EB, Kapfinger E, and Geis J. The structural architecture of adult mammalian cartilage evolves by a synchronized process of tissue resorption and neoformation during postnatal development. *Osteoarthritis Cartilage* **15**, 403–413 (2007); (e,f) from Miller D and Benedek G. *Intraocular Light Scattering*. Charles C Thomas, Springfield, IL, 1973, p. 20; (g) courtesy of Dr Lin Han; (h) from Canty EG, Lu Y, Meadows RS, et al. Coalignment of plasma membrane channels and protrusions (fibripositors) specifies the parallelism of tendon. *J. Cell Biol.* **165**, 553–563 (2004); (j) from Center for Mechanics, J Helfenstein, ETH, Zurich.)

weight of human femoral head cartilage and the nucleus of the intervertebral disk, compared with about 0.1–1% in tendon and the loose connective tissue of umbilical cord (Wharton's jelly). Therefore, cartilaginous tissue would be expected to have a higher osmotic swelling pressure. However, when excised specimens of adult human articular cartilage, disk, and umbilical cord are placed in identical physiological saline, cartilage swells by a few percent, the disk may swell by 100% (annulus fibrosus) to 250% (nucleus pulposus), and native umbilical cord strips may swell 50–100% beyond their *in vivo* wet weight. Thus, the content of fixed-charge groups alone is only one of the factors that control tissue swelling. The size and arrangement of collagen fibrils in the extracellular matrix [2], the content and type of chemical crosslinkages, mechanical entanglements, and other factors all play a role in determining the swollen volume of a particular tissue. *In vitro* swelling experiments are often performed under physiological as well as nonphysiological conditions to probe the role of individual molecular constituents in the tissue's overall structure and rheological behavior.

Loose Irregular Connective Tissues

Wharton's jelly, named for the English anatomist Thomas Wharton (1614–1673), is a gelatinous mucus-like tissue within the umbilical cord, consisting mainly of chondroitin sulfate and hyaluronan GAG chains synthesized by resident fibroblasts. This tissue functions to protect and insulate umbilical cord blood vessels. When exposed to temperature changes, Wharton's jelly collapses structures within the umbilical cord and thus provides a physiological clamping of the cord within about five minutes after childbirth. Investigators long ago observed that swelling of umbilical cord tissue depended on bath pH and salt concentration. Analog synthetic gels of collagen and hyaluronic acid were made and mechanically characterized by ultracentrifugation and viscometry. From these experiments, it was postulated that a mixture of collagen fibers, polysaccharides, and water could form a structure that would withstand compression by hindering water flow. Umbilical cord has been used as a model for studying properties of loose connective tissues, which constitute the predominant part of the extravascular space of the microcirculation. This tissue has also gained interest as a source of adult stem cells.

More recently, Espinosa et al. [3] studied the swelling of cervical mucus granules, an important process that affects human reproduction. Cervical cells secrete mucin glycoprotein granules that swell to become a hydrogel after exocytosis. The extent of swelling is dramatically changed by alterations in local pH and calcium ion concentration, which can vary by orders of magnitude within the cervix, *in vivo*, resulting in marked changes in the rheological properties of cervical mucus pre- and post-ovulation. Espinosa et al. quantified the kinetics of swelling using an experimental procedure developed by Tanaka and Fillmore [4] to study the contributions of gel-network elasticity and intra-gel fluid flow (see Section 7.5). It was determined that the characteristic time constant for the swelling of a spherical mucin granule increased linearly with the square of the gel radius, a hallmark of poroelastic behavior based on a "gel diffusivity" equal to the product of the elastic modulus and fluid permeability of the gel. Mucin gels secreted by cells of the gastrointestinal and esophageal tracts also exhibit marked swelling regulated by local pH, owing to the high concentration of O-linked and N-linked oligosaccharides (Figure 7.2).

Polyelectrolyte Hydrogels

Hydrogels made from natural or synthetic polymeric networks have received great attention for decades as materials for drug delivery, as scaffolds for tissue engineering, and in a wide variety of medical and biological applications. These gels are also used as model systems for tissue swelling behavior. Of particular relevance here is the class of polyelectrolyte hydrogels made from crosslinked or entangled macromolecular networks containing ionizable fixed-charge groups (positive, negative, or both). Swelling and shrinking of polyelectrolyte gels can be induced by modulating the fixed-charge density or by screening electrostatic and osmotic interactions within the network by controlling electrolyte pH or ionic strength. As a result, the effective pore size of the gel can be modified and active control of gel permeability to proteins, drugs, and other solutes can be achieved. An entire field has emerged

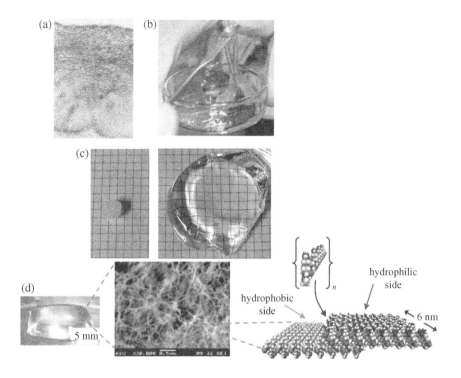

Figure 7.2 (a) Normal human esophageal mucosa, cryostat-sectioned and stained with PAS/Alcian blue. (b) Photomicrograph of gel formed from purified pig gastrointestinal mucins. (c) Polyelectrolyte gels bearing tetraalkylammonium tetraphenylborate as a lipophilic and bulky ionic group swell up to 500 times their dry size, with applications as absorbents for waste oil and spills in the environment. (d) Self-assembling peptide hydrogel made from peptide monomers having the sequence Ac–(RADA)$_4$–CONH$_2$, which first assemble into peptide nanofibers on addition of salt and then further assemble into a hydrogel for use in tissue engineering and drug delivery. ((a) from Dixon J, Strugala V, Griffin SM, et al. Esophageal mucin: an adherent mucus gel barrier is absent in the normal esophagus but present in columnar-lined Barrett's esophagus. *Am. J. Gastroenterol.* **96**, 2575–2583 (2001); (b) courtesy of Professor Katharina Ribbeck; (c) from Ono T, Sugimoto T, Shinkai S, and Sada K. Lipophilic polyelectrolyte gels as super-absorbent polymers for nonpolar organic solvents. *Nature Mater.* **6**, 429–433 (2007); (d) from Koutsopoulos S, Unsworth LD, Nagai Y, and Zhang S. Controlled release of functional proteins through designer self-assembling peptide nanofiber hydrogel scaffold. *Proc. Natl Acad. Sci. USA* **106**, 4623–4628, 2009.)

in this regard, focused on chemical, electrical, pH, and temperature-controlled changes in "smart" gel rheology and permeability to fluids and solutes.

Corneal Stroma: Regular Dense Connective Tissue

The central 90% of the cornea (the corneal stroma) is an approximately 0.5 mm-thick lamellar structure of dense regular connective tissue. Each approximately 2 μm-thick lamella (Figure 7.1(e)) contains a highly oriented array of collagen fibrils; the fibril orientation of each successive lamella is staggered at various angles. This ultrastructure gives the cornea its classification as a "regular dense connective tissue." The mechanism governing the arrangement and spacing of the fibrils is still under investigation, but it is thought that members of the family of small leucine-rich proteoglycans (Section 1.4) may play a major role as they bind to outer surfaces of the fibrils. Thus, collagen fibril size and the regularity of spacing are essential to corneal transparency. Under certain pathological conditions, the cornea will swell irregularly (edema) and the presence and size of the water pools between fibrils produces opacification (Figure 7.1(f)). The swelling of the stroma has been linked to the Donnan osmotic swelling pressure of the small proteoglycans and their constituent chondroitin and dermatan sulfate GAG chains. It has been hypothesized that this swelling tendency is dynamically balanced *in vivo* by the cornea's endothelial cell layers, which actively transport water out of the cornea. These phenomena have led to detailed studies of the physicochemical and electromechanical properties of the stroma and the biological, metabolic, and electrophysiological properties of the cornea's outer linings.

Tendon: Regular Oriented Connective Tissue

Tendons and ligaments are examples of "regular, ordinary, oriented connective tissues" in the same family as corneal stroma. They form the connections from bone to muscle and from bone to bone, respectively. These tissues have extraordinary tensile strength due to the high proportion of oriented collagen fibers. Tendon fibroblasts are responsible for synthesis, secretion, and assembly of collagen into oriented fibrils in health and disease, and enable injured or severed tendons to regenerate under certain conditions, a property not shared by all connective tissues. (For example, the chondrocytes of cartilage cannot offset the degradative process that leads to osteoarthritis in synovial joints.) *In vitro* swelling experiments under nonphysiological conditions have led to important discoveries concerning tendon ultrastructure and crosslinking. Glimcher et al. [5] found that insoluble Achilles tendon collagen swelled markedly to an almost gel-like state when exposed to mild acetic acid solution. However, decalcified chicken bone collagen (also insoluble) failed to swell at all. This behavior was ascribed to the differences in the number, chemical nature, distribution, and stability of intra- and intermolecular crosslinkages in these two collagens. Tendon collagen fibrils are thought to be assembled and maintained extracellularly following post-translational modifications of secreted procollagen molecules. However, recent discoveries suggest that during embryonic development, procollagen processing and collagen fibrillogenesis are initiated in Golgi-to-plasma membrane carriers, and that 28 nm-diameter fibrils are released by "fibripositors" parallel to the tendon axis and projected into parallel channels between cells (Figure 7.1(h)) [6].

Cartilages: Specialized Dense Connective Tissue

Articular cartilage, a specialized, dense skeletal connective tissue, is the whitish covering of the bony ends of synovial joints that functions as a bearing material. It is composed of aqueous electrolyte (about 80%) and cells that synthesize an extracellular matrix of collagens, proteoglycans, and glycoproteins (Figure 7.1(a–d)). The relatively high content of aggrecan proteoglycans in normal articular cartilage gives this tissue a large swelling pressure due to electrostatic repulsive interactions. Cartilage is very resilient to deformation and can resist tensile, compressive, and shear forces. In humans, the peak forces across the hip and knee joints may reach several times body weight. Cartilage contact deformations in the knee and ankle can quickly reach as high as 10–40% under dynamic loading conditions *in vivo* [7]. Osteoarthritis can lead to degradation of cartilage, resulting in its eventual destruction and joint dysfunction. During the past decades, many investigators have focused on the relation between the biomechanical, cell biological, and matrix biochemical events that are associated with this degenerative process. In particular, the role of the proteoglycan, glycoprotein, and collagen components and the interactions between these components in normal and pathological cartilage have received much attention. In adult articular cartilage, the denser network of collagen fibrils is prestressed *in vivo* by the swelling pressure of the aggrecan proteoglycans held within the network. When specimens of normal cartilage and osteoarthritic cartilage were placed in physiological saline, the degenerated cartilage swelled easily while the normal tissue did not, even though the degenerated cartilage had significantly less aggrecan charge groups. This was interpreted as the result of weakening of the tensile restraining forces of the collagen network of degenerated cartilage [2].

7.2.2 Nonequilibrium Swelling Measurements Reveal Governing Kinetics

While equilibrium swelling experiments can illuminate the ultrastructure of biological tissues and biomaterials, *nonequilibrium* experiments and associated theoretical models are essential to understanding the *mechanism(s) of swelling* and the chemical, mechanical, and electrical rate processes that regulate swelling. Nonequilibrium swelling behavior has been studied extensively in many tissues and synthetic hydrogels, including cornea, tendon, cartilage, muscle, intervertebral disk, epithelium, blood vessel walls, mucin granules, and a variety of synthetic gels.

Experimental techniques have included measurements of changes in tissue thickness or volume caused by gradients in chemical concentration, mechanical pressure, or electrical potential, and measurements of changes in swelling forces at constant compressive or tensile strain (e.g., isometric compression or tension). When tissue or gel bath composition (pH, salt concentration) is altered, several nonequilibrium rate processes may occur simultaneously. The kinetics of each rate process can be approximated by a characteristic time constant associated with a linearized model of the process of interest. Important rate processes and their associated time constants include the following:

(1) **Diffusion of mobile ions within the matrix (Chapter 1):** $\tau_{\text{diff}} \sim L^2/D_i$. Chemical diffusion kinetics can be characterized by a time constant proportional to the square of a characteristic specimen dimension L and inversely proportional to ionic diffusivity D_i.

(2) **Diffusion-limited binding of ions or other solutes to matrix macromolecular sites (Chapter 1):** $\tau_{\text{dr}} \equiv \tau_{\text{diff}}(1+\beta)$. Diffusion-limited chemical reactions can significantly impede the transport of mobile species into the matrix, as represented by the diffusion–reaction parameter β, which is generally a function of the binding site density and binding rate constants (e.g., $\beta = n/K_d$ in Example 1.6.2).

(3) **Electrodiffusion (Chapter 3):** $\tau_{\text{ed}} \equiv \tau_{\text{diff}}[1 + (E_0 L/2\pi V_T)^2]^{-1}$. When an electric field E_0 is applied across a tissue or a membrane of thickness L, τ_{ed} is the time needed for the establishment of a new equilibrium profile of mobile ion concentration within the tissue. The new concentration profile represents a competition between ion migration caused by the applied voltage drop $E_0 L$ and ion diffusion represented by the "thermal voltage" $V_T \equiv RT/F$ (assuming that no binding reactions occur). If the ionic solute can additionally bind to matrix sites, the electrodiffusion–reaction time constant takes the form $\tau_{\text{edr}} \equiv \tau_{\text{ed}}(1+\beta)$.

(4) **Readjustment of local (double layer) electric fields and forces (Chapters 2 and 4):** $\tau_{\text{ch.rel.}} \sim \epsilon/\sigma$. Readjustment of local electric fields within the matrix, after an instantaneous change in the ionic content of the interstitial fluid, is proportional to the ratio of the interstitial fluid's permittivity ϵ to its conductivity σ [8]. $\tau_{\text{ch.rel.}}$ varies from 10^{-6}–10^{-5} s in bone, with its relatively low fluid content, to 10^{-9} s in tendon, cartilage, blood vessels, and cell cytoskeleton. Charge relaxation times are so short that this process

will never be rate-limiting for swelling of macroscopic tissues having dimensions L much greater than a Debye length (about 1 nm under physiological conditions).

(5) **Mechanical readjustment (swelling) of the tissue matrix (Chapter 5 and below):** $\tau_{matrix} \sim L^2/Hk$. This process involves elastic reconfiguration of the matrix molecules (characterized by an elastic modulus H) simultaneously with relative fluid flow into or out of the matrix (characterized by the matrix hydraulic permeability k). Neglecting solid matrix-associated viscoelastic processes, the mechanical swelling of polymer gels [4] and many biological tissues has been described in the small-strain limit by a poroelastic time constant inversely proportional to the hydraulic permeability k and the elastic modulus of the matrix. When the intrinsic viscoelastic behavior of the solid matrix is important, yet another time constant would be appropriate for these poro-viscoelastic processes.

A comparison of the magnitudes of the above time constants can give valuable insights into the *mechanisms* that govern swelling in different tissues and porous hydrated biomaterials. Based on a review of numerous published experimental and theoretical studies, several conclusions can be drawn. In relatively high-modulus soft tissues such as tendon and cartilage, it was found that $\tau_{matrix} \ll \tau_{diff}$. That is, electromechanical and osmotic swelling (deswelling) induced by changes in bath neutral salt concentration occur at least as rapidly as chemical diffusion. Thus, ionic diffusion appears to be the rate-limiting mechanism for swelling in such experiments. In much softer tissues and hydrogels having a lower modulus (e.g., the nucleus pulposus of the disk and gel scaffolds for tissue engineering) and/or lower fluid hydraulic permeability, measurements have shown that $Hk \ll D_i$ ($\tau_{diff} \ll \tau_{matrix}$). For such materials, changes in external bath concentration will chemically equilibrate with the interstitial fluid much faster than subsequent mechanical swelling (deswelling). In these cases, the swelling kinetics could provide a direct measure of the product of the tissue's elastic modulus and hydraulic permeability, as long as matrix viscoelasticity is not important. Conversely, for tissues in which $\tau_{matrix} \ll \tau_{diff}$, free swelling kinetics are dominated by chemical processes (i.e., the slowest, rate-limiting process). In such cases, the measured swelling kinetics may be invaluable as a quantitative, nondestructive measure of certain biochemical properties, such as diffusion-limited binding (τ_{dr}) of Ca^{2+}, H^+, or other ions to tissues and gels. The kinetics of swelling brought about by electrodiffusion-induced changes in intra-tissue salt concentration should be governed by τ_{ed} as long as $\tau_{matrix} \ll \tau_{ed}$.

7.3 EQUILIBRIUM ELASTIC BEHAVIOR

The complex ultrastructure and biochemical composition of biological tissues is directly reflected in their biomechanical behavior. For example, the connective tissues described in Section 7.2 are generally inhomogeneous and anisotropic, and their deformational behavior can be markedly nonlinear. Nevertheless, these tissues can exhibit near-elastic behavior in equilibrium within limited regimes of strain. Thus, the approach embodied in the texts of Fung [9, 10] and others to first model the equilibrium elastic behavior of tissues at the macro-level and to compare these models with experimental evidence is the starting point for our treatment. This approach sets the stage for further investigation of the coupling of chemical, electrical, and mechanical interactions at the micro (cell) and nano (molecular) scales.

In Section 7.3.1, we review definitions and our notation for the stress and strain tensors for an elastic medium. Then, in Section 7.3.2, the generalized Hooke's constitutive laws relating stress to strain are summarized in the context of experimental configurations that have been used to measure tissue-level equilibrium elastic moduli that can describe biomechanical behavior under certain limiting conditions. Problems 7.1–7.5 at the end of the chapter provide additional examples relating experimental observations to models of equilibrium elastic behavior.

7.3.1 Stress and Strain Tensors for an Elastic Medium

We can now take full advantage of previous sections in this text. General relationships between force, stress, and force density have been discussed in detail in Section 4.2, in which the traction force vector acting on an arbitrary surface is represented by the components of a stress tensor (Figures 4.2 and 4.3). In addition, the force density acting on a material is then represented as the divergence of the stress tensor of interest (equation (4.8)). For forces of electrical origin, the Maxwell stress tensor is derived (Section 4.3), incorporating forces associated with mobile charge carriers as well as fixed or induced dipoles associated with electrical polarization forces. In Chapter 5, viscous forces are defined in terms of a fluid viscous stress tensor; the flow-associated strain rate tensor is derived (equation (5.27)), and constitutive relations between viscous stress and fluid strain rate are described. Experimental configurations for measuring the coefficients of viscosity are then delineated. By analogy to these sections of Chapter 5, the elastic stress and strain tensors and constitutive laws relating them are now summarized.

Stress in a Continuous Elastic Medium

With reference to Figure 4.2, the stress tensor σ_{ij} is represented as an array of nine stress components:

$$\sigma_{ij} = \begin{bmatrix} \sigma_{11} & \sigma_{12} & \sigma_{13} \\ \sigma_{21} & \sigma_{22} & \sigma_{23} \\ \sigma_{31} & \sigma_{32} & \sigma_{33} \end{bmatrix} \tag{7.1}$$

where i is the direction of the normal to the plane on which the stress acts, j is the direction of the stress, $i = j$ correspond to normal stresses, and $i \neq j$ are shear stress components. In the absence of body torques in equilibrium, the net torque is zero and the stress tensor is thereby symmetric: $\sigma_{ij} = \sigma_{ji}$.

Strain in an Elastic Continuum

To examine the types of solid deformations that can give rise to elastic stresses, we focus on material elements initially located at the undeformed positions $P(x_1, x_2, x_3)$ and $Q = (x_1 + \Delta x_1, x_2 + \Delta x_2, x_3 + \Delta x_3)$ in Figure 7.3. Upon deformation of the material, these elements are located, respectively, at the positions $(x_1 + u_1, x_2 + u_2, x_3 + u_3)$ and $(x_1 + \Delta x_1 + u_1 + \Delta u_1, x_2 + \Delta x_2 + u_2 + \Delta u_2, x_3 + \Delta x_3 + u_3 + \Delta u_3)$. The Δx_i and Δu_i are infinitesimal in length, and the Δu_i represent relative deformations between the two deformed positions. Writing the 1-component

Figure 7.3 Solid deformations resulting in elastic stress.

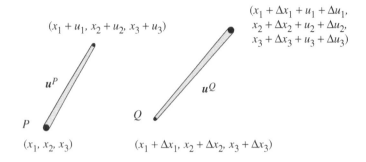

of u^Q gives

$$u_1^Q = u_1^P + \left(\frac{\partial u_1}{\partial x_1} \Delta x_1 + \frac{\partial u_1}{\partial x_2} \Delta x_2 + \frac{\partial u_1}{\partial x_3} \Delta x_3 \right) \tag{7.2}$$

where the term in parentheses on the right-hand side of (7.2) corresponds to Δu_1. More generally, using index notation to summarize all three components,

$$u_i^Q = u_i^P + \frac{\partial u_i}{\partial x_j} \Delta x_j$$

$$= \underbrace{u_i^P}_{\text{TRANSLATION}} + \underbrace{\frac{1}{2} \left(\frac{\partial u_i}{\partial x_j} - \frac{\partial u_j}{\partial x_i} \right) \Delta x_j}_{\text{ROTATION}} + \underbrace{\frac{1}{2} \left(\frac{\partial u_i}{\partial x_j} + \frac{\partial u_j}{\partial x_i} \right) \Delta x_j}_{\text{DEFORMATION}} \tag{7.3}$$

By analogy with (5.26) and Figure 5.2, the first two terms on the right-hand side of (7.3) correspond to rigid body translation and rigid body rotation, respectively. Only the rightmost term corresponds to the combination of material displacements that result in solid deformation, giving rise to the definition of the elastic strain tensor, strictly valid for infinitesimal displacements (i.e., the Cauchy infinitesimal strain tensor),

$$\boxed{\varepsilon_{ij} = \frac{1}{2} \left(\frac{\partial u_i}{\partial x_j} + \frac{\partial u_j}{\partial x_i} \right)} \tag{7.4}$$

From the form of the strain tensor (7.4), we can picture simple examples of normal and shear strain components as shown schematically in two dimensions in Figure 7.4.

Figure 7.4 Pure strain in two dimensions: (a) uniaxial extension; (b) pure shear; (c) simple shear.

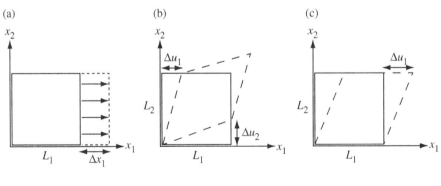

$$\varepsilon_{11} = \frac{\partial u_1}{\partial x_1} = \frac{\Delta x_1}{L_1} \qquad \varepsilon_{12} = \varepsilon_{21} = \frac{1}{2} \left(\frac{\Delta u_1}{L_2} + \frac{\Delta u_2}{L_1} \right) \qquad \varepsilon_{12} = \frac{1}{2} \left(\frac{\Delta u_2}{L_1} \right)$$

7.3.2 **Generalized Hooke's Law for Homogeneous, Isotropic Elastic Materials**

Having introduced the stress and strain tensors, we now relate them by means of the generalized form of Hooke's law. This law states that the components of stress are linearly related to the components of strain, which can be written concisely as a tensor equation

$$\sigma_{ij} = c_{ijkl}\varepsilon_{ij} \tag{7.5}$$

In matrix form, (7.5) becomes,

$$\begin{bmatrix} \sigma_{11} & \sigma_{12} & \sigma_{13} \\ \sigma_{21} & \sigma_{22} & \sigma_{23} \\ \sigma_{31} & \sigma_{32} & \sigma_{33} \end{bmatrix} = c_{ijkl} \begin{bmatrix} \varepsilon_{11} & \varepsilon_{12} & \varepsilon_{13} \\ \varepsilon_{21} & \varepsilon_{22} & \varepsilon_{23} \\ \varepsilon_{31} & \varepsilon_{32} & \varepsilon_{33} \end{bmatrix} \tag{7.6}$$

from which it is clear that the fourth-rank tensor c_{ijkl} stands for 81 constants or "elastic moduli" that relate each component of stress to strain. At this point, we can parallel the discussion of the viscous stress–strain rate relations in Section 5.5, where we used arguments of symmetry, orientation, and isotropy to reduce the number of constants (coefficients of viscosity) from 81 to 2. First, since the elastic stress and strain tensors are symmetric, the maximum number of independent elastic constants is reduced immediately from 81 to 36. We can further reason that normal stresses produce only normal strains, and shear stresses produce only shear strains, and that the materials of initial interest are isotropic (i.e., the elastic properties are identical in all directions). Along with symmetry, this will reduce the number of independent moduli to two (for a detailed discussion, see [11]). Taken together, the generalized Hooke's law then takes the form,

$$\boxed{\sigma_{ij} = 2G\varepsilon_{ij} + \lambda\delta_{ij}\varepsilon_{kk}} \tag{7.7}$$

where G and $\lambda = K - \frac{2}{3}G$ are called the elastic Lamé constants and δ_{ij} is the Kronecker delta function. In terms of elastic moduli, G is the shear modulus and K is the bulk modulus, the latter being related to the change in volume of an elastic material subjected to hydrostatic compression. The generalized Hooke's law can also be written in the inverse form

$$\boxed{\varepsilon_{ij} = \left(\frac{1+\nu}{E}\right)\sigma_{ij} - \frac{\nu}{E}\delta_{ij}\sigma_{kk}} \tag{7.8}$$

where ν is the Poisson's ratio and E is the Young's modulus. We will see from the examples and problems involving experiments to measure the various elastic constants that any given modulus can be expressed in terms of two other moduli.

EXAMPLE

Example 7.3.1 Simple Shear Referring to Figure 7.4(c), a shear stress σ_{21} is applied to the upper surface at $x_2 = L_2$, causing a simple shear deformation as shown by the dashed lines. Experimental measurement reveals a linear relation between the applied shear stress and the ratio $\Delta u_1/L_2$:

$$\sigma_{21} = G\left(\frac{\Delta u_1}{L_2}\right) = G\frac{\partial u_1}{\partial x_2} \tag{7.9}$$

The strain tensor component ε_{21} is

$$\varepsilon_{21} = \frac{1}{2}\left(\frac{\partial u_2}{\partial x_1} + \frac{\partial u_1}{\partial x_2}\right) = \frac{1}{2}\frac{\partial u_1}{\partial x_2} \tag{7.10}$$

Combining (7.9) and (7.10) gives a relation between the shear stress and strain components in terms of the shear modulus G:

$$\sigma_{21} = 2G\varepsilon_{21} \tag{7.11}$$

We note that this is consistent with the generalized Hooke's law (7.7) for $i = 2$ and $j = 1$, for which $\delta_{ij} = 0$.

EXAMPLE

Example 7.3.2 Moduli in Uniaxial Confined and Unconfined Compression A cylindrical disk specimen (e.g., biological tissue or cell-seeded hydrogel scaffold) is first held within a cylindrical confining chamber (Figure 7.5(a)) and subjected to uniaxial compression. The bottom and side walls of the chamber are impermeable to fluid flow, but the upper compression platen itself is porous to enable fluid exudation caused by compression of the hydrated specimen. Upon re-equilibration after compression, fluid flow ceases and only the solid matrix provides compressive stiffness. For a simple elastic model of the equilibrium compressive modulus, Hooke's law (7.7) can be used to relate the applied stress σ_{11} to the resulting strain ε_{11}. According to convention, outward stresses are defined as positive (see, e.g., Figure 4.2); however, σ_{11} and ε_{11} are defined here as positive in compression for simplicity:

$$\sigma_{11} = 2G\varepsilon_{11} + \lambda(\varepsilon_{11} + \varepsilon_{22} + \varepsilon_{33}) \tag{7.12}$$

Figure 7.5 (a) Uniaxial confined compression of a cylindrical disk of tissue via a porous platen at $x_1 = 0$. (b) Unconfined compression of the same tissue disk using impermeable platens at $x_1 = 0$ and $x_1 = L$.

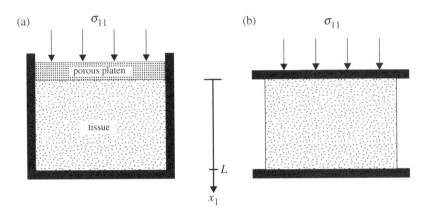

Since the confining chamber constrains deformations such that $\varepsilon_{22} = \varepsilon_{33} = 0$, axial compression leads to the relation

$$\sigma_{11} = (2G + \lambda)\varepsilon_{11} \tag{7.13}$$

where the confined compression modulus $2G + \lambda$ is sometimes referred to as the aggregate modulus H or, in the acoustics literature, as the longitudinal modulus M.

In the uniaxial unconfined compression configuration of Figure 7.5(b), Hooke's law (7.8) gives

$$\varepsilon_{11} = \left(\frac{1 + v}{E}\right)\sigma_{11} - \frac{v}{E}(\sigma_{11} + \sigma_{22} + \sigma_{33}) \tag{7.14}$$

With the radial periphery of the disk being a free surface ($\sigma_{22} = \sigma_{33} = 0$), and assuming that there is no friction at the platen–specimen interfaces at $x_1 = 0$ and $x_1 = L$, (7.14) reduces to a measurement of the Young's modulus in compression, E:

$$\sigma_{11} = E\varepsilon_{11} \tag{7.15}$$

Example 7.3.3 Measurement of Poisson's Ratio A tissue specimen modeled as homogeneous and isotropic is subjected to a uniaxial tensile stress σ_{11} as in Figure 7.6. After equilibration in the strained configuration, a noticeable necking-in is observed in the x_2 and x_3 directions. To measure Poisson's ratio, we note the zero-stress condition at the free surfaces, $\sigma_{22} = \sigma_{33} = 0$, and use (7.8) to obtain

$$\varepsilon_{11} = \left(\frac{1 + v}{E}\right)\sigma_{11} - \frac{v}{E}\sigma_{11} = \frac{\sigma_{11}}{E}$$

$$\varepsilon_{22} = \varepsilon_{33} = -\frac{v}{E}\sigma_{11}$$

From these relations, Poisson's ratio is defined as

$$v = -\frac{\varepsilon_{22}}{\varepsilon_{11}} = -\frac{\varepsilon_{33}}{\varepsilon_{11}} \tag{7.16}$$

In this configuration, Poisson's ratio is related to the "necking-in" (in the 2- and 3-directions) caused by tensile strain in the 1-direction. The associated changes in volume are addressed in Problems 7.1 and 7.2. For most isotropic materials, v has values in the range $0 < v < 0.5$, where $v \rightarrow 0.5$ corresponds to the limit of incompressibility. For real biological tissues, such deformation may lead to exudation and loss of intra-tissue fluid; the resulting loss of tissue volume must be interpreted carefully (Problem 7.1).

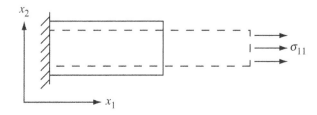

Figure 7.6 Experiment to measure the Poisson's ratio of a homogeneous, isotropic elastic material in uniaxial tension.

Example 7.3.4 Measurement of the Bulk Modulus Referring to Problem 7.3 and the configuration of Figure 7.25(a), a tissue specimen is modeled as a single (solid) phase homogeneous, isotropic elastic material having bulk modulus K. The material is subjected to a uniform compressive (normal) hydrostatic stress, and we use the generalized Hooke's law (7.7) to find K in terms of the Lamé constants G and λ. The normal stresses $\sigma_{11} = \sigma_{22} = \sigma_{33} = P$ are

$$\sigma_{11} = 2G\varepsilon_{11} + \lambda(\varepsilon_{11} + \varepsilon_{22} + \varepsilon_{33})$$

$$\sigma_{22} = 2G\varepsilon_{22} + \lambda(\varepsilon_{11} + \varepsilon_{22} + \varepsilon_{33})$$

$$\sigma_{33} = 2G\varepsilon_{33} + \lambda(\varepsilon_{11} + \varepsilon_{22} + \varepsilon_{33})$$

The sum of the three normal stresses can be represented using index notation (where an index repeated twice such as "kk" means sum on k):

$$\sigma_{kk} = (2G + 3\lambda)\varepsilon_{kk} = 3P \tag{7.17}$$

Dividing both sides of (7.17) by a factor of 3 and defining P as the mean normal stress gives,

$$P = \frac{\sigma_{kk}}{3} = \left(\lambda + \tfrac{2}{3}G\right)\varepsilon_{kk}$$

and we arrive at the desired relation,

$$K = \lambda + \tfrac{2}{3}G \tag{7.18}$$

When (7.18) is combined with the confined compression modulus $2G + \lambda$, we also find that $H = K + \tfrac{4}{3}G$.

Example 7.3.5 Wave Propagation in Isotropic Elastic Media The equations of motion for propagation of waves in elastic media can be derived using the fact that a force density can be written as the divergence of a stress tensor (i.e., equation (4.8)), and the stress tensor of interest here is that from Hooke's law 7.7. Longitudinal (dilatational) waves in an infinite elastic medium, corresponding to displacements u_1 in the direction of propagation, would thereby satisfy a wave equation having the form

$$\rho\frac{\partial^2 u_1}{\partial t^2} = (2G + \lambda)\frac{\partial^2 u_1}{\partial x_1^2} \tag{7.19}$$

while shear (transverse) waves have displacements satisfying a wave equation of a form such as

$$\rho\frac{\partial^2 u_2}{\partial t^2} = G\frac{\partial^2 u_2}{\partial x_1^2} \tag{7.20}$$

The corresponding wave propagation speeds are $[(2G + \lambda)/\rho]^{1/2}$ and $(G/\rho)^{1/2}$, respectively, where ρ is the mass density of the material. These concepts are pursued further in the context of waves in the gelatinous tectorial membrane of the inner ear (Problem 7.7).

EXAMPLE

EXAMPLE

7.4 VISCOELASTIC BEHAVIOR

7.4.1 Experimental Evidence

Studies of the stress–strain properties of biological materials in their native wet state often reveal time-dependent behavior. Such behavior has been referred to, generically, as "viscoelasticity," a descriptor that incorporates features of "solid-like" and "liquid-like" contributions to the overall rheological response. Classic examples of time-dependent behaviour include the creep and stress relaxation and frequency-dependent oscillatory stress–strain responses shown in Figure 7.7. When a sudden constant compressive load is applied to an intact disk explant (Figure 7.7(a)), the tissue compresses rapidly at first and then slowly continues to creep until a new equilibrium is reached. In contrast, a rapid increase in displacement to a new fixed tensile strain, such as that applied to the mouse tail tendon fascicle of Figure 7.7(b), causes an immediate increase in load (stress) followed by a slow stress relaxation towards a final nonzero equilibrium value that depends on the stiffness of the tendon's solid extracellular matrix. When a low-amplitude sinusoidal torsional strain is applied to a cylindrical specimen of cartilage, a sinusoidal shear stress results, but offset by a phase angle that leads the applied strain by an amount that varies with frequency (Figure 7.7(c)).

While these particular examples focus on tissue-level behavior, similar time-dependent responses have been observed with individual cells and systems of intracellular and extracellular molecular networks using state-of-the-art biophysical and rheological measurement technologies (e.g., atomic force microscopy, optical tweezers, and micro- and nano-indenters).

Based on such observations, investigators have developed a range of phenomenological models (continuum and molecular) formulated

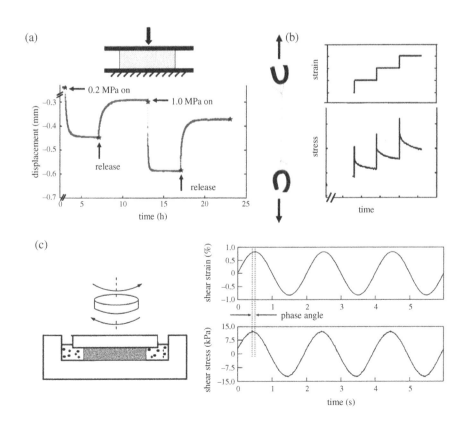

Figure 7.7 (a) Creep displacements of rat disk caused by an applied stress of 0.2 MPa and then 1.0 MPa, with incomplete recovery after the load is removed. (b) Stress relaxation of mouse tail tendon fascicle caused by successive 0.5% increments in tensile strain. (c) Sinusoidal torsional shear strain applied to a cylindrical disk specimen of articular cartilage resulting in a shear stress sinusoid that leads strain by the indicated phase angle (dotted lines) that varies with frequency. ((a) adapted from MacLean JJ, Owen JP, and Iatridis JC. Role of endplates in contributing to compression behaviors of motion segments and intervertebral discs. *J. Biomech.* **40**, 55–63 (2007); (b) adapted from Robinson PS, Huang T, Kazam E, et al. Influence of decorin and biglycan on mechanical properties of multiple tendons in knockout mice. *J. Biomech. Eng.* **127**, 181–185 (2005); (c) adapted from Jin M and Grodzinsky A. Effect of electrostatic interactions between glycosaminoglycans on the shear stiffness of cartilage: a molecular model and experiments. *Macromolecules* **34**, 8330–8339 (2001).)

as constitutive laws to describe creep and stress relaxation. In addition to representing the experimental data in a convenient manner, these models attempt to explain the biophysical and physicochemical mechanisms responsible for energy dissipation and stress–strain time dependence. Historically, such approaches were motivated by experimental and theoretical studies of solid (single-phase) polymers [12, 13]. These methods were subsequently applied to the characterization of porous, hydrated biomaterials [9]. In this section, we summarize simple models of linear viscoelastic behavior, and in Section 7.5, we focus on poroelastic descriptions that emphasize the role of fluid–solid frictional interactions in determining the deformational behavior of hydrated biomaterials.

7.4.2 Lumped Element Models of Creep and Stress Relaxation

The behavior of viscoelastic materials is often modeled mathematically using the general theory of linear systems via Fourier and Laplace transforms and the superposition (convolution) integral. Using methods similar to those for analysis of electrical circuits, lumped element models provide a convenient tool to represent time-dependent stress–strain relationships. The simplest such models include a single spring connected to a single dashpot in parallel or series. The stress–strain constitutive laws for the individual spring and dashpot elements are shown in Figure 7.8: stress T is linearly related to strain e for the spring (E), and to the first derivative of strain for the dashpot (η). We use the nomenclature T and e for stress and strain to distinguish the lumped element model "terminal variables" from true stress and strain (point by point) within a material.

Example 7.4.1 Voigt Model for Creep   To model the creep strain caused by an applied step in stress T_0 at $t = 0^+$, given initial rest ($e = 0$ for $t < 0$), we first try the simple parallel spring–dashpot model of Figure 7.9(a), called the Voigt model.

The strain variable e across the complete Voigt element is related to the strain across each individual element by

$$e(t) = e_1(t) = e_2(t) \qquad (7.21)$$

and the stress T is related to T_1 and T_2 by

$$T(t) = T_1 + T_2 \qquad (7.22)$$

Combining (7.21) and (7.22) with the stress–strain constitutive laws for the individual spring and dashpot elements (Figure 7.8) gives the overall differential equation for the system:

$$\eta \frac{de}{dt} + Ee = T(t) \qquad (7.23)$$

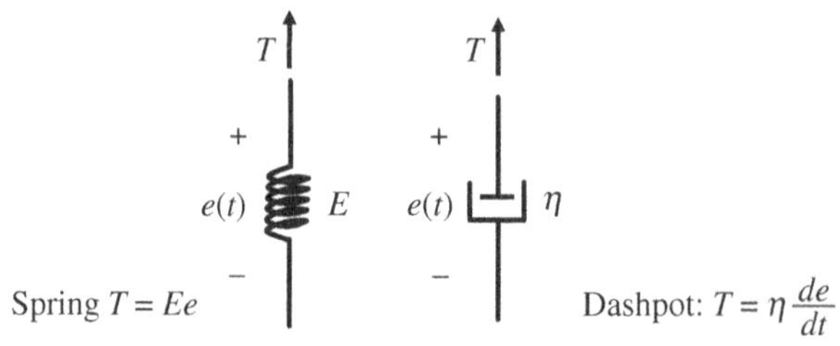

Figure 7.8 Stress–strain constitutive laws for spring and dashpot elements.

Spring $T = Ee$

Dashpot: $T = \eta \dfrac{de}{dt}$

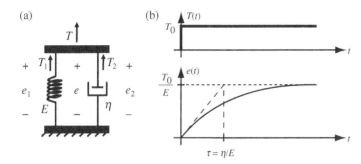

Figure 7.9 (a) Voigt model. (b) Applied step in stress and resulting creep strain response.

where $T(t)$ is the applied unit step in stress (Figure 7.9(b)). This differential equation has the solution

$$e(t) = Ae^{-t/\tau} + B, \qquad \tau = \frac{\eta}{E} \qquad (7.24)$$

The constants A and B are found from the initial rest condition and by noting that as $t \to \infty$, the creep strain is taken up entirely by the spring at the final equilibrium value, $e = T_0/E$:

$$e(t) = \frac{T_0}{E}(1 - e^{-t/\tau}) \qquad (7.25)$$

The creep response of this two-element Voigt model (Figure 7.9(b)) has some of the important qualitative features of the experimental data segments of Figure 7.7(a). However, closer inspection would reveal that the initial displacement after application of the step in load is much more rapid than that given by the exponential curve of Figure 7.9(b). To account for this observed behavior, additional lumped elements would need to be added to the model (see, e.g., Example 7.4.3). In addition, the Voigt model cannot predict stress relaxation, which leads us to the next (Maxwell) model.

Example 7.4.2 Maxwell Model for Stress Relaxation To represent the stress response of a biomaterial subjected to an applied step in strain e_0 at $t = 0^+$, given initial rest ($T = 0$ for $t < 0$), we try the simple configuration of a spring and dashpot connected in series, referred to as the Maxwell model (Figure 7.10(a)). Following the method of Example 7.4.1, the stress and strain variables T and e are first related to the stresses and strains across the individual elements by

$$e(t) = e_1(t) + e_2(t), \qquad \frac{de(t)}{dt} = \frac{de_1(t)}{dt} + \frac{de_2(t)}{dt} \qquad (7.26)$$

$$T(t) = T_1(t) = T_2(t) = \eta \frac{de_2}{dt} \qquad (7.27)$$

Equations (7.26) and (7.27) are combined with the constitutive law for the spring to give the governing differential equation

$$\frac{dT(t)}{dt} + \frac{E}{\eta}T(t) = E\frac{de}{dt} \qquad (7.28)$$

Figure 7.10 (a) Maxwell model. (b) Applied step in strain and resulting stress relaxation response.

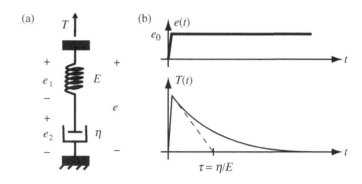

Since $e(t)$ is a step function, the drive on the right-hand side of (7.28) is its derivative, the unit impulse; thus, the solution to (7.28) subject to the condition that $T(t) \to 0$ at $t \to \infty$ is

$$T(t) = Ee_0 e^{-t/\tau}, \qquad \tau = \frac{\eta}{E} \tag{7.29}$$

The stress relaxation response of the two-element Maxwell model has some of the attributes of the data in Figure 7.7(b). However, there are two key problems with this idealized model. First, the theoretical stress response (7.29) relaxes to zero, while in many experiments, the stress relaxes to a new finite equilibrium value. In addition, the Maxwell model cannot predict creep.

Example 7.4.3 Standard Three-Element Linear Solid The limitations of the two-element Voigt and Maxwell models can be overcome in part by adding additional elements. As an example, the standard linear solid of Figure 7.11 displays both creep and stress relaxation responses. Show that the stress–strain relation has the form

$$T(t) + \alpha \frac{dT(t)}{dt} = E_1 e(t) + \beta \frac{de(t)}{dt} \tag{7.30}$$

where $\alpha = \tau$, $\beta = (E_1 + E_2)\tau$, and $\tau = \eta/E_2$.

The three-element model provides a first-order representation of the observed behavior of many materials within their viscoelastic range. This model is examined in the context of a force relaxation experiment from the literature in Problem 7.9, and the response of this model to a sinusoidal drive is also explored there.

Extensions of this approach involve the use of multiple springs and dashpots in series and/or parallel circuit configurations. Of course, the use of additional elements provides more adjustable coefficients, which can enable a better fit to experimental data. The challenge is then to identify the specific material property or molecular mechanism that may be associated with each element in the model. In terms of representing experimental data, the use of an infinite number of elements leads to the concept of the so-called relaxation- and retardation-time spectra. This approach involves the use of superposition or convolution integrals

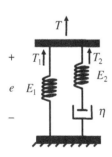

Figure 7.11 Standard linear solid.

EXAMPLE

[9, 12, 13]. It is assumed that the viscoelastic material behaves as a linear time-invariant (LTI) system, and that an arbitrary input to the system (stress or strain) can be represented in terms of a summation of impulse or step functions. The system response can then be computed as the sum (convolution integral) of the impulse or step responses. For example, the stress relaxation exponential of (7.29) can be considered as the response to a step in strain applied to the Maxwell element of Figure 7.10. With the general strain input represented by a series of incremental steps in strain, the stress for a linear viscoelastic material subjected to this input can be computed from the convolution integral

$$T(t) = \int_{-\infty}^{t} G(t-\tau)\frac{de}{d\tau}\,d\tau \qquad (7.31)$$

where $G(t-\tau)$ can be thought of as the response to the impulse function $de/d\tau$.

7.4.3 Dynamic Measurements and the Complex Modulus

In the experiment of Figure 7.7(c), a tissue specimen is subjected to a sinusoidal strain (in this case, shear) having small enough amplitude that the resultant stress is found to be a pure sinusoid shifted by a phase angle δ. The phase angle is observed to change with frequency, but the stress always leads the strain. (This latter requirement is consistent with energy dissipation being a positive-definite number.) Complex notation is now used to derive the form of the frequency-dependent modulus of such a linear viscoelastic material, called the complex modulus. (This form of the complex modulus is also useful when the frequency dependence derives from other physical mechanisms such as poroelasticity, as will be shown in Sections 7.5 and 7.6.)

The applied strain is first written as the real part of the sinusoidal steady state form

$$e(t) = e_0 \cos \omega t \equiv \text{Re}\left[e_0 e^{j\omega t}\right] \qquad (7.32)$$

where the term $e^{j\omega t}$ is defined by the Euler formula for complex numbers as

$$e^{j\omega t} = \cos \omega t + j \sin \omega t \qquad (7.33)$$

The resultant stress has the phase-shifted form

$$T(t) = T_0 \cos(\omega t + \delta) = \text{Re}\left[T_0 e^{j(\omega t + \delta)}\right] = \text{Re}\left[\hat{T}(\omega)e^{j\omega t}\right] \qquad (7.34)$$

where $\hat{T}(\omega)$ is the complex amplitude of the stress, equal to $T_0(\omega)e^{j\delta}$.

The complex modulus $\hat{E}(\omega)$ is then defined as

$$\hat{E}(\omega) \equiv \frac{\hat{T}(\omega)}{e_0} = \frac{|T_0(\omega)|\,e^{j\delta}}{e_0} = \left(\frac{|T_0(\omega)|\,\cos\delta}{e_0}\right) + j\left(\frac{|T_0(\omega)|\sin\delta}{e_0}\right) \qquad (7.35)$$

or, more concisely,

$$\hat{E}(\omega) = E'(\omega) + jE''(\omega) \qquad (7.36)$$

where $E'(\omega)$ is the storage modulus and $E''(\omega)$ is the loss modulus. E' is in phase with the applied strain and accounts for the elastic energy storage in the material; E'' is 90° out of phase with the strain and accounts for energy dissipation due to specific molecular and/or viscous drag-associated mechanisms [12, 13]. (In various texts, G or other symbols may be used to represent the complex modulus rather than E.)

In general, the frequency-dependent behavior of the material can be described by the magnitude and phase of the complex modulus, $|\hat{T}(\omega)|$ and δ, or, equivalently, by the combination of the storage and loss moduli, E' and E''. Some investigators report $\tan \delta = E''(\omega)/E'(\omega)$ rather than δ.

Example 7.4.4 Dynamic Behavior of the Three-Element Linear Solid   The standard linear solid of Figure 7.11 is described by the differential equation (7.30). The complex modulus describing this three-element spring–dashpot model can be derived directly from (7.30) by inspection, noting that we can convert from the time domain to the frequency domain using $d/dt \rightarrow j\omega$ and replacing $T(t)$ with $\hat{T}(\omega)$ and $e(t)$ with $\hat{e}(\omega)$:

$$\hat{T}(\omega) + j\omega\alpha\,\hat{T}(\omega) = E_1\hat{e}(\omega) + j\omega\beta\hat{e}(\omega) \tag{7.37}$$

$$(1 + j\omega\alpha)\hat{T}(\omega) = (E_1 + j\omega\beta)\hat{e}(\omega) \tag{7.38}$$

$$\frac{\hat{T}(\omega)}{\hat{e}(\omega)} = \hat{E}(\omega) = \frac{E_1 + j\omega\beta}{1 + j\omega\alpha} = \left(\frac{E_1 + \omega^2\alpha\beta}{1 + \omega^2\alpha^2}\right) + j\left(\frac{\omega(\beta - \alpha E_1)}{1 + \omega^2\alpha^2}\right)$$

$$= E'(\omega) + jE''(\omega) \tag{7.39}$$

Thus, the complex modulus for the three-element viscoelastic solid can be written as a function of the spring and dashpot parameters, $\hat{E}(\omega)(E_1, E_2, \eta)$ (see Problem 7.9(d)). Note that the physical constraint $\beta > \alpha E_1$ in (7.39) to ensure positive-definite energy dissipation.

7.4.4 Molecular Interpretations of Solid Phase Viscoelastic Behavior

The very extensive literature on the chemistry and physics of solid polymers and solutions of complex macromolecules has provided the basis for theoretical and experimental interpretations of the molecular origins of viscoelastic behavior. In particular, statistical thermodynamic theories of the rubber-like state have focused on polymer chain configurations and have provided the basis for molecular theories of an elastic network. Models incorporating linear flexible random coils and bead–spring models in the presence and absence of hydrodynamic effects have then been used to characterize certain modes of molecular interactions in the short-time region around the glass-to-rubber transition zone (e.g., the Rouse and Zimm models [12, 13]). In the long-time flow regime of viscoelastic interactions, entanglement and reptation provide widely accepted molecular models. For detailed discussions, the reader is referred to several treatises on the subject [12–15]. Briefly, regarding the time dependence associated with solid polymeric materials, deformations can cause changes in the orientation and configuration of polymers. On very short timescales, the polymer network does not have time to reorient and regain initial intermolecular distances and configurations. The distortions produced may be of high energy, resulting in an effectively higher modulus at these short timescales. At longer times, the molecules of a crosslinked polymer network can reorient and chain segments can translate to their initial configuration so as to relieve the initially high stress. Relaxation to lower-energy configurations thereby results in a decrease in the effective modulus of the material.

While the above discussion focuses primarily on solid polymer networks, the role of fluid motion within and through biological tissues and gels has received great attention with regard to time- and frequency-dependent behavior, which can deviate substantially from the spring–dashpot models described in this section (see Section 7.5). In addition, the constitutive equations based on the theory of linear viscoelasticity assume small strains. However, many soft tissues may undergo strains that are large enough to violate this assumption. Here, the reader is referred to Fung's theory of quasilinear viscoelasticity, which is linear with respect to relaxation but accounts for larger deformations [9].

7.5 POROELASTIC BEHAVIOR OF BIOMATERIALS: THEORIES AND EXPERIMENTS

The nonequilibirum rheological behavior of porous, hydrated tissues and molecular networks is the result of complex interactions involving both the interstitial fluid and the solid matrix constituents. In some cases, single-phase viscoelastic models (e.g., lumped element spring–dashpot models) may be able to characterize such behavior within limited ranges of stress, strain, and time (frequency). For example, cyclic testing of tendons and ligaments in tension has been described by the quasi-linear viscoelastic theory of Fung. Tensile testing of connective tissues is sometimes dominated by the oriented collagen fiber constituents, and therefore flow of water may be somewhat less important. Similarly, the dynamic shear behavior of soft tissues may involve little or no flow of interstitial fluid relative to the solid matrix, especially at very small strains. In these cases, time-dependent behavior may be ascribed to the properties of the crosslinked, entangled deforming solid network of the tissue.

However, the swelling and compressional behavior of tissues and hydrogels necessarily involves flow of fluid within and through the matrix. The resulting frictional interactions between fluid and solid phases can lead to remarkably different rheological properties, which are generally categorized as "poroelastic." We will see that frequency (time) dependence and energy dissipation associated with poroelastic deformations involve spatial as well as temporal phase delays between stress and strain at neighboring positions within the material. Mathematically, the displacement profile within the material is found to be described by the solution of a partial differential equation in space and time.

The theory of poroelasticity has been successfully applied to the deformational behavior of soft connective tissues, gels, soils, rocks, and many other materials composed of a multiple of solid and fluid electrolyte phases. The following discussion will focus on the fundamental laws that are combined to formulate a simple linear poroelastic model for the mechanics of hydrated biomaterials. In addition, the influence of chemical and electrical interactions on stress–strain constitutive laws and material properties will be mentioned here and in more detail in Section 7.6.

7.5.1 Brief Historical Background

The early concepts underlying flows through consolidating porous media are generally credited to Terzaghi and his studies of soil mechanics [16]. The generalization of these concepts to fully three-dimensional analyses for any loading variable in an arbitrary porous medium is attributed to Biot [17], who is thought of as the father of poroelastic mechanics. The most general treatment considers an anisotropic, compressible viscoelastic skeleton containing a compressible viscous fluid, capable of sustaining quasistatic consolidation as well as the propagation of shear and longitudinal elastic waves and of acoustic waves [18]. Detailed discussions of the various formulations of poroelastic mechanics may be found in review articles [17–21]. Biot [17, 18] and Rice and Cleary [19], for example, utilize a formulation involving the pore pressure and the total stress acting on the fluid-filled medium as basic state variables. Other approaches have utilized a mixture theory [21, 22] in which both fluid and solid phases are mixed and assumed to co-exist simultaneously at every point in space; individual equations of motion are written for the fluid and solid phases. During the past decades, investigators have applied these approaches to the swelling and deformation of hydrogels [23], a wide range of hard and soft biological tissues [22, 24–26], and, more recently, motions of the intracellular cytoskeleton [27] (see Problem 7.13).

In their seminal paper on swelling of gels, Tanaka and Fillmore [28] provided direct experimental evidence (Figure 7.12) that gel swelling occurred with kinetics that were well predicted by a simple model involving two material properties: the elastic modulus and the hydraulic permeability of the gel. They derived an equation of motion for the displacement field of spherical polyacrylamide gel beads undergoing free swelling in water by combining an elastic stress–strain constitutive law for the gel fiber network, a constitutive law for the relative motion of fluid within the swelling gel, and conservation of momentum for the deforming gel:

$$\frac{\partial u}{\partial t} = D_{\text{gel}} \frac{\partial}{\partial r} \left[\frac{1}{r^2} \left(\frac{\partial}{\partial r} (r^2 u) \right) \right] \tag{7.40}$$

This is a linear diffusion equation in spherical coordinates for the gel displacement u in terms of the gel diffusivity [23] $D_{\text{gel}} = H/f$, with H the longitudinal modulus (Section 7.3) and $1/f$ essentially the Darcy hydraulic permeability of the gel. A comparison of theory and experiment (Figure 7.12) verified the applicability of a gel swelling time constant τ_{gel}, obtained from a variable-separable solution to (7.40):

$$\tau_{\text{gel}} = \frac{L^2}{\pi^2 D_{\text{gel}}} \tag{7.41}$$

We will see that this characteristic time constant is a hallmark of poroelastic behavior. (In this particular paper [28], Tanaka and Fillmore did not incorporate pore pressure terms in their constitutive equation. Thus, while solution of (7.40) successfully describes the free swelling displacement field of the solid fiber network, further extensions of the model would be needed to calculate internal pressure distributions and fluid velocity profiles caused by deformation, as we discuss more generally below.)

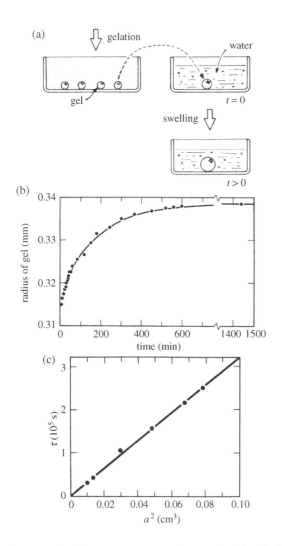

Figure 7.12 (a) Swelling of a 5% polyacrylamide spherical gel bead in water. (b) Radius of a typical gel bead as it swells in water, measured optically. (c) Characteristic time for the swelling of gel beads plotted as a function of the square of the final bead radius a. (From Tanaka T and Fillmore DJ, Kinetics of swelling of gels. *J. Chem. Phys.* **70**, 1214–1218 (1979), Figures 1, 5, and 6.)

7.5.2 Linear, Isotropic, Homogeneous Poroelastic Media

In this section, a simple model for linear poroelastic materials is developed by combining a stress–strain constitutive law with Darcy's law for fluid flow within the material, conservation of mass, and conservation of linear momentum. We assume that (1) the material is composed of a solid interconnected elastic network filled with fluid; (2) the fluid and the fibers of the solid network are each intrinsically incompressible; (3) the deformable solid network can be modeled as a linear, isotropic, homogeneous (Hookean) network (thereby subject to the limiting condition of small strain); and (4) time- (frequency-) dependent deformations cause dissipation that is associated solely with fluid–solid frictional interactions (i.e., the fluid itself can be thought of as intrinsically inviscid, but percolation of the fluid through the solid causes dissipation). In addition, we first treat examples in which there are no gradients in chemical concentration or electrical potential. Together, these simplifying assumptions will enable us to identify key physical processes that differentiate poroelastic behavior from the solid phase viscoelastic description of Section 7.4.

Total Stress–Strain Poroelastic Constitutive Law

Using the total stress and pore pressure formulation [18, 19, 29], we define the total (Cauchy) stress acting on an element of the material as the sum of the elastic stress on the solid network, (7.7), and the fluid pore pressure p:

$$\sigma_{ij} = 2G\varepsilon_{ij} + \delta_{ij}\lambda\varepsilon_{kk} - \delta_{ij}p \qquad (7.42)$$

where ε_{ij} is the infinitesimal strain tensor of (7.4), and p accounts for the effects of deformation on pressurization of the fluid phase under nonequilibrium conditions, which would be measurable, for example, by a sensor attached to the upper porous platen of the confined compression chamber of Figure 7.5(a). (The direction of the total stress components is defined by convention as shown previously in Figure 4.2.)

For the simple one-dimensional configuration of uniaxial confined compression, (7.42) becomes

$$\sigma = H\varepsilon - p \qquad (7.43)$$

Constitutive Law for Fluid Flow Relative to the Solid Network

Frictional interactions between the fluid and solid matrix within the material are described by Darcy's law, which relates the fluid flow to the gradient in fluid pressure at any point within the material. The fluid velocity $\boldsymbol{U}$ is related to the gradient in p by

$$\boldsymbol{U} = -k\nabla p \qquad (7.44)$$

where $\boldsymbol{U}$ is the total-area-averaged velocity of the fluid relative to the solid at each position and k is the tissue hydraulic permability. Implicit to the fluid flow law (7.44) is that inertial effects are negligible, a reasonable assumption for the known fluid mass density, time rates of change (frequencies), and frictional damping forces that are of interest. For cases in which fluid inertia is important, Darcy's law would be replaced by a more general statement of conservation of momentum for the fluid.

Conservation of Mass

To further define the relative fluid velocity $\boldsymbol{U}$ in Darcy's law (7.44) within the context of conservation of mass, we refer to the differential element of fluid-saturated solid network shown in Figure 7.13. The porosity of the network, ϕ, is defined in terms of the fractional volumes of fluid, V_f, and solid, V_s, as

$$\phi = \frac{V_f}{V_f + V_s} = \frac{V_f}{V_{tot}} \equiv \frac{A_f}{A_T} \qquad (7.45)$$

where A_f is the fraction of the total surface area A_T occupied by the fluid, and the last equality on the right is valid for isotropic materials. The local velocities of fluid and solid elements within the network are defined as v_f and v_s, respectively (Figure 7.13). Then $\boldsymbol{U}$ is defined with respect to these local velocities as

$$\boldsymbol{U} \equiv (v_f - v_s)\frac{A_f}{A_T} = \phi(v_f - v_s) \qquad (7.46)$$

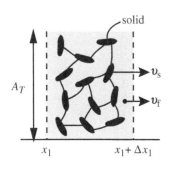

Figure 7.13 Differential element of fluid-saturated solid network.

where $(v_f - v_s)A_f$ can be thought of as the local volume flux of fluid with respect to the solid, and $\boldsymbol{U}$ is this volume flux normalized to the total area A_T.

Referring again to Figure 7.13, the x_1 component of conservation of mass can be derived by noting that a net flux of fluid into the Δx_1-thick element should give rise to an increase in total volume of the element:

$$\underbrace{\left[A_f(v_f - v_s)\Big|_{x_1} - A_f(v_f - v_s)\Big|_{x_1+\Delta x_1} \right] \Delta t}_{\text{fluid accumulation into element in } \Delta t}$$

$$= A_T \underbrace{\left[\Delta x_1 \, (1 + \varepsilon_{11})\Big|_{t_0+\Delta t} - \Delta x_1 \, (1 + \varepsilon_{11})\Big|_{t_0} \right]}_{\text{net increase in volume of element}} \tag{7.47}$$

Dividing both sides of (7.47) by A_T, Δx_1, and Δt, taking appropriate limits to obtain derivatives with respect to x_1 and t, and using the definition of $\boldsymbol{U}$ from (7.46), gives

$$-\frac{\partial U_1}{\partial x_1} = \frac{\partial \varepsilon_{11}}{\partial t} = \frac{\partial}{\partial t}\frac{\partial u_1}{\partial x_1} \tag{7.48}$$

$$-\frac{\partial U_1}{\partial x_1} = \frac{\partial}{\partial x_1}\left(\frac{\partial u_1}{\partial t} \right) \tag{7.49}$$

In three dimensions, (7.49) becomes:

$$\boxed{\nabla \cdot \boldsymbol{U} = \nabla \cdot [\phi(v_f - v_s)] = -\nabla \cdot v_s} \tag{7.50}$$

In one-dimension, integration of (7.48) gives

$$\boxed{U_1 = -\frac{\partial u_1}{\partial t} + U_0} \tag{7.51}$$

where the integration constant U_0 corresponds to the possibility of a steady state velocity through the material.

Conservation of Momentum

In the absence of inertial effects, conservation of momentum requires that

$$\nabla \cdot \underline{\underline{\sigma}} = 0 \tag{7.52}$$

This is an excellent approximation for frequencies and strain rates of physiological interest. In uniaxial geometry, (7.52) becomes

$$\frac{\partial \sigma_{11}}{\partial x_1} = 0 \tag{7.53}$$

7.5.3 Poroelastic Diffusion Equation

For the case of chemically homogeneous systems (no chemical transport or gradients in chemical species), (7.42), (7.44), (7.50), and (7.52) constitute a complete description of tissue stress σ, strain ε, displacement u, fluid velocity U, and pore pressure p. In the one-dimensional uniaxial geometry of Figure 7.14, these laws combine to give a diffusion equation for each variable; for example, the displacement $u_1(x, t)$ is a solution of a diffusion equation having the "diffusivity" Hk:

$$\frac{\partial u_1(x_1, t)}{\partial t} = Hk\frac{\partial^2 u_1(x_1, t)}{\partial x_1^2} \tag{7.54}$$

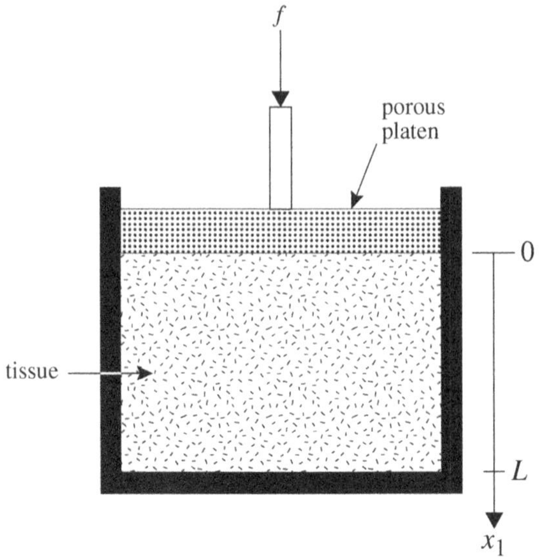

Figure 7.14 Uniaxial confined compression configuration.

Application of (7.54) for the cases of stress relaxation and creep are given in Examples 7.5.1 and 7.5.2, respectively. When the material properties (e.g., H and k) depend on the chemical environment, constitutive laws for their concentration dependence must be derived or measured as exemplified in Section 7.5.4. Electrical gradients and electrokinetic interactions in poroelastic media are discussed in Section 7.6.

Example 7.5.1 Stress Relaxation in Uniaxial Confined Compression With the tissue specimen initially at rest, a rapid small compressive displacement u_0 is applied at time $t = 0^+$ at the position $x_1 = 0$ (Figure 7.14). The confining chamber prevents material bulging at the sides and forces the displacement to be zero at $x_1 = L$. Using the method of separation of variables, directly analogous to the diffusion problem of Figure 1.18 with constant-concentration boundary conditions, the solution of (7.54) subject to fixed displacement boundary conditions at $x_1 = 0, L$ is

$$u_1(x_1, t) = u_0 \left(1 - \frac{x_1}{L}\right) - \sum_{n=1}^{\infty} \left(\frac{2u_0}{n\pi}\right) \sin\left(\frac{n\pi x_1}{L}\right) e^{-t/\tau_n} \quad (7.55)$$

where the relaxation time constant for the nth term in the series is

$$\tau_n = \frac{L^2}{n^2 \pi^2 Hk} \quad (7.56)$$

The characteristic *stress relaxation time constant* is defined from the slowest ($n = 1$) term in the series solution, $\tau = L^2/(\pi^2 Hk)$. The space–time evolution of the displacement $u_1(x_1, t)$ is shown schematically in Figure 7.15(a). At early times, the local strain is steepest near the porous platen, since fluid has exuded from the surface of the tissue during compression. In contrast, the fluid pressure is zero at the surface and highest at $x_1 = L$. During relaxation, fluid no longer leaves the tissue but redistributes within the specimen. At any position x_1, the total stress is the sum of the elastic stress and fluid pressure at that position; this sum is independent of x_1 and decays with the

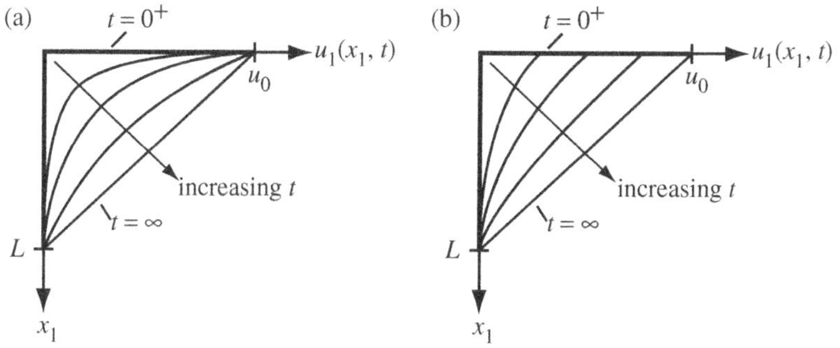

(a) $t = 0^+$ $u_1(x_1, t)$ u_0 increasing t $t = \infty$ L x_1

(b) $t = 0^+$ $u_1(x_1, t)$ u_0 increasing t $t = \infty$ L x_1

Figure 7.15 (a) Evolution of the displacement profile $u_1(x_1, t)$ in a linear poroelastic material during stress relaxation following a rapid step compressive displacement $u_1 = u_0$ at the surface $x_1 = 0$ in uniaxial confined compression. (b) Displacement profile $u_1(x_1, t)$ during creep compression following application of a constant compressive stress σ_0 at the surface in the uniaxial confined compression configuration of Figure 7.14.

characteristic time constant τ_1 from (7.56). The final displacement profile at equilibrium ($t = \infty$) is linear and corresponds to a uniform strain field; thus, the final relaxed stress is the product of this final strain and the known longitudinal modulus of the material, $H = 2G + \lambda$. We note that the result (7.55) is the response to an idealized step in applied displacement. In practice, the initial displacement occurs over a rapid but finite time, and the partial relaxation that could occur during this initial compression can also be evaluated from a separate solution of (7.54).

Example 7.5.2 Poroelastic Creep in Uniaxial Confined Compression In the configuration of Figure 7.14, a constant load is applied at $x_1 = 0$ corresponding to a stress σ_0. At $x_1 = L$, the displacement is zero. Using separation of variables, find the displacement profile $u_1(x_1, t)$ and show that the characteristic creep time is

$$\tau_1 = \frac{4L^2}{\pi^2 Hk} \tag{7.57}$$

Interestingly, while scaling analyses alone would suggest that the "diffusion times" for both stress relaxation and creep are of order L^2/Hk, the exact solutions show that a creep experiment takes four times longer to perform than stress relaxation for the same poroelastic material and geometric configuration. This is a result of the constant-displacement boundary condition for the stress relaxation experiment versus the constant-stress condition at $x_1 = 0$ for creep. (This is directly analogous to constant-concentration versus constant-flux conditions for chemical diffusion.) The space–time evolution of the displacement $u_1(x_1, t)$ during creep is shown schematically in Figure 7.15(b). We note that the slope of the displacement (i.e., the strain) at $x_1 = 0$ is invariant with time, consistent with the pressure being zero at the surface and the condition of constant stress at $x_1 = 0$. As with the stress relaxation example above, once $u_1(x_1, t)$ is known, $\varepsilon_{11}(x_1, t)$, $p(x_1, t)$, and $U_1(x_1, t)$ can be derived.

7.5.4 Revised Stress–Strain Constitutive Law Including Electrical and Ionic Interactions

Distinct macromolecular constituents within biological tissues contribute to material properties such as the elastic moduli and the total swelling stress. For example, in extracellular matrices that have a high concentration of charged proteoglycans and proteins, repulsive interactions between the charge groups can significantly increase tissue swelling stress. These intermolecular forces also affect tissue moduli, resulting in added resistance to deformation. Since intra-tissue pH and ionic strength can directly modulate these electrical interaction forces, the chemical environment also plays an important role in determining poroelastic material behavior [30–35]. As an example, we describe in this section a modification of the equilibrium stress–strain constitutive law [30, 31] that includes such effects along with experimental measurements of model parameters.

Increasing the salt concentration of a tissue's bathing solution screens electrical repulsive forces between macromolecular charge groups by decreasing the electrical interaction distance (Debye length) within the tissue (Chapter 4). In a confined compression test in which the tissue thickness is held constant (Figure 7.16), changes in nanoscale electrical

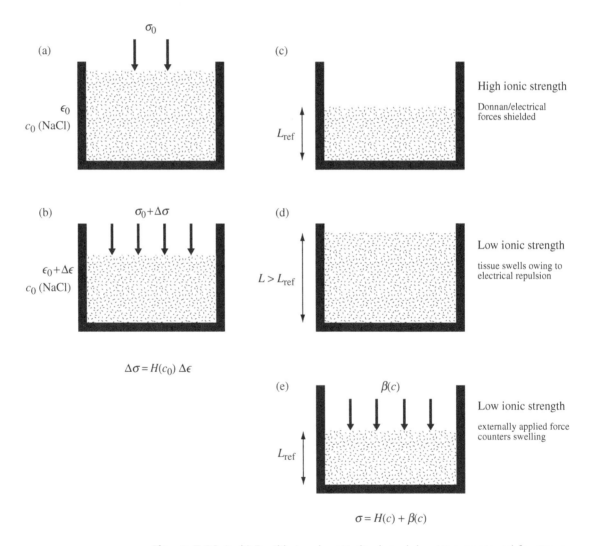

Figure 7.16 (a, b) Equilibrium longitudinal modulus $H(c)$ measured for tissue in buffer at pH_0 and ionic strength given by NaCl concentration c_0. (c–e) Measurement of chemical stress $\beta(c)$: (c) free swelling thickness measured at high ionic strength such that all electrical repulsive and osmotic forces are shielded; (d) with bath changed to lower ionic strength c, tissue swells; (e) an externally applied compressive stress equal to $\beta(c)$ must be applied to compress the tissue back to the inital reference thickness of (c).

repulsive forces give rise to changes in the macroscopic swelling stress and a concomitant change in the applied stress needed to keep the strain constant in equilibrium. Conversely, when the stress on the tissue is held constant, a change in salt concentration alters intermolecular repulsion forces and intermolecular spacing, thereby causing a change in macroscopic tissue thickness.

To account for these effects, the stress–strain constitutive law (7.42) can be modified to take the form

$$\sigma_{ij} = 2G(c)\varepsilon_{ij} + [\lambda(c)\varepsilon_{kk} - \beta(c)]\,\delta_{ij} \tag{7.58}$$

where the Lamé constants $G(c)$ and $\lambda(c)$ are now dependent on ionic concentrations, and the chemical stress $\beta(c)$ represents the chemical analog of the thermal stress in the thermoelasticity. In equilibrium, the fluid pressure in (7.42) is zero (ambient), and for a homogeneous, isotropic, charged poroelastic material, (7.58) states that three concentration-dependent material properties are required to completely describe the chemical modulation of swelling behavior. (To account for tissue anisotropy, (7.58) must be generalized and β can vary with direction.) For the uniaxial confined compression geometry of Figure 7.16, with stress σ_{11} and strain ε_{11} defined as positive in compression, (7.58) takes the form

$$\sigma_{11} = [2G(c) + \lambda(c)]\varepsilon_{11} + \beta(c) = H(c)\varepsilon + \beta(c) \tag{7.59}$$

For a homogeneous sample in equilibrium with bathing NaCl concentration c, a mechanical stress equal to the swelling stress $[H(c)\varepsilon + \beta(c)]$ is required to keep the tissue at a given thickness in the configuration of Figure 7.16. The modulus H can be measured at any concentration c_0 by applying an increment in compressive stress, $\Delta\sigma$ (strain $\Delta\varepsilon$). The origin of the concentration-dependent chemical stress β can be understood in the following way. A tissue sample in equilibrium with a bath whose salt concentration is large enough to shield electrical interactions is used to define the reference thickness for the sample (Figure 7.16(c)). When the bath salt concentration is decreased, electrical repulsive forces cause tissue swelling in the thickness direction (Figure 7.16(d)). Equilibrium is reached when these repulsive forces are balanced by the constraining force of the stretched network. The stress σ required to compress the sample back to its initial strain-free state (Figure 7.16(e)) defines the chemical stress $\beta(c)$, i.e., the stress that would be measured in equilibrium if the sample were held at its reference thickness while the salt concentration was decreased.

Figure 7.17 shows data for the NaCl-concentration dependence of H and β, measured for adult bovine articular cartilage [30]. The decrease in the modulus with increased concentration is due to electrostatic shielding of aggrecan–GAG charge groups by counter-ions; electrostatic forces are shielded by 1.0 M NaCl. By comparing the modulus at 0.15 M NaCl with the modulus at 1.0 M, we therefore conclude that electrostatic repulsive interactions account for at least half of the modulus at physiological ionic strength in these cartilage specimens. The chemical stress shows a more rapid decrease with concentration, and the electrostatic shielding effects of increased NaCl concentration result in less stress being required to keep the tissue at its reference volume.

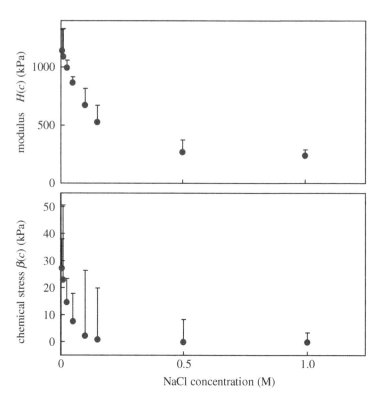

Figure 7.17 Equilibrium modulus $H(c)$ and chemical stress $\beta(c)$ of bovine cartilage as a function of bath NaCl concentration. (Adapted from Eisenberg SR and Grodzinsky AJ. Swelling of articular cartilage and other connective tissues: electromechanochemical forces. *J. Orthop. Res.* **3**, 148–159 (1985).)

7.6 ELECTROKINETIC TRANSDUCTION IN POROELASTIC MEDIA

Chapter 6 focused on electrokinetic transduction in biological tissues, hydrogels, and membranes modeled as porous materials having ionized charge groups on the inner pore surfaces. Flows of fluid and electric current across such materials caused by gradients in pressure and electrical potential were modeled in a macrocontinuum sense by the phenomenological coupling relations shown in Figure 6.18. A nano-continuum representation of electrokinetic transduction was introduced in the context of a cylindrical pore model, pictured in Figure 6.4(c) and described by (6.17). This pore model provided a convenient way of introducing the phenomena of electroosmosis and streaming potential and relating these electrokinetic effects to macroscale systems, including microfluidics. In those examples, the biomaterials were essentially modeled as *rigid* porous media that could support relative flow of interstitial fluid.

In this section, we extend these electrokinetic concepts for the case of *deformable* charged molecular networks, i.e., poroelastic materials in which the molecular network has built-in fixed-charge groups. In such systems, nonuniform deformations of the matrix can induce local fluid flows, which, in turn, generate nonuniform streaming potential gradients. Conversely, applied electric currents can cause both electroosmosis of interstitial fluid and oppositely directed electrophoretic motions of the charged matrix itself. Together, these electrokinetic effects result in nonuniform matrix deformations, swelling, or current-induced swelling stresses consistent with the mechanical and electrical boundary conditions imposed on the system.

We restrict the discussion to linear, isotropic, homogeneous poroelastic biomaterials in which there are no applied or induced chemical gradients, thereby highlighting linear electrokinetic effects in deformable media. For mathematical simplicity, we focus on the one-dimensional system of Figure 7.18; extensions to three dimensions will be evident. The simplest continuum model combines the stress–strain constitutive law ((7.60) below), conservation of mass (7.61) and conservation of momentum (7.62) from Section 7.5, together with expanded constitutive laws for fluid flow (7.63) and electric current density (7.64) and a statement of conservation of charge (7.65):

$$\sigma_{11}(x_1, t) = H\varepsilon_{11}(x_1, t) - p(x_1, t) \tag{7.60}$$

$$U_1(x_1, t) = -\frac{\partial u_1(x_1, t)}{\partial t} + U_0(t) \tag{7.61}$$

$$\frac{\partial \sigma_{11}(x_1, t)}{\partial x_1} = 0 \tag{7.62}$$

$$U_1(x_1, t) = -k_{11}\frac{\partial p(x_1, t)}{\partial x_1} + k_{12}\frac{\partial V(x_1, t)}{\partial x_1} \tag{7.63}$$

$$J_1(x_1, t) = k_{21}\frac{\partial p(x_1, t)}{\partial x_1} - k_{22}\frac{\partial V(x_1, t)}{\partial x_1} \tag{7.64}$$

$$\frac{\partial J_1(x_1, t)}{\partial x_1} = -\frac{\partial \rho_e(x_1, t)}{\partial t} \simeq 0 \tag{7.65}$$

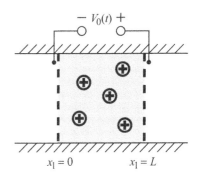

Figure 7.18 Flow-induced streaming potential measured using a high-impedance amplifier such that the electrical current $i = 0$.

In (7.61), U_0 corresponds to the possibility of a constant fluid velocity across the material of Figure 7.18 (i.e., an integration constant from the differential form of conservation of mass (7.50). Equation (7.63) is Darcy's law with the addition of a term for electroosmotically induced fluid flow. Similarly, (7.64) is Ohm's law including a term for pressure-induced current density. Conservation of charge (7.65) states that in biological systems of interest, charge relaxation times are so rapid (of the order of 10^{-9} s) that $\partial \rho_e / \partial t \simeq 0$ (Chapter 2) and therefore the current density has no divergence. This is not a limitation to steady state behavior; rather, "quasistatic" forces and flows that may involve relatively rapid (higher frequency) processes are still described by (7.60)–(7.65) even though $J_1(x_1, t)$ must be solenoidal.

The electrokinetic coupling coefficients k_{ij} have been described in Chapter 6, and are intrinsic material properties that depend on the electrical, mechanical, and chemical properties of the tissue matrix. These coefficients describe the streaming potential $(-k_{21}/k_{22})$, first and second electroosmotic flows (k_{12}/k_{22} and k_{12}, respectively), and the streaming current (k_{21}/k_{11}), as summarized by Katchalsky and Curran [36]. k_{22} is the electrokinetic conductivity, which may have both Ohmic and convective contributions (see (6.17)). $k_{12} = k_{21}$ by reciprocity in the linear system, and they are equal to $\rho_m k'$, the product of the macro-continuum fixed-charge density ρ_m and the hydraulic permeability $k' = k_{11} - k_{12}k_{21}/k_{22}$ (see Problem 6.3). This k' is the "open-circuit" Darcy permeability that would be measured in the configuration of Figure 7.18 with the electrodes unconnected. In contrast, k_{11}, is the "short-circuit" Darcy permeability, measured with the electrodes shorted to each other. The open-circuit permeability can be derived from (7.63) and (7.64) with $J = 0$. The backflow term ($k_{12}k_{21}/k_{22}$) represents the electrical force exerted by the streaming potential field ($\partial V/\partial x_1$) on the space charge entrained within the fluid phase, which reduces the fluid flow and hence lowers the effective hydraulic permeability. The streaming potential field can be suppressed by increasing the salt concentration of the interstitial fluid. This fact provides an experimental method for testing the significance of the backflow term, which can be quite appreciable in magnitude [37]. We note finally that (7.63) and (7.64) are point-by-point relations within the material that are reminiscent of the one-dimensional "transmembrane" jump relations from Chapter 6:

$$\begin{bmatrix} U_1 \\ J_1 \end{bmatrix} = \begin{bmatrix} L_{11} & L_{12} \\ L_{21} & L_{22} \end{bmatrix} \begin{bmatrix} \Delta P \\ \Delta V \end{bmatrix} \tag{7.66}$$

Equations (7.60)–(7.65) can now be combined to arrive at the poro-elastic electrokinetic diffusion equation in terms of the displacement $u_1(x_1, t)$:

$$\frac{\partial u_1(x_1, t)}{\partial t} = Hk' \frac{\partial^2 u_1(x_1, t)}{\partial x_1^2} + U_0(t) + \left(\frac{k_{12}}{k_{22}} \right) J_0(t) \tag{7.67}$$

where J_0 represents the possibility of a finite uniform current density if the electrodes are short-circuited. Equation (7.67) can be solved subject to appropriate mechanical and electrical boundary conditions at $x_1 = 0$ and $x_1 = L$.

Example 7.6.1 Compression-Induced Oscillatory Streaming Potential

A cylindrical disk of hydrated tissue having negative fixed-charge groups is placed in the confined compression configuration of Figure 7.19(a) and allowed to come to equilibrium in a bathing electrolyte of known pH and ionic strength. The potential between the upper and lower electrodes is measured using a high-input-impedance op-amp, and therefore the terminal current is always nearly zero ($i = 0$). An oscillatory displacement is applied by the upper porous platen/electrode surface at $x_1 = 0$ having the form, $u_1 = u_0 \cos \omega t$. The fluid pressure is zero (ambient) at $x_1 = 0$. At the fluid-impermeable bottom surface $x_1 = L$, $u_1 = 0$, and the fluid velocity (in (7.61)) $U_0 = 0$. After sinusoidal steady state is reached, show that the complex displacement amplitude within the tissue specimen has the form [38, 39]

$$\hat{u}(x_1, \omega) = u_0 \frac{\sinh \gamma (L - x_1)}{\sinh \gamma L} \tag{7.68}$$

$$\gamma = \sqrt{\frac{j\omega}{Hk'}} \tag{7.69}$$

Equation (7.68) describes the envelope of the peak displacement amplitude as a function of x_1. At any instant in time, a snapshot of $u_1(x_1, t)$ would show an oscillatory diffusion wave having a peak amplitude that decreases with depth in a manner bounded by the hyperbolic sine function having the value

Figure 7.19 Compression-induced streaming potential in uniaxial confined compression.

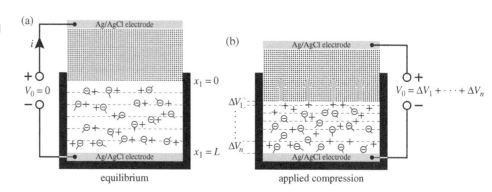

zero at $L = x_1$. The penetration depth δ of this diffusion wave is

$$\delta = \sqrt{\frac{2Hk'}{\omega}} \tag{7.70}$$

At any instant in time, the local strain amplitude is highest at the surface. We can picture the resultant streaming potential to be the sum of increments in potential ΔV_n produced by the local compression of each element of tissue in the depth direction, Δx_1 (Figure 7.19(b)). The local fluid velocity and streaming potential in each element are out of phase in space and in time from those in the neighboring element. Details for experiments performed using adult articular cartilage in this configuration are described in the literature [38, 39]. Typical waveforms of the sinusoidal streaming potential and load produced by an applied sinusoidal displacement having constant amplitude over a wide range of frequency are shown in Figure 7.20.

In order to model the streaming potential produced by the displacement profile (7.68), we note that the current density J in (7.64) is negligible when a high-input-impedance amplifier is used to measure the potential. Thus, $\partial p / \partial x_1$ can be written in terms of $\partial V / \partial x_1$, and (7.63) becomes

$$U_1 = -k' \left(\frac{k_{22}}{k_{21}} \right) \frac{\partial V}{\partial x_1} \tag{7.71}$$

To find the streaming potential measured between the positive electrode at $x_1 = 0$ (articular surface) and the negative electrode at $x_1 = L$, (7.71) is intergrated from L to 0:

$$\hat{V}(\omega) = - \int_L^0 \frac{k_{21}}{k' \, k_{22}} \hat{U}(x_1, \omega) \, dx_1 \tag{7.72}$$

where $\hat{V}(\omega)$ and $\hat{U}(x_1, \omega)$ are the complex amplitudes corresponding to the sinusoidal steady state. If the velocity profile $\hat{U}(z, \omega)$ can be measured or calculated from $\hat{u}_1(x_1, \omega)$, then the streaming potential $\hat{V}(\omega)$ can be computed and compared with experimental results.

The theoretical model for the streaming potential, (7.72), was compared with deformation-induced potentials measured across plugs of bovine articular cartilage in uniaxial confined

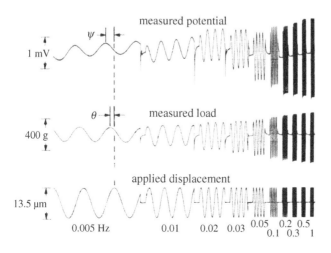

Figure 7.20 Dynamic load and streaming potential response to applied dynamic displacements at various frequencies.

compression [38, 39]. The mechanical stiffness, the mechanical phase angle between stress and displacement, and the electrical potential and its phase angle were compared with the corresponding theory. By fitting the mechanical theory to data, H and k were computed. This is important since the electrical phase angle ψ is an extra variable that contains no other adjustable parameters and thereby provided independently measured confirmation that the underlying mechanism for potential generation was electrokinetic and not, for example, piezoelectric. (A significant piezoelectric response in wet soft tissues would be highly unlikely owing to macroscopic symmetry considerations and the fast charge relaxation time constants (10^{-7} to 10^{-9} s) compared with the much slower fluid-velocity-dependent relaxations that are rate-limiting for the streaming potential.) Hence, the simultaneous measurement of streaming potential and stiffness can provide important information on velocity and deformation fields within charged poroelastic materials.

7.7 PROBLEMS

PROBLEM

Problem 7.1 Poisson's Ratio; Incompressibility A volume element has initial unstrained volume $V_0 = X_1 X_2 X_3$ (Figure 7.21). A tensile stress σ_{11} results in a deformation of the volume element such that $X_1 \rightarrow X_1 + \Delta x_1$, $X_2 \rightarrow X_2 + \Delta x_2$, etc., where the strain $\varepsilon_{11} \equiv (\Delta x_1)/X_1$. The tissue can be modeled as a linear, homogeneous, isotropic, elastic solid, having a Poisson's ratio ν.

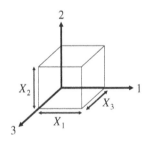

Figure 7.21 Volume element for Problem 7.1.

(a) Find expressions for ε_{22} and ε_{33} in terms of σ_{11} and ε_{11}.

(b) Find an expression for the final deformed volume V_f of the volume element in terms of X_1, X_2, X_3, ε_{11}, and ν.

(c) Show that $V_f = V_0$ only if $\nu = 0.5$.

(d) A strip of tendon 20 mm long and 1 mm×1 mm in cross section is subjected to tensile loading in the configuration shown in Figure 7.22. The tissue's extracellular matrix is composed primarily of a dense, hydrated, oriented type I collagen fiber network. A tensile strain ε_{11} is applied in the x_1 direction. The specimen's dimensions in the final equilibrium (strained) state are length 22 mm and cross section 0.8 mm×0.8 mm.

   You first model the specimen as a linear, homogeneous, elastic solid having Poisson's ratio ν and Young's modulus E. Using your expressions from part (b), calculate the specimen's Poisson's ratio ν from the change in specimen dimensions.

(e) Does your answer to part (d) make physical sense? Justify your answer and discuss in the context of real tissue versus the assumed model of a linear homogeneous elastic solid.

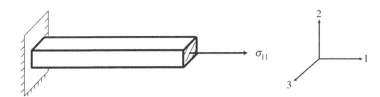

Figure 7.22 Tensile loading on strip of tendon.

PROBLEM

Problem 7.2 Elastic Behavior of Intervertebral Disk A schematic of the intervertebral disk is shown in Figure 7.23(a). Experiments in confined and unconfined compression are used to characterize the equilibrium elastic behavior of the disk as well as a tissue-engineered gel (Figure 7.23(b)) that is to be used as a disk replacement material.

(a) *Homogeneous, isotropic behavior*: Assume, first, that both the intervertebral disk and the gel can be modeled as identical uniform, isotropic, linear elastic cylinders of radius R and thickness h.

  (i) For both structures, find an expression for the confined compression modulus H in terms of the Young's modulus E and Poisson's ratio ν.

  (ii) For what value(s) of ν are H and E equal?

(b) *Inhomogeneous and anisotropic behavior*: For the disk of Figure 7.24 the NP can be modeled as an isotropic gel composed of proteoglycans and randomly oriented type II collagen. However, the AF has highly oriented type I collagen fibrils that are oriented on average in the circumferential direction, as indicated. In equilibrium, three different Young's moduli can be ascribed to AF to characterize its anisotropy: circumferential (E_c), radial (E_r), and axial (E_z). Thin strip specimens are used to measure E_c and E_r in tension, while the little disk is used to measure E_z in compression.

  (i) E_z is found to be 0.5 MPa. Give qualitative estimates for E_c and E_r, and state your reasoning.

  (ii) Assume that the unconfined compression modulus of the gel of Figure 7.23(b) is 0.5 MPa. Is the effective unconfined compression modulus of the whole disk of Figure 7.23(a) greater than or less than that of the gel? State your reasoning.

(c) Find an expression for the shear modulus of the gel of Figure 7.23(b) in terms of the unconfined compression modulus and the Poisson's ratio. Is the effective shear modulus of the whole disk (Figure 7.23(a)) greater than or less than that of the gel? State your reasoning.

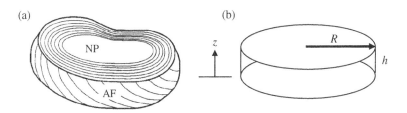

(a)   (b)

Figure 7.23 (a) Schematic of intervertebral disk (AF, annulus fibrosus; NP, nucleus pulposis). (b) Schematic of disk replacement.

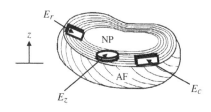

Figure 7.24 Thin specimens used to measure the Young's moduli E_c, E_r, and E_z.

Problem 7.3 Measurement of Bulk Modulus

(**a**) A fluid containing a specimen of connective tissue is pictured in Figure 7.25(a). In equilibrium, the tissue is modeled as a linear, homogeneous, isotropic, elastic solid. The fluid is pressurized so that only normal stresses are applied to the tissue. We have used the generalized Hooke's law (7.7) relating σ_{ij} to ε_{ij} and found that $K = \lambda + \frac{2}{3}G$, where K is the bulk modulus, G the shear modulus, and λ the second elastic Lamé coefficient.

Use the alternative form of the generalized Hooke's Law (7.8) to find K in terms of the Young's modulus E and Poisson's ratio v. Find K in the limit $v = 0.5$.

(**b**) The specimen of tissue in Figure 7.25(a) is now modeled as a porous, fluid-filled cross-linked network of intrinsically incompressible macromolecules having an intra-tissue water content of 75%. The specimen is again immersed in a pressurized fluid, which can communicate with the network's intra-tissue fluid. When the fluid bath is pressurized, does the specimen deform (i.e., change shape or volume)? If so, by what mechanism and by how much?

(**c**) Repeat part (b) for the case of a cell, as shown in Figure 7.25(b). When pressurized, does the cell deform and, if so, by what mechanisms?

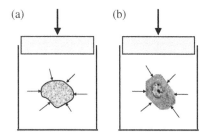

Figure 7.25 Application of hydrostatic (normal) stress.

Problem 7.4 Shear Stress A shear stress σ_{21} is applied to the tissue specimen of Figure 7.26(a). The applied shear stress produces a displacement of the upper surface ΔL that is 5% of its initial height h. In equilibrium, this tissue can be modeled as a linear, homogeneous, isotropic elastic solid.

(**a**) Find an expression for the equilibrium shear modulus G in terms of σ_{21}, ΔL, and h, and calculate the value of G.

(**b**) Find an expression for the Young's modulus E in terms of the shear modulus G and the Poisson's ratio, v, by applying both forms of Hooke's law to the shear configuration of Figure 7.26(a).

(**c**) Assume now that the equilibrium Poisson's ratio can take on values between 0 and 0.5 (as has been observed for a variety of tissues). Find expressions for E in terms of G in the two limits $v = 0$ and $v = 0.5$. For which case is E greater? Give a physical justification for your answer.

Figure 7.26 Shear modulus.

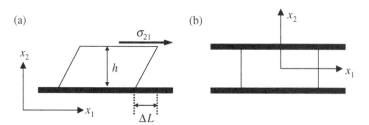

(**d**) The tissue is now tested in the unconfined compression configuration of Figure 7.26(b). Find expressions for σ_{11} and σ_{33} in terms of σ_{22}.

Problem 7.5 Effects of Tissue Anisotropy on Tissue and Cell Deformation Corneal stroma consists of type I collagen molecules co-assembled along with those of type V collagen into 4 nm "microfibrils" that form right-handed helices, and then into larger bundles of 36 nm fibrils [40]. The approximately 0.5–1 mm full thickness of the stroma ($h \simeq 1$ mm in Figure 7.27) consists of a stack of approximately 2 μm-thick sheets, or lamellae. The fibrils are highly oriented within the plane of each lamella, but the fibril orientation changes from lamella to lamella (Figure 7.27); thus, macroscopically, collagen orientation is random but within the plane of the cornea. The ability of corneal stroma to remain transparent is thought to depend on the precise spacing between fibrils, which is maintained, in part, by the presence of a variety of additional fibrillar and non-fibrillar collagen types, along with several members of the small, leucine-rich proteoglycan family (including decorin, lumican, keratocan, and osteoglycin (also known as mimecan)). Knockout mice in which one or more of these proteoglycans have been deleted develop corneal opacity, which mimics certain human diseases. The microfibril is known to be the principal tensile element within the radially oriented fibrils, and the swelling and biomechanical properties of the corneal stroma are thus dominated by these fibrils.

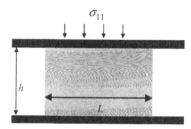

Figure 7.27 Cylindrical disk of corneal stroma.

(**a**) Figure 7.27 pictures a cylindrical disk of corneal stroma tissue subjected to unconfined compression. As a zeroth-order, simplified model, assume that the tissue can be modeled as an isotropic, homogeneous, linear elastic solid in equilibrium. A combination of experiments showed that the bulk modulus K was related to the Young's modulus E by $E = 3K = 1$ MPa.

Assume that there is no friction between the tissue and the compression platens (no shear stress at the interfaces). The tissue is compressed to 50% of its original height h (assume that this is still in the linear strain range).

Sketch the final equilibrium compressed state of the tissue. Show the final dimensions of the tissue disk by calculating any changes in the disk diameter. (State all reasoning and show all your work.)

(**b**) Repeat part (a) for the case in which the tissue is found to have $E = 3G$ (where G is the shear modulus).

(**c**) Given the anisotropic microstructure of the collagen fibril orientation within real corneal stroma, as pictured in Figure 7.27, would application of unconfined compression lead to the final deformed state of (a) or (b) as you calculated it above? (Use qualitative reasoning; no math here, but justify your answer.)

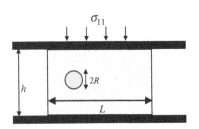

Figure 7.28 Cylindrical disk of tissue made of the extracellular medium of Figure 7.27, showing one of many "spherically shaped" cells within the matrix (cell size is greatly exaggerated).

(**d**) Figure 7.28 pictures a homogeneous, isotropic, elastic extracellular medium subjected to unconfined compression. In addition, a spherical cell within the extracellular medium is shown. The cell can be modeled in equilibrium as a porous elastic solid having a Young's modulus of 1 kPa. The tissue in Figure 7.28 is again compressed to 50% of its original height h.

Use physical reasoning (no math needed). For both cases, tissue $E = 3K$ and $E = 3G$, sketch the approximate final equilibrium compressed shape of the cell within the compressed tissue, denoting appropriate final dimensions of the cell. State all your reasoning.

Problem 7.6 General Relation Between Force Density and Stress Tensor In fluid mechanics, conservation of momentum leads to the Navier–Stokes equation for the fluid velocity v, which has the form

$$\rho \frac{\partial v}{\partial t} = \mu \nabla^2 v + \text{(other terms)}$$

where ρ is the mass density of the fluid and μ is the viscosity. The left-hand term represents inertia, the first term on the right-hand side is the force density associated with viscous stresses, and other terms on the right represent other relevant force densities that may be acting on an element of fluid (e.g., hydrostatic pressure ∇p, the electrical force density $\rho_e E$, where ρ_e is the electrical charge density, etc.).

In general, all such force densities, including that associated with elastic deformation, can be written as the divergence of a stress tensor.

Show that a vector force density F can, in general, be written as the divergence of its associated stress tensor $\underline{\underline{\sigma}}$, in vector notation,

$$F = \nabla \cdot \underline{\underline{\sigma}}$$

or, in index notation for the j component of the force,

$$F_j = \frac{\partial \sigma_{ij}}{\partial x_i}$$

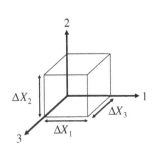

Figure 7.29 Cubic element for Problem 7.6.

One approach is to start with index notation and calculate the net force in one particular direction in terms of stress components acting on the differential faces of the cubic element shown in Figure 7.29, then extrapolate to all three directions.

Problem 7.7 Propagating Waves in Biological Tissues: the Elastic Limit This problem is motivated by a paper by Ghaffari et al. [41] on experiments to measure propagating waves in the mammalian tectorial membrane of the cochlea *in vitro* (Figure 7.30(a, b)).

(a)

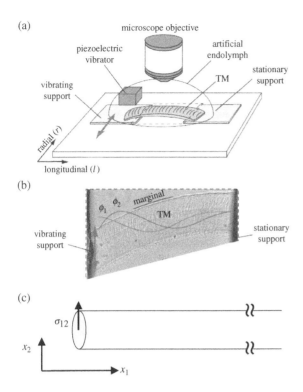

(b)

(c)

Figure 7.30 (a) Schematic of tectorial membrane (TM) segment suspended between two supports. (b) Light-microscope image of TM. ϕ_1 and ϕ_2 are schematic waveforms representing displacement snapshots at sequential instants. (c) Idealized cylindrical tube model. ((a) and (b) from Ghaffari R, Aranyosi AJ, and Freeman DM. Longitudinally propagating traveling waves of the mammalian tectorial membrane. *Proc. Natl Acad. Sci. USA* **104**, 16510–16515 (2007).)

The tectorial membrane (TM) is a highly hydrated gel-like connective tissue composed of "heteropolymeric" collagen fibrils (i.e., fibrils containing types II, IX, and XI collagen, similar to that in articular cartilage) as well as two major glycoproteins, α- and β-tectorin, and other proteoglycans. It is known that sound in the 20 Hz to 20 kHz range of human hearing results in the propagation of waves down the TM, at the same frequency as the input, in the "longitudinal" direction (that is, the x_1 direction in Figure 7.30(c)). In mouse models, gene deletion of one or more of the above matrix molecules can dramatically alter the wave propagation properties of the TM. In general, the viscoelastic properties of the TM result in decay of the wave amplitude along the direction of propagation (as shown in Figure 7.30(b)). In this problem, however, we focus on the elastic properties of the TM that enable wave propagation.

To simplify the problem, the TM is modeled here as a cylindrical tube of connective tissue that behaves as a linear, homogeneous, isotropic material having moduli G, E, λ, etc., and mass density ρ. (We neglect complex boundary conditions at the surface that may contribute to dissipation and other viscoelastic-like losses.) An oscillatory stress σ_{12} (caused by impinging sound waves) is applied to the left-hand face of the "elastic tube" of Figure 7.30(c). We assume that a shear wave of displacement having the form $u_2(x_1, t)$ is induced within the tube, which propagates in the x_1 direction.

(a) By analogy to the statement of Problem 7.6, use index notation to write the general form of the wave equation for the elastic solid by writing conservation of momentum, but here in terms of material displacements u_2 and the stress σ_{12}.

Figure 7.31 Charge titration behavior of collagen and proteoglycan components of TM.

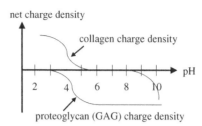

Then assume that the stress σ_{12} can be obtained from the generalized Hooke's law, and convert the equation to a one-dimensional wave equation for u_2 in terms of ρ and the appropriate elastic modulus.

(**b**) Experiments using the actual configuration of Figure 7.30(a, b) are accomplished by applying sinusoidal displacements u_2 to the left-hand edge of the TM (rather than applying a stress). It was observed that a displacement amplitude of about 100 nm at an imposed frequency of 15 kHz resulted in a shear wave having a wavelength of about 350 μm. What is the estimated value of the shear modulus of the TM?

(**c**) If the Poisson's ratio of the TM gel is approximately 0.5, what is its Young's modulus, E?

Problem 7.8 Material Moduli Are Affected by ECM Charge Density Assume that the collagen component of the tectorial membrane of Problem 7.7 has the "charge titration" behavior shown in Figure 7.31. In addition, the proteoglycans and their associated glycosaminoglycan (GAG) chains have a net negative charge for pH > 3.

For the three sketches below, the important result is the relative magnitude, not the exact magnitude.

(**a**) Sketch the net charge of the tissue versus bath pH over the entire pH range 2–10.

(**b**) Sketch the dependence of the equilibrium Young's modulus E on pH for the pH range 2–10. State your reasoning.

(**c**) Repeat part (b) for the case in which proteolytic enzymes (e.g., caused by disease or purposely in animal models of disease) have removed the proteoglycan constituents, leaving the collagen network intact.

Problem 7.9 Standard Three-Element Model for Viscoelastic Tissue and Cell Behavior

(**a**) A form of the "standard three-element model" for linear viscoelastic solids (Figure 7.32(a)) is used by Darling et al. [42] to represent the stress relaxation behavior (Figure 7.32(c)) of a cell as tested via atomic force microscopy as shown schematically in Figure 7.32(b). This three-element model

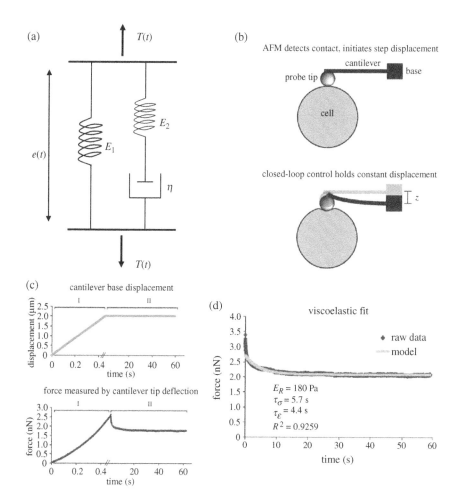

Figure 7.32 Standard three-element model for viscoelastic tissue and cell behavior. (From Darling EM, Zauscher S, and Guilak F. Viscoelastic properties of zonal articular chondrocytes measured by atomic force microscopy. *Osteoarthritis Cartilage* **14**, 571–579 (2006).)

can be represented by an ordinary differential equation of the form

$$T + \alpha \frac{dT}{dt} = E_1 e + \beta \frac{de}{dt} \qquad (7.73)$$

Show that this form is correct, and find the algebraic expressions for α and β in terms of the element values E_1, E_2, and η.

(b) Based on the form of (7.73), find expressions for the stress relaxation time constant τ_e (the relaxation time at constant strain) and the creep time constant τ_c (the relaxation time under constant load) in terms of E_1, E_2, and η.

(c) In Figure 7.32(d), the stress relaxation data of Figure 7.32(c) (region II of the "force" curve) are compared with the predictions of the three-element model of Figure 7.32(a) by obtaining best fit values of the three model parameters. Describe qualitatively how you would change the model in an attempt to improve the fit to the data even further at very early times (i.e., within the first 2 s) on the time scale of Figure 7.32(d). In the frequency domain, this would correspond to testing frequencies above about 1 Hz in part (d) below.

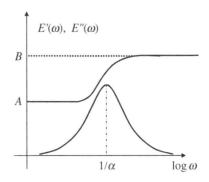

Figure 7.33 Frequency dependences of $E'(\omega)$ and $E''(\omega)$ in Problem 7.9(d).

(**d**) Based on the differential equation (7.73), derive an expression for the complex modulus that describes the frequency behavior of the three-element model of Figure 7.32(a) having the form

$$\hat{E}(\omega) = E'(\omega) + jE''(\omega)$$

Show that $E'(\omega)$ and $E''(\omega)$ have the frequency dependences shown qualitatively in Figure 7.33 by reasoning the low- and high-frequency limits. Find the constants A and B in terms of the element values E_1 and E_2 based on physical (and/or mathematical) arguments.

Problem 7.10 Poroelastic Behavior: Linear, Isotropic, Homogeneous Tissue A cylindrical disk of tissue is placed in the testing chamber shown in Figure 7.14 and subjected to uniaxial confined compression. Tissue behavior is to be modeled using a simple poroelastic description for a uniform, isotropic, linear material.

(**a**) Show that the displacement $u_1(x_1, t)$ is described by a partial differential equation having the form of a diffusion equation with equivalent "diffusivity" equal to Hk, the product of the confined compression modulus $H = 2G + \lambda$ and the hydraulic permeability k.

(**b**) A step in displacement is applied at $x_1 = 0$ having amplitude u_0. State the boundary conditions on $u_1(x_1 = 0, t)$ and $u_1(x_1 = L, t)$ and the initial condition $u_1(x_1, t = 0)$ that would be used to solve for the displacement $u_1(x_1, t)$ occurring during this "stress relaxation." (Do not solve.)

(**c**) A step in stress of amplitude σ_0 is applied at $x_1 = 0$. State the boundary conditions on the displacement (or its slope) and the initial condition on $u_1(x_1, t = 0)$ that would be used to solve for the creep displacement $u_1(x_1, t)$. (Do not solve.)

(**d**) For the stress relaxation example of part (b), solve the diffusion equation for the displacement $u_1(x_1, t)$ for all (x_1, t) given the initial and boundary conditions you provided above. (*Hint:* Using separation of variables, your answer will involve a sum of terms that are each periodic in space and

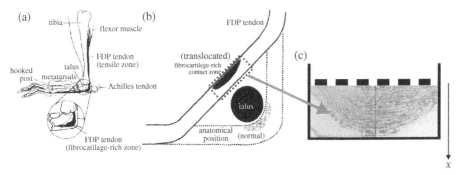

Figure 7.34 (a, b) Schematic of loading of rabbit flexor tendon (FDP, flexor digitorum profundus). (c) Confined compression experiment. ((a) and (b) from Malaviya P, Butler DL, Boivin GP, et al. An *in vivo* model for load-modulated remodeling in the rabbit flexor tendon. *J. Orthop. Res.* **18**, 116–125 (2000).)

exponentially decaying in time. Find the Fourier coefficients and the relaxation times that are given in (7.55).)

(**e**) Compare the relaxation time constant that you found in part (d) with the stress relaxation time for the viscoelastic model in Section 7.4. In both poroelastic and viscoelastic models, describe the state of strain (i.e., the spatial distribution of the displacement profile) that is present in each model at a time corresponding to one stress relaxation time constant.

Problem 7.11 *In Vivo* Model for Load-Modulated Remodeling in the Rabbit Flexor Tendon A study by Malaviya et al. [43] tested the hypothesis that eliminating *in vivo* compression of the wraparound, fibrocartilage-rich zone of the rabbit flexor tendon results in rapid depletion of fibrocartilage and changes in its mechanical properties and extracellular matrix composition. Flexor tendons normally wrap around the talus (heel) in the rabbit. The normal and shear forces exerted on the tendon by the bone result in cell synthesis of a fibrocartilage region within the tendon (the shaded region in Figure 7.34(b)). Malaviya et al. translocated the tendon away from the heel bone to eliminate *in vivo* compression and shear of the fibrocartilage zone and, at 4 weeks after surgery, cored cylindrical disks of tissue (dotted area in Figure 7.34(b)), and measured mechanical properties in uniaxial confined compression (Figure 7.34(c)). Cores from translocated tendons were compared with untreated control tendons and tendons that had sham surgery.

The data obtained from a variety of oscillatory (sinusoidal steady state) mechanical testing protocols can be compared with theoretical predictions based on idealized models that approximate the tissue as homogeneous, linear, and isotropic. Models (a), (b), and (c), shown in Figure 7.35, are linear viscoelastic models. Model (d) is described by

$$\frac{\partial u}{\partial t} = Hk\frac{\partial^2 u}{\partial x^2}$$

and represents the model for a uniform, isotropic, linear poroelastic material having longitudinal modulus $H = 2G + \lambda$ and hydraulic permeability k; here u is the displacement in the (axial) x direction.

Figure 7.35 Idealized models (a)–(c) in Problem 7.11.

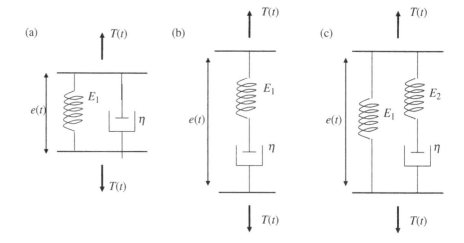

For models (a), (b), (c), and (d), sketch the magnitude of the frequency-dependent modulus versus frequency (call it $|\hat{G}(\omega)| = |T|/|e|$ for models (a)–(c) and $|\hat{H}(\omega)| = |\sigma|/|\varepsilon|$ for model (d)). Justify your answers for each model: state physical reasoning for the behavior at very low and very high frequencies. You do not have to derive the mathematical formula for the modulus for each model if you can justify it based on physical reasoning. Remember that the magnitude $|G|$ is different than G' or G'' alone; $|G|$ is a function of both G' and G''.

Using the poroelastic approach of model (d), you should now attempt to approximate the depth-dependent nonuniformity of the tissue in Figure 7.34(c) using the "two-layer" model shown in Figure 7.36, where the confined compression modulus H and hydraulic permeability k take on different values for the upper and lower tissue. A step in displacement having amplitude u_0 is applied at $x = 0$, $t = 0$.

(**i**) State the partial differential equation that governs the displacement $u(x, t)$ in each region (do not solve).

(**ii**) State the boundary conditions on u at $x = 0$, $x = w$, and $x = L$ for all time, and the initial condition $u(x, t = 0^+)$ that would be used to solve for the displacement $u(x, t)$ everywhere in the tissue during a stress relaxation experiment (do not solve.)

(**iii**) State the relation between the fluid pressure in the upper and lower regions (1 and 2) at the position $x = w$ during the

Figure 7.36 "Two-layer" model.

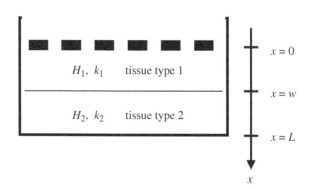

stress relaxation process. Repeat for the total stress σ and fluid velocity U (relative to the solid matrix) in regions 1 and 2 at $x = w$.

(**iv**) If $k_1 = k_2$, but $H_1 < H_2$, does the stress relaxation process in the upper region 1 occur more quickly, more slowly, or at the same rate as stress relaxation in region 2? State your reasoning.

Problem 7.12 Convective Augmentation of Nutrient Transport into Tissue-Engineered Constructs This problem emphasizes the effects of fluid flow into and across soft, gel-like tissue-engineered constructs sometimes used for augmenting nutrient transport during tissue growth. Such fluid flow is induced by applying a pressure gradient across the construct (see Figure 7.37). The question is whether such fluid flow can cause inadvertent deformation (e.g., consolidation or compaction) of the tissue that may be harmful to the cells. This issue is important in state-of-the-art microelectromechanical systems applications (in which cells and neo-tissues are grown within microfluidic channels), as well as more macroscopic-scale applications. A cylindrical disk of porous, hydrated tissue has thickness d and diameter D, and is held within a chamber that confines the disk at its radial periphery. The tissue is supported by a rigid, porous filter located at the position $z = 0$. A constant pressure drop P_0 is applied across the tissue from left to right, resulting in a constant fluid flow velocity U_0 with respect to the tissue (the rigid filter prevents the tissue from moving, but does not impede fluid flow).

(**a**) Assuming a one-dimensional model, write expressions for (1) conservation of momentum, (2) Darcy's law, (3) total stress versus strain (including hydrostatic pressure), and (4) conservation of mass. In your conservation-of-mass equation, you should include a term $U = U_0$ that corresponds to the possibility of a constant fluid flow in the steady state (when $\partial/\partial t = 0$).

(**b**) Combine your equations in part (a) to find a differential equation for u_z in terms of the constant velocity U_0, the hydraulic permeability k, and the confined compression modulus $H = 2G + \lambda$.

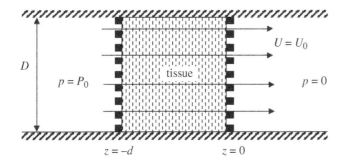

Figure 7.37 Effects of fluid flow on scaffold compaction.

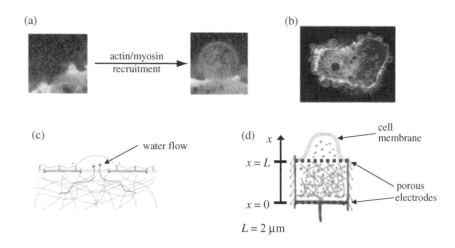

Figure 7.38 Intracellular poroelasticity. ((a) and (c) from Charras GT, Yarrow JC, Horton MA, et al. Non-equilibration of hydrostatic pressure in blebbing cells. *Nature* **435**, 365–369 (2005).)

(**c**) For the case of steady flow ($\partial/\partial t = 0$), integrate your differential equation to find an expression for the displacement u_z in terms of two integration constants. Find the two constants from the boundary conditions: (i) zero displacement at $z = 0$; (ii) zero strain at $z = -d$.

(**d**) Derive expressions for $u_z(z)$ and $e_{zz}(z)$ within the tissue. Sketch u_z and e_{zz} as functions of z within the tissue ($-d < z < 0$). Do you find compaction or not?

Problem 7.13 Intracellular Poroelasticity The recruitment of the actin/myosin cytoskeleton to form protrusive blebs at the cell membrane (Figure 7.38(a, b)) has been hypothesized to be a key step in motility for certain cell types. Contraction of the cell cortex is thought to induce pressure gradients within the cell, giving rise to fluid flow within the cytoskeleton and local nonuniform deformations, which are especially pronounced in the region of the bleb (Figure 7.38(c)). Local rupture of the cell membrane may further facilitate outward flow of fluid. This problem explores this hypothesis in terms of a one-dimensional model (Figure 7.38(d)). Assume, first, that the cytoskeleton in the region of interest behaves as a linear, homogeneous, isotropic, negatively charged ($k_{12} = k_{21}$) poroelastic network having a longitudinal modulus H and a short-circuit Darcy hydraulic permeability k_{11}. Two nano-wire-mesh porous electrodes with an enmeshed pressure sensor have been designed to enable measurement of the streaming potential drop across the $L = 2\,\mu m$ height of the bleb region in Figure 7.38(d), as well as the pressure drop across the bleb region generated by cytoskeletal contraction. The electrodes constrain the displacement of the cytoskeletal network at $x = 0$ and $x = L$ to be zero.

(**a**) Derive the partial differential equation that can be used to solve for the space-time distribution of pressure $p(x, t)$ and

PROBLEM

relative fluid velocity $U(x, t)$ within the one-dimensional bleb of Figure 7.38(d), including the effect of the charge of the actin/myosin network.

(**b**) Assume that local cell motion results in an instantaneous step in pressure $p = P_0$ at the base of the bleb, $x = 0$, in Figure 7.38(d), while the ambient pressure outside the cell in the medium ($x = L$) is always $p = 0$. State the initial condition as well as any appropriate boundary conditions on p, U, the displacement u, and stress σ needed to solve the equations of part (a). Assume that the electrical potential V between the electrode at $x = 0$ with respect to the reference ground at $x = L$ is measured using a high-input-impedance electrometer amplifier.

(**c**) Find an analytical expression for the final steady state relative fluid velocity $U(x)$ within the cytoskeletal network in the region $0 < x < L$. Then solve for the corresponding spatial dependence of the electrical potential $V(x)$. (Note that this part can be done independently of part (a).)

(**d**) What is the time constant for the transient change in fluid velocity and electrical potential after application of the step in pressure? Write your expression in terms of the material properties (including appropriate k_{ij}) and appropriate geometrical constants.

(**e**) Much recent research has suggested that the cytoskeletal network behaves like a viscoelastic solid. Assume that the solid phase of the cytoskeletal network of Figure 7.38(d) can be modeled as a standard three-element viscoelastic solid. For oscillatory sinusoidal steady state behavior, show that the viscoelastic description of the solid network can be embedded within the sinusoidal steady state form of the poroelastic equation of motion to give a one-dimensional "poro-viscoelastic" description. Describe how you would formulate an expression for the complex modulus of the material in terms of the poroelastic material properties, including the viscoelastic behavior of the solid network.

(**f**) Find the displacement $u(x, t)$ caused by a step in current applied at the electrodes, leading to a step in current density $J_0(t)$ (see (7.67)).

7.8 REFERENCES

[1] Buschmann MD and Grodzinsky AJ (1995) A molecular model of proteoglycan associated forces in cartilage mechanics. *J. Biomech. Eng.* **117**, 179–192.

[2] Maroudas A (1976) Balance between swelling pressure and collagen tension in normal and degenerate cartilage. *Nature* **260**, 808–809.

[3] Espinosa M, Noé G, Troncoso C, et al. (2002) Acidic pH and increasing [Ca^{2+}] reduce the swelling of mucins in primary cultures of human cervical cells. *Hum. Reprod.* **17**, 1964–1972.

[4] Tanaka T and Fillmore DJ (1979) Kinetics of swelling of gels. *J. Chem. Phys.* **70**, 1214–1218.

[5] Glimcher MJ and Krane SM (1968) The organization and structure of bone and the mechanism of calcification. In *Treatise on Collagen*, Volume 2B (Gould BS, ed.). Academic Press, New York, pp. 68–251.

[6] Canty EG, Lu Y, Meadows RS, et al. (2004) Coalignment of plasma membrane channels and protrusions (fibripositors) specifies the parallelism of tendon. *J. Cell Biol.* **165**, 553–563.

[7] Van de Velde SK, Bingham JT, Hosseini A, et al. (2009) Increased tibiofemoral cartilage contact deformation in patients with anterior cruciate ligament deficiency. *Arthritis Rheumatism* **60**, 3693–3702.

[8] Stratton JA (1941) *Electromagnetic Theory.* McGraw-Hill, New York (reprinted 2007 by IEEE Press/Wiley, Hoboken, NJ).

[9] Fung YC (1993) *Biomechanics: Mechanical Properties of Living Tissues*, 2nd ed. Springer-Verlag, New York.

[10] Fung YC (1990) *Biomechanics: Motion, Flow, Stress, and Growth.* Springer-Verlag, New York.

[11] Long RR (1961) *Mechanics of Solids and Fluids.* Prentice Hall, Englewood Cliffs, NJ.

[12] Ferry JD (1980) *Viscoelastic Properties of Polymers*, 3rd ed. Wiley, New York.

[13] Ward IM (1983) *Mechanical Properties of Solid Polymers*, 2nd ed. Wiley-Interscience, London.

[14] Flory PJ (1953) *Principles of Polymer Chemistry.* Cornell University Press, Ithaca, NY.

[15] Doi M and Edwards SF (1986) *The Theory of Polymer Dynamics.* Oxford University Press, Oxford.

[16] Terzaghi K (1943) *Theoretical Soil Mechanics.* Wiley, New York.

[17] Biot MA (1941) General theory of three-dimensional consolidation. *J. Appl. Phys.* **12**, 155–164.

[18] Biot MA (1962) Mechanics of deformation and acoustic propagation in porous media. *J. Appl. Phys.* **33**, 1482–1498.

[19] Rice JR and Cleary MP (1976) Some basic stress diffusion solutions for fluid-saturated elastic porous media with compressible constituents. *Rev. Geophys. Space Phys.* **14**, 227–241.

[20] Detournay E and Cheng AHD (1993) Fundamentals of poroelasticity. In *Comprehensive Rock Engineering: Principles, Practice and Projects,* Volume II: *Analysis and Design Method* (Fairhurst C, ed.). Pergamon Press, Oxford, pp. 113–171.

[21] Bowen RM (1976) Theory of mixtures. In *Continuum Physics,* Volume III: *Mixtures and EM Field Theories* (Eringen AC, ed.). Academic Press, New York, pp. 1–127.

[22] Mow VC, Kuei SC, Lai WM, and Armstrong CG (1980) Biphasic creep and stress relaxation of articular cartilage in compression: theory and experiments. *J. Biomech. Eng.* **102**, 73–84.

[23] Tanaka T, Hocker LO, and Benedek GB (1973) Spectrum of light scattered from a viscoelastic gel. *J. Chem. Phys.* **59**, 5151–5159.

[24] Lanir Y (1987) Biorheology and fluid flux in swelling tissues. I. Bicomponent theory for small deformations, including concentration effects. *Biorheology* **24**, 173–187.

[25] Cowin SC (1999) Bone poroelasticity. *J. Biomech.* **32**, 217–238.

[26] Huyghe JM, Molenaar MM, and Baajens FPT (2007) Poromechanics of compressible charged porous media using the theory of mixtures. *J. Biomech. Eng.* **129**, 776–785.

[27] Charras GT, Yarrow JC, Horton MA, et al. (2005) Non-equilibration of hydrostatic pressure in blebbing cells. *Nature* **435**, 365–369.

[28] Tanaka T and Fillmore DJ (1979) Kinetics of swelling of gels. *J. Chem. Phys.* **70**, 1214–1218.

[29] Ateshian GA, Ellis BJ, and Weiss JA (2007) Equivalence between short-time biphasic and incompressible elastic material responses. *J. Biomech. Eng.* **129**, 405–412.

[30] Eisenberg SR and Grodzinsky AJ (1985) Swelling of articular cartilage and other connective tissues: electromechanochemical forces. *J. Orthop. Res.* **3**, 148–159.

[31] Eisenberg SR and Grodzinsky AJ (1987) The kinetics of chemically induced nonequilibrium swelling of articular cartilage and corneal stroma. *J. Biomech. Eng.* **109**, 79–89.

[32] Frank EH, Grodzinsky AJ, Phillips SL, and Grimshaw PE (1990) Physicochemical and bioelectrical determinants of cartilage material properties. In *Biomechanics of Diarthrodial Joints* (Mow VC, Ratcliffe A, and Woo SL-Y, eds.). Springer-Verlag, New York, pp. 261–282.

[33] Lai WM, Hou JS, and Mow VC (1991) A triphasic theory for the swelling and deformation behaviors of articular cartilage. *J. Biomech. Eng.* **113**, 245–258.

[34] Huyghe JM and Janssen JD (1997) Quadriphasic mechanics of swelling incompressible porous media. *Int. J. Eng. Sci.* **35**, 793–802.

[35] Loret B and Simões FMF (2010) Effects of pH on transport properties of articular cartilages. *Biomech. Model. Mechanobiol.* **9**, 45–63.

[36] Katchalsky A and Curran PF (1965) *Nonequilibrium Thermodynamics in Biophysics.* Harvard University Press, Cambridge, MA.

[37] Mattern KJ, Nakornchai C, and Deen WM (2008) Darcy permeabilty of agarose–glycosaminoglycan gels analyzed using fiber-mixture and Donnan models. *Biophys. J.* **95**, 648–656.

[38] Frank EH and Grodzinsky AJ (1987) Cartilage electromechanics I: electrokinetic transduction and the effects of pH and ionic strength. *J. Biomech.* **20**, 615–627.

[39] Frank EH and Grodzinsky AJ (1987) Cartilage electromechanics II: a continuum model of cartilage electrokinetics and correlation with experiments. *J. Biomech.* **20**, 629–639.

[40] Holmes DF, Gilpin CJ, Baldock C, et al. (2001) Corneal collagen fibril structure in three dimensions: structural insights into fibril assembly, mechanical properties, and tissue organization. *Proc. Natl Acad. Sci. USA* **98**, 7307–7312.

[41] Ghaffari R, Aranyosi AJ, and Freeman DM (2007) Longitudinally propagating traveling waves of the mammalian tectorial membrane. *Proc. Natl Acad. Sci. USA* **104**, 16510–16515.

[42] Darling EM, Zauscher S, and Guilak F (2006) Viscoelastic properties of zonal articular chondrocytes measured by atomic force microscopy. *Osteoarthritis Cartilage* **14**, 571–579.

[43] Malaviya P, Butler DL, Boivin GP, et al. (2000) An *in vivo* model for load-modulated remodeling in the rabbit flexor tendon. *J. Orthop. Res.* **18**, 116–125.

Appendix A
Integral Theorems

In formulating laws of continuum mechanics, it is often convenient to consider a volume or surface always made up of the same particles. Such volumes and surfaces are therefore time-dependent. To convert a law stated in terms of integrals into a differential law, it is necessary to be able to reverse the order in which integration and differentiation are carried out. It is to this end that the following theorems are derived. These can be termed generalized forms of the Leibnitz rule for differentiating an integral.

TIME RATE OF CHANGE OF A SURFACE INTEGRAL

Consider the differentiation of a surface integral having not only a time-dependent integrand, but a time varying surface of integration as well. By definition,

$$\frac{d}{dt} \int_{S(t)} \boldsymbol{A} \cdot \boldsymbol{n}\, da = \lim_{\Delta t \to 0} \frac{1}{\Delta t} \left[\int_{S(t+\Delta t)} \boldsymbol{A}(t + \Delta t) \cdot \boldsymbol{n}\, da - \int_{S(t)} \boldsymbol{A}(t) \cdot \boldsymbol{n}\, da \right] \quad (A.1)$$

The surface $S(t)$ moves with velocity v_s so that in the time interval from t to $t + \Delta t$, S assumes a new position as depicted in Figure A.1. With the objective of evaluating the right-hand side of (A.1), it is only necessary to know the quantity to the term linear in Δt. Hence, observe that

$$\boldsymbol{A}(t + \Delta t) = \boldsymbol{A}(t) + \left. \frac{\partial \boldsymbol{A}}{\partial t} \right|_t \Delta t \quad (A.2)$$

so that, to first order in Δt, (A.1) becomes

$$\frac{d}{dt} \int_{S(t)} \boldsymbol{A} \cdot \boldsymbol{n}\, da = \lim_{\Delta t \to 0} \frac{1}{\Delta t} \left\{ \int_{S(t)} \frac{\partial \boldsymbol{A}}{\partial t} \Delta t \cdot \boldsymbol{n}\, da \right.$$
$$\left. + \left[\int_{S(t+\Delta t)} \boldsymbol{A}(t) \cdot \boldsymbol{n}\, da - \int_{S(t)} \boldsymbol{A}(t) \cdot \boldsymbol{n}\, da \right] \right\} \quad (A.3)$$

The integrals in this expression are over surfaces at different instants in time, but what is being integrated in each case is the same function. This makes it possible to exploit the conventional Gauss' theorem applied to a volume V swept out by the surface S during the interval Δt:

$$\int_V \nabla \cdot \boldsymbol{A}\, dV = \oint_S \boldsymbol{A} \cdot \boldsymbol{i}_n\, da \quad (A.4)$$

With $\boldsymbol{n}$ defined as the unit normal to the surfaces S composing the top and bottom of V in Figure A.1, it is seen that $\boldsymbol{n} = \boldsymbol{i}_n$ on the top surface but $\boldsymbol{n} = -\boldsymbol{i}_n$ on the bottom surface. On the sides, a differential area

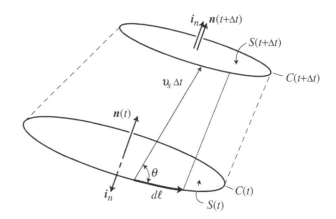

Figure A.1 Surface $S(t)$ with periphery $C(t)$ moves at velocity v_s.

element defined as having a direction normal to the surface enclosing V is

$$\boldsymbol{i}_n \, da \approx d\boldsymbol{\ell} \times v\Delta t \tag{A.5}$$

as can be seen by recognizing that $|d\boldsymbol{\ell} \times v| = |d\boldsymbol{\ell}||v_s| \sin\theta$, so that $\boldsymbol{i}_n \, da$ is the area of the differential element of area joining the top and bottom surfaces. Still simply evaluating (A.5), we can write the surface integral as a sum of integrations over the top, bottom, and sides, so that, to first order in Δt,

$$\int_V \nabla \cdot \boldsymbol{A} \, dV \longrightarrow \int_{S(t)} (\nabla \cdot \boldsymbol{A}) v_s \Delta t \cdot \boldsymbol{n} \, da = \int_{S(t+\Delta t)} \boldsymbol{A} \cdot \boldsymbol{n} \, da$$

$$- \int_{S(t)} \boldsymbol{A} \cdot \boldsymbol{n} \, da + \oint_{C(t)} \boldsymbol{A} \cdot d\boldsymbol{\ell} \times v_s \Delta t \tag{A.6}$$

The two terms after the equality are those required to evaluate the expression in square brackets in (A.3). Recalling that $\boldsymbol{A} \cdot d\boldsymbol{\ell} \times v_s = -\boldsymbol{A} \times v_s \cdot d\boldsymbol{\ell}$, it follows that

$$\frac{d}{dt} \int_{S(t)} \boldsymbol{A} \cdot \boldsymbol{n} \, da = \int_{S(t)} \left[\frac{\partial \boldsymbol{A}}{\partial t} + (\nabla \cdot \boldsymbol{A}) v_s \right] \cdot \boldsymbol{n} \, da + \oint_C \boldsymbol{A} \times v_s \cdot d\boldsymbol{\ell} \tag{A.7}$$

where Δt has been cancelled out and terms omitted that are of higher order in Δt and would tend to zero in the limit $\Delta t \to 0$.

In using (A.7) to take the time derivative inside the surface integral, remember that v_s is defined as the velocity of the surface, a geometric relation between $S(t)$ and v_s. In applying (A.7) to a physical problem, $S(t)$ is usually identified with particle motions, and hence v_s assumes a physical interpretation not implied by the theorem of (A.7).

TIME RATE OF CHANGE OF A VOLUME INTEGRAL

To convert the theorem of (A.7) to one replacing the surface integration on the left by a volume integration, picture the surface of Figure A.2. The contour C is the "drawstring" on a bag that is the surface S. As this string is pulled tight, $C \to 0$ and the surface S becomes closed. Then,

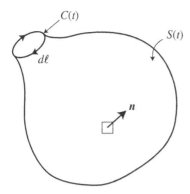

Figure A.2 Surface S and contour C. As $C \to 0$, S becomes closed with enclosed volume V.

(A.7) states that

$$\frac{d}{dt} \oint_S \boldsymbol{A} \cdot \boldsymbol{n}\, da = \oint_S \left[\frac{\partial \boldsymbol{A}}{\partial t} + (\nabla \cdot \boldsymbol{A}) v_s \right] \cdot \boldsymbol{n}\, da \qquad (A.8)$$

By Gauss' theorem, the closed surface integral on the left is the integral over the volume of $\zeta = \nabla \cdot \boldsymbol{A}$. A similar application to the first term on the right together with a permutation of time derivative and divergence operation then makes it possible to write (A.8) as the desired theorem

$$\frac{d}{dt} \int_V \zeta\, dV = \int_V \frac{\partial \zeta}{\partial t}\, dV + \oint_S \zeta v_s \cdot \boldsymbol{n}\, da \qquad (A.9)$$

Appendix B
Differential Operators in
Various Coordinate Systems

Table B.1 Differential operators in Cartesian coordinates. γ is a scalar function and $\boldsymbol{C}$ is a vector given by $\boldsymbol{C} = C_x \boldsymbol{i}_x + C_y \boldsymbol{i}_y + C_z \boldsymbol{i}_z$.

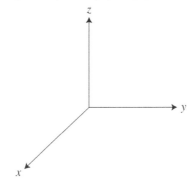

Cartesian Coordinates

Scalar gradient

$$\nabla \gamma = \frac{\partial \gamma}{\partial x} \boldsymbol{i}_x + \frac{\partial \gamma}{\partial y} \boldsymbol{i}_y + \frac{\partial \gamma}{\partial z} \boldsymbol{i}_z$$

Divergence

$$\nabla \cdot \boldsymbol{C} = \frac{\partial C_x}{\partial x} + \frac{\partial C_y}{\partial y} + \frac{\partial C_z}{\partial z}$$

Curl

$$\nabla \times \boldsymbol{C} = \left(\frac{\partial C_z}{\partial y} - \frac{\partial C_y}{\partial z} \right) \boldsymbol{i}_x$$
$$+ \left(\frac{\partial C_x}{\partial z} - \frac{\partial C_z}{\partial x} \right) \boldsymbol{i}_y$$
$$+ \left(\frac{\partial C_y}{\partial x} - \frac{\partial C_x}{\partial y} \right) \boldsymbol{i}_z$$

Scalar Laplacian

$$\nabla^2 \gamma = \frac{\partial^2 \gamma}{\partial x^2} + \frac{\partial^2 \gamma}{\partial y^2} + \frac{\partial^2 \gamma}{\partial z^2}$$

Vector Laplacian

$$\nabla^2 \boldsymbol{C} = \nabla^2 C_x \boldsymbol{i}_x + \nabla^2 C_y \boldsymbol{i}_y + \nabla^2 C_z \boldsymbol{i}_z$$

Table B.2 Additional differential operations in cartesian coordinates. v is a vector given by $v = v_x \boldsymbol{i}_x + v_y \boldsymbol{i}_y + v_z \boldsymbol{i}_z$ and τ is a symmetric tensor.

$$v \cdot \nabla v = \left(v_x \frac{\partial v_x}{\partial x} + v_y \frac{\partial v_x}{\partial y} + v_z \frac{\partial v_x}{\partial z} \right) \boldsymbol{i}_x$$

$$+ \left(v_x \frac{\partial v_y}{\partial x} + v_y \frac{\partial v_y}{\partial y} + v_z \frac{\partial v_y}{\partial z} \right) \boldsymbol{i}_y$$

$$+ \left(v_x \frac{\partial v_z}{\partial x} + v_y \frac{\partial v_z}{\partial y} + v_z \frac{\partial v_z}{\partial z} \right) \boldsymbol{i}_z$$

$$\tau : \nabla v = \tau_{xx} \frac{\partial v_x}{\partial x} + \tau_{yy} \frac{\partial v_y}{\partial y} + \tau_{zz} \frac{\partial v_z}{\partial z}$$

$$+ \tau_{xy} \left(\frac{\partial v_x}{\partial y} + \frac{\partial v_y}{\partial x} \right) + \tau_{yz} \left(\frac{\partial v_y}{\partial z} + \frac{\partial v_z}{\partial y} \right) + \tau_{zx} \left(\frac{\partial v_z}{\partial x} + \frac{\partial v_x}{\partial z} \right)$$

$$\nabla \cdot \tau = \left(\frac{\partial \tau_{xx}}{\partial x} + \frac{\partial \tau_{xy}}{\partial y} + \frac{\partial \tau_{xz}}{\partial z} \right) \boldsymbol{i}_x$$

$$+ \left(\frac{\partial \tau_{xy}}{\partial x} + \frac{\partial \tau_{yy}}{\partial y} + \frac{\partial \tau_{yz}}{\partial z} \right) \boldsymbol{i}_y$$

$$+ \left(\frac{\partial \tau_{xz}}{\partial x} + \frac{\partial \tau_{yz}}{\partial y} + \frac{\partial \tau_{zz}}{\partial z} \right) \boldsymbol{i}_z$$

Table B.3 Differential operators in cylindrical coordinates. γ is a scalar function and $\boldsymbol{C}$ is a vector given by $\boldsymbol{C} = C_r \boldsymbol{i}_r + C_\phi \boldsymbol{i}_\phi + C_z \boldsymbol{i}_z$.

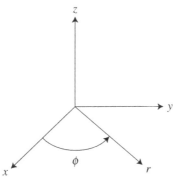

Cylindrical Coordinates

Scalar gradient

$$\nabla \gamma = \frac{\partial \gamma}{\partial r} \boldsymbol{i}_r + \frac{1}{r} \frac{\partial \gamma}{\partial \phi} \boldsymbol{i}_\phi + \frac{\partial \gamma}{\partial z} \boldsymbol{i}_z$$

Divergence

$$\nabla \cdot \boldsymbol{C} = \frac{1}{r} \frac{\partial}{\partial r}(rC_r) + \frac{1}{r} \frac{\partial C_\phi}{\partial \phi} + \frac{\partial C_z}{\partial z}$$

Curl

$$\begin{aligned} \nabla \times \boldsymbol{C} = & \left(\frac{1}{r} \frac{\partial C_z}{\partial \phi} - \frac{\partial C_\phi}{\partial z} \right) \boldsymbol{i}_r \\ & + \left(\frac{\partial C_r}{\partial z} - \frac{\partial C_z}{\partial r} \right) \boldsymbol{i}_\phi \\ & + \left(\frac{1}{r} \frac{\partial}{\partial r}(rC_\phi) - \frac{1}{r} \frac{\partial C_r}{\partial \phi} \right) \boldsymbol{i}_z \end{aligned}$$

Scalar Laplacian

$$\nabla^2 \gamma = \frac{1}{r} \frac{\partial}{\partial r} \left(r \frac{\partial \gamma}{\partial r} \right) + \frac{1}{r^2} \frac{\partial^2 \gamma}{\partial \phi^2} + \frac{\partial^2 \gamma}{\partial z^2}$$

Vector Laplacian

$$\begin{aligned} \nabla^2 \boldsymbol{C} = & \left(\nabla^2 C_r - \frac{C_r}{r^2} - \frac{2}{r^2} \frac{\partial C_\phi}{\partial \phi} \right) \boldsymbol{i}_r \\ & + \left(\nabla^2 C_\phi - \frac{C_\phi}{r^2} + \frac{2}{r^2} \frac{\partial C_r}{\partial \phi} \right) \boldsymbol{i}_\phi \\ & + \nabla^2 C_z \boldsymbol{i}_z \end{aligned}$$

Table B.4 Additional differential operations in cylindrical coordinates. v is a vector given by $v = v_r \boldsymbol{i}_r + v_\phi \boldsymbol{i}_\phi + v_z \boldsymbol{i}_z$ and τ is a symmetric tensor.

$$
v \cdot \nabla v = \left(v_r \frac{\partial v_r}{\partial r} + \frac{v_\phi}{r} \frac{\partial v_r}{\partial \phi} - \frac{v_\phi^2}{r} + v_z \frac{\partial v_r}{\partial z} \right) \boldsymbol{i}_r
$$
$$
+ \left(v_r \frac{\partial v_\phi}{\partial r} + \frac{v_\phi}{r} \frac{\partial v_\phi}{\partial \phi} + \frac{v_r v_\phi}{r} + v_z \frac{\partial v_\phi}{\partial z} \right) \boldsymbol{i}_\phi
$$
$$
+ \left(v_r \frac{\partial v_z}{\partial r} + \frac{v_\phi}{r} \frac{\partial v_z}{\partial \phi} + v_z \frac{\partial v_z}{\partial z} \right) \boldsymbol{i}_z
$$

$$
\tau : \nabla v = \tau_{rr} \frac{\partial v_r}{\partial r} + \tau_{\phi\phi} \left(\frac{1}{r} \frac{\partial v_\phi}{\partial \phi} + \frac{v_r}{r} \right) + \tau_{zz} \frac{\partial v_z}{\partial z}
$$
$$
+ \tau_{r\phi} \left(r \frac{\partial (v_\phi/r)}{\partial r} + \frac{1}{r} \frac{\partial v_r}{\partial \phi} \right) + \tau_{\phi z} \left(\frac{\partial v_\phi}{\partial z} + \frac{1}{r} \frac{\partial v_z}{\partial \phi} \right) + \tau_{rz} \left(\frac{\partial v_z}{\partial r} + \frac{\partial v_r}{\partial z} \right)
$$

$$
\nabla \cdot \tau = \left(\frac{1}{r} \frac{\partial (r\tau_{rr})}{\partial r} + \frac{1}{r} \frac{\partial \tau_{r\phi}}{\partial \phi} - \frac{\tau_{\phi\phi}}{r} + \frac{\partial \tau_{rz}}{\partial z} \right) \boldsymbol{i}_r
$$
$$
+ \left(\frac{\partial \tau_{r\phi}}{\partial r} + \frac{2\tau_{r\phi}}{r} + \frac{1}{r} \frac{\partial \tau_{\phi\phi}}{\partial \phi} + \frac{\partial \tau_{\phi z}}{\partial z} \right) \boldsymbol{i}_\phi
$$
$$
+ \left(\frac{1}{r} \frac{\partial (r\tau_{rz})}{\partial r} + \frac{1}{r} \frac{\partial \tau_{\phi z}}{\partial \phi} + \frac{\partial \tau_{zz}}{\partial z} \right) \boldsymbol{i}_z
$$

Table B.5 Differential operators in spherical coordinates. γ is a scalar function and $\boldsymbol{C}$ is a vector given by $\boldsymbol{C} = C_r \boldsymbol{i}_r + C_\theta \boldsymbol{i}_\theta + C_\phi \boldsymbol{i}_\phi$.

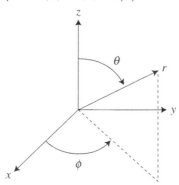

Spherical Coordinates

Scalar gradient
$$\nabla \gamma = \frac{\partial \gamma}{\partial r} \boldsymbol{i}_r + \frac{1}{r} \frac{\partial \gamma}{\partial \theta} \boldsymbol{i}_\theta + \frac{1}{r \sin \theta} \frac{\partial \gamma}{\partial \phi} \boldsymbol{i}_\phi$$

Divergence
$$\nabla \cdot \boldsymbol{C} = \frac{1}{r^2} \frac{\partial}{\partial r} \left(r^2 C_r \right) + \frac{1}{r \sin \theta} \frac{\partial}{\partial \theta} \left(C_\theta \sin \theta \right) + \frac{1}{r \sin \theta} \frac{\partial C_\phi}{\partial \phi}$$

Curl
$$\nabla \times \boldsymbol{C} = \left(\frac{1}{r \sin \theta} \frac{\partial}{\partial \theta} (C_\phi \sin \theta) - \frac{1}{r \sin \theta} \frac{\partial C_\theta}{\partial \phi} \right) \boldsymbol{i}_r$$
$$+ \left(\frac{1}{r \sin \theta} \frac{\partial C_r}{\partial \phi} - \frac{1}{r} \frac{\partial}{\partial r} (r C_\phi) \right) \boldsymbol{i}_\theta$$
$$+ \left(\frac{1}{r} \frac{\partial}{\partial r} (r C_\theta) - \frac{1}{r} \frac{\partial C_r}{\partial \theta} \right) \boldsymbol{i}_\phi$$

Scalar Laplacian
$$\nabla^2 \gamma = \frac{1}{r^2} \frac{\partial}{\partial r} \left(r^2 \frac{\partial \gamma}{\partial r} \right) + \frac{1}{r^2 \sin \theta} \frac{\partial}{\partial \theta} \left(\sin \theta \frac{\partial \gamma}{\partial \theta} \right) + \frac{1}{r^2 \sin^2 \theta} \frac{\partial^2 \gamma}{\partial \phi^2}$$

Vector Laplacian
$$\nabla^2 \boldsymbol{C} = \left(\nabla^2 C_r - \frac{2 C_r}{r^2} - \frac{2}{r^2} \frac{\partial C_\theta}{\partial \theta} - \frac{2 C_\theta \cos \theta}{r^2 \sin \theta} - \frac{2}{r^2 \sin \theta} \frac{\partial C_\phi}{\partial \phi} \right) \boldsymbol{i}_r$$
$$+ \left(\nabla^2 C_\theta + \frac{2}{r^2} \frac{\partial C_r}{\partial \theta} - \frac{C_\theta}{r^2 \sin^2 \theta} - \frac{2 \cos \theta}{r^2 \sin^2 \theta} \frac{\partial C_\phi}{\partial \phi} \right) \boldsymbol{i}_\theta$$
$$+ \left(\nabla^2 C_\phi + \frac{2}{r^2 \sin \theta} \frac{\partial C_r}{\partial \phi} + \frac{2 \cos \theta}{r^2 \sin^2 \theta} \frac{\partial C_\theta}{\partial \phi} - \frac{C_\phi}{r^2 \sin^2 \theta} \right) \boldsymbol{i}_\phi$$

Table B.6 Additional differential operations in spherical coordinates. v is a vector given by $v = v_r \boldsymbol{i}_r + v_\theta \boldsymbol{i}_\theta + v_\phi \boldsymbol{i}_\phi$, and τ is a symmetric tensor.

$$v \cdot \nabla v = \left(v_r \frac{\partial v_r}{\partial r} + \frac{v_\theta}{r} \frac{\partial v_r}{\partial \theta} + \frac{v_\phi}{r \sin\theta} \frac{\partial v_r}{\partial \phi} - \frac{v_\theta^2 + v_\phi^2}{r} \right) \boldsymbol{i}_r$$

$$+ \left(v_r \frac{\partial v_\theta}{\partial r} + \frac{v_\theta}{r} \frac{\partial v_\theta}{\partial \theta} + \frac{v_\phi}{r \sin\theta} \frac{\partial v_\theta}{\partial \phi} + \frac{v_r v_\theta}{r} - \frac{v_\phi^2 \cos\theta}{r \sin\theta} \right) \boldsymbol{i}_\theta$$

$$+ \left(v_r \frac{\partial v_\phi}{\partial r} + \frac{v_\theta}{r} \frac{\partial v_\phi}{\partial \theta} + \frac{v_\phi}{r \sin\theta} \frac{\partial v_\phi}{\partial \phi} + \frac{v_\phi v_r}{r} + \frac{v_\theta v_\phi \cos\theta}{r \sin\theta} \right) \boldsymbol{i}_\phi$$

$$\tau : \nabla v = \tau_{rr} \frac{\partial v_r}{\partial r} + \tau_{\theta\theta} \left(\frac{1}{r} \frac{\partial v_\theta}{\partial \theta} + \frac{v_r}{r} \right) + \tau_{\phi\phi} \left(\frac{1}{r \sin\theta} \frac{\partial v_\phi}{\partial \phi} + \frac{v_r}{r} + \frac{v_\theta \cos\theta}{r \sin\theta} \right)$$

$$+ \tau_{r\theta} \left(\frac{1}{r} \frac{\partial v_r}{\partial \theta} + \frac{\partial v_\theta}{\partial r} - \frac{v_\theta}{r} \right) + \tau_{r\phi} \left(\frac{\partial v_\phi}{\partial r} + \frac{1}{r \sin\theta} \frac{\partial v_r}{\partial \phi} - \frac{v_\phi}{r} \right)$$

$$+ \tau_{\theta\phi} \left(\frac{1}{r} \frac{\partial v_\phi}{\partial \theta} + \frac{1}{r \sin\theta} \frac{\partial v_\theta}{\partial \phi} - \frac{v_\phi \cos\theta}{r \sin\theta} \right)$$

$$\nabla \cdot \tau = \left(\frac{1}{r^2} \frac{\partial (r^2 \tau_{rr})}{\partial r} + \frac{1}{r \sin\theta} \frac{\partial (\tau_{r\theta} \sin\theta)}{\partial \theta} + \frac{1}{r \sin\theta} \frac{\partial \tau_{r\phi}}{\partial \phi} - \frac{\tau_{\theta\theta} + \tau_{\phi\phi}}{r} \right) \boldsymbol{i}_r$$

$$+ \left(\frac{1}{r^2} \frac{\partial (r^2 \tau_{r\theta})}{\partial r} + \frac{1}{r \sin\theta} \frac{\partial (\tau_{\theta\theta} \sin\theta)}{\partial \theta} + \frac{1}{r \sin\theta} \frac{\partial \tau_{\theta\phi}}{\partial \phi} + \frac{\tau_{r\theta}}{r} - \frac{\tau_{\phi\phi} \cos\theta}{r \sin\theta} \right) \boldsymbol{i}_\theta$$

$$+ \left(\frac{1}{r^2} \frac{\partial (r^2 \tau_{r\phi})}{\partial r} + \frac{1}{r} \frac{\partial \tau_{\theta\phi}}{\partial \theta} + \frac{1}{r \sin\theta} \frac{\partial \tau_{\phi\phi}}{\partial \phi} + \frac{\tau_{r\phi}}{r} + \frac{2\tau_{\theta\phi} \cos\theta}{r \sin\theta} \right) \boldsymbol{i}_\phi$$

Table B.7 Solutions of Laplace's equation in two dimensions.

$$\nabla^2 \Phi = 0$$

Rectangular Coordinates
(independent of z)

e^{kx} and e^{-kx} may be replaced by $\sinh kx$ and $\cosh kx$.

$$\Phi = e^{kx}(A_1 \sin ky + A_2 \cos ky) + e^{-kx}(B_1 \sin ky + B_2 \cos ky)$$

$$\Phi = Axy + Bx + Cy + D; \qquad (k = 0)$$

Cylindrical Coordinates
(independent of z)

$$\Phi = r^n(A_1 \sin n\phi + A_2 \cos n\phi) + r^{-n}(B_1 \sin n\phi + B_2 \cos n\phi)$$

$$\Phi = (A_1\phi + A_2) \ln \frac{R}{r} + B_1\phi + B_2; \qquad (n = 0)$$

Spherical Coordinates
(independent of ϕ):

$$\Phi = A r \cos\theta + \frac{B}{r^2} \cos\theta + \frac{C}{r} + D$$

Appendix C
Vector Identities

In the following identities, ϕ and ψ are arbitrary scalar functions, and $\boldsymbol{A}$, $\boldsymbol{B}$, and $\boldsymbol{C}$ are arbitrary vectors.

$$\boldsymbol{A} \times \boldsymbol{B} \cdot \boldsymbol{C} = \boldsymbol{A} \cdot \boldsymbol{B} \times \boldsymbol{C}$$

$$\boldsymbol{A} \times (\boldsymbol{B} \times \boldsymbol{C}) = \boldsymbol{B}(\boldsymbol{A} \cdot \boldsymbol{C}) - \boldsymbol{C}(\boldsymbol{A} \cdot \boldsymbol{B})$$

$$\nabla(\phi + \psi) = \nabla\phi + \nabla\psi$$

$$\nabla \cdot (\boldsymbol{A} + \boldsymbol{B}) = \nabla \cdot \boldsymbol{A} + \nabla \cdot \boldsymbol{B}$$

$$\nabla \times (\boldsymbol{A} + \boldsymbol{B}) = \nabla \times \boldsymbol{A} + \nabla \times \boldsymbol{B}$$

$$\nabla(\phi\psi) = \phi\nabla\psi + \psi\nabla\phi$$

$$\nabla \cdot (\psi\boldsymbol{A}) = \boldsymbol{A} \cdot \nabla\psi + \psi\nabla \cdot \boldsymbol{A}$$

$$\nabla \cdot (\boldsymbol{A} \times \boldsymbol{B}) = \boldsymbol{B} \cdot \nabla \times \boldsymbol{A} - \boldsymbol{A} \cdot \nabla \times \boldsymbol{B}$$

$$\nabla \cdot \nabla\phi = \nabla^2\phi$$

$$\nabla \cdot \nabla \times \boldsymbol{A} = 0$$

$$\nabla \times \nabla\phi = 0$$

$$\nabla \times (\nabla \times \boldsymbol{A}) = \nabla(\nabla \cdot \boldsymbol{A}) - \nabla^2\boldsymbol{A}$$

$$(\nabla \times \boldsymbol{A}) \times \boldsymbol{A} = (\boldsymbol{A} \cdot \nabla)\boldsymbol{A} - \tfrac{1}{2}\nabla(\boldsymbol{A} \cdot \boldsymbol{A})$$

$$\nabla(\boldsymbol{A} \cdot \boldsymbol{B}) = (\boldsymbol{A} \cdot \nabla)\boldsymbol{B} + (\boldsymbol{B} \cdot \nabla)\boldsymbol{A} + \boldsymbol{A} \times (\nabla \times \boldsymbol{B}) + \boldsymbol{B} \times (\nabla \times \boldsymbol{A})$$

$$\nabla \times (\phi\boldsymbol{A}) = \nabla\phi \times \boldsymbol{A} + \phi\nabla \times \boldsymbol{A}$$

$$\nabla \times (\boldsymbol{A} \times \boldsymbol{B}) = \boldsymbol{A}(\nabla \cdot \boldsymbol{B}) - \boldsymbol{B}(\nabla \cdot \boldsymbol{A}) + (\boldsymbol{B} \cdot \nabla)\boldsymbol{A} - (\boldsymbol{A} \cdot \nabla)\boldsymbol{B}$$

Appendix D
System of Units

INTERNATIONAL SYSTEM OF UNITS (SI)

The SI is founded on seven SI base units for seven base quantities assumed to be mutually independent:

Quantity	Name	Symbol
Length	meter	m
Mass	kilogram	kg
Time	second	s
Electric current	ampere	A
Temperature	kelvin	K
Amount of substance	mole	mol
Luminous intensity	candela	cd

DERIVED UNITS AND UNITS WITH SPECIAL NAMES

Derived quantity	Name	Symbol	Equivalent units
Area	square meter	m^2	
Volume	cubic meter	m^3	
Volume	liter	L	$1\,L = 10^{-3}\,m^3$
Velocity	meter per second	$m\,s^{-1}$	
Acceleration	meter per second squared	$m\,s^{-2}$	
Force	newton	N	$m\,kg\,s^{-2}$
Pressure, Stress	pascal	Pa	$N\,m^{-2} = m^{-1}\,kg\,s^{-2}$
Pressure	bar	bar	$1\,bar = 100\,kPa$
Energy	joule	J	$N\,m^{-2} = m^2\,kg\,s^{-2}$
Power, radiant flux	watt	W	$J\,s^{-1} = m^2\,kg\,s^{-3}$
Plane angle	radian	rad	$m\,m^{-1} = 1$
Solid angle	steradian	sr	$m^2\,m^{-2} = 1$
Mass density	kilogram per cubic meter	$kg\,m^{-3}$	
Concentration	mole per cubic meter	$mol\,m^{-3}$	
Specific volume	cubic meter per kilogram	$m^3\,kg^{-1}$	
Catalytic activity	katal	kat	$s^{-1}\,mol$
Catalytic concentration	katal per cubic meter	$kat\,m^{-3}$	
Dynamic viscosity	pascal second	Pa s	$1\,poise = 0.1\,Pa\,s$
Kinematic viscosity	meter squared per second	$m^2\,s^{-1}$	$1\,stoke = 10^{-4}\,m^2\,s^{-1}$
Moment of force	newton meter	N m	
Surface tension	newton per meter	$N\,m^{-1}$	
Angular velocity	radian per second	$rad\,s^{-1}$	
Angular acceleration	radian per second squared	$rad\,s^{-2}$	
Celsius temperature	degree Celsius	°C	K
Wavenumber	reciprocal meter	m^{-1}	
Frequency	hertz	Hz	s^{-1}

Derived quantity	Name	Symbol	Equivalent units
Current density	ampere per square meter	$A\,m^{-2}$	
Electric charge	coulomb	C	$s\,A$
Electric potential	volt	V	$W\,A^{-1} = m^2\,kg\,s^{-3}\,A^{-1}$
Capacitance	farad	F	$C\,V^{-1} = m^{-2}\,kg^{-1}\,s^4\,A^2$
Electric resistance	ohm	Ω	$V\,A^{-1} = m^2\,kg\,s^{-3}\,A^{-2}$
Electric conductance	siemens	S	$A\,V^{-1} = m^{-2}\,kg^{-1}\,s^3\,A^2$
Magnetic field strength	ampere per meter	$A\,m^{-1}$	
Magnetic flux	weber	Wb	$V\,s = m^2\,kg\,s^{-2}\,A^{-1}$
Magnetic flux density	tesla	T	$Wb\,m^{-2} = kg\,s^{-2}\,A^{-1}$
Inductance	henry	H	$Wb\,A^{-1} = m^2\,kg\,s^{-2}\,A^{-2}$
Electric field strength	volt per meter	$V\,m^{-1}$	
Electric charge density	coulomb per cubic meter	$C\,m^{-3}$	
Electric flux density	coulomb per square meter	$C\,m^{-2}$	
Dielectric permittivity	farad per meter	$F\,m^{-1}$	
Magnetic permeability	henry per meter	$H\,m^{-1}$	
Luminous flux	lumen	lm	$cd\,sr = m^2\,m^{-2}\,cd = cd$
Illuminance	lux	lx	$lm\,m^{-2} = m^{-2}\,cd$
Radioactive decay	becquerel	Bq	s^{-1}
Absorbed energy	gray	Gy	$J\,kg^{-1} = m^2\,s^{-2}$
Dose equivalent (d)	sievert	Sv	$J\,kg^{-1} = m^2\,s^{-2}$
Luminance	candela per square meter	$cd\,m^{-2}$	
Radioactive decay	curie	Ci	$1\,Ci = 3.7 \times 10^{10}\,Bq$
	roentgen	R	$1\,R = 2.58 \times 10^{-4}\,C\,kg^{-1}$
	rad	rad	$1\,rad = 1\,cGy = 10^{-2}\,Gy$
	rem	rem	$1\,rem = 1\,cSv = 10^{-2}\,Sv$
Heat flux density	watt per square meter	$W\,m^{-2}$	
Heat capacity, entropy	joule per kelvin	$J\,K^{-1}$	
Specific heat capacity	joule per kilogram kelvin	$J\,kg^{-1}\,K^{-1}$	
Specific energy	joule per kilogram	$J\,kg^{-1}$	
Thermal conductivity	watt per meter kelvin	$W\,m^{-1}\,K^{-1}$	
Energy density	joule per cubic meter	$J\,m^{-3}$	
Molar energy	joule per mole	$J\,mol^{-1}$	
Molar heat capacity	joule per mole kelvin	$J\,mol^{-1}\,K^{-1}$	
Exposure	coulomb per kilogram	$C\,kg^{-1}$	
Absorbed dose rate	gray per second	$Gy\,s^{-1}$	
Radiant intensity	watt per steradian	$W\,sr^{-1}$	

Appendix E
Physical Constants

Quantity	Symbol	Numerical value
Avogadro constant	N_A	$6.022 \times 10^{23}\,\mathrm{mol}^{-1}$
Molar gas constant	R	$8.314\,\mathrm{J\,mol}^{-1}\,\mathrm{K}^{-1}$
Boltzmann constant	$k_B = R/N_A$	$1.381 \times 10^{-23}\,\mathrm{J\,K}^{-1}$
Elementary charge	e	$1.602 \times 10^{-19}\,\mathrm{C}$
Faraday constant	$F = N_A\,e$	$96\,485\,\mathrm{C\,mol}^{-1}$
Speed of light in vacuum	c	$2.998 \times 10^{8}\,\mathrm{m\,s}^{-1}$
Magnetic permeability in vacuum	μ_0	$4\pi \times 10^{-7}\,\mathrm{H\,m}^{-1}$
Electric permittivity in vacuum	$\epsilon_0 = 1/(\mu_0 c^2)$	$8.854 \times 10^{-12}\,\mathrm{F\,m}^{-1}$
Atomic mass constant	$m_u = \frac{1}{12} m(^{12}\mathrm{C}) = 1\,\mathrm{u}$	$1.661 \times 10^{-27}\,\mathrm{kg}$
Standard gravity	g	$9.80665\,\mathrm{m\,s}^{-2}$
Gravitational constant	G	$6.674 \times 10^{-11}\,\mathrm{m}^3\,\mathrm{kg}^{-1}\,\mathrm{s}^{-2}$

Index

Aggrecan, 18, 20, 20F
Aggregate modulus, 251
Agrin, 19
Ambipolar
  diffusion coefficient, 110
  diffusion equation, 110
  mobility, 111
Ampère's law, 35
Amino acids, 19, 19F
Amino groups, ionizable, 18, 19F
Aneurism, 197
Annulus fibrosus, 241
Arrhenius, 11
Asporin, 19

Bamacan, 19
Bernoulli's equation, 175, 197
Biglycan, 19
Biharmonic equation, 185
Binary electrolyte
  transport, 71, 126
Binding site density, 15
Boltzmann
  constant, 8
  distribution, 89
  equilibrium, 91
  factor, 151
  model, 90
  statistics, 66, 90
Boundary conditions
  chemical, 21–23
  electroquasistatic, 51, 53
  magnetoquasistatic, 54
Boundary layer, 183, 184F
  diffusion, 191–194, 192F, 195F
  viscous, 192–194, 192F, 195F
Boundary value problems
  chemical, 23
  EQS and MQS, 51
Brevican, 18
Brownian motion, 6
Bulk modulus, 249, 252, 274
Bullard's equation, 194

Capacitive coupling, 59
Carboxyl groups, ionizable, 18, 19F
Cartilage, articular, 241F, 243, 244
  specialized dense connective
    tissue, 244
  equilibrium modulus, ionic and
    electrical effects, 268F
  chemical stress, ionic and electrical
    effects, 268F
Cell
  bleb, 285
  deformation, 275
  electrokinetic interactions, 235F
  osmotically swollen, 234
  stress relaxation, 279
Cell surface, 15
  impedance, 234
  receptors, IGF-1, 14
  surface charge by electrophoresis,
    229, 229F
Cervical mucus granules, 242

Charge density
  volume, free, 34, 142
  vs. pH, 106
Charge groups, ionizable
  extracellular matrix, 15–21
  dissociation, 95F
Charge relaxation, 53, 75, 245
  in charged porous media, 111
  time constant, 108
Chemical potential, 87
Chemical reaction, 10–21, 24
  diffusion-limited, 24
Chemical stress, 266F, 267
Chemotaxis, 9, 9F
Chondroadherin, 19
Chondroitin sulfate, 20, 20F, 242
Co-ion (minority carrier) transport,
  81, 111, 112
Collagens, 16–19, 16F, 17F, 277
Collagen charge groups, 17
Collagen fibrils, 241, 241F
Collagen titration, 17, 18F, 106F
Collagen transmembrane potential,
  105, 105F
Complex modulus, 257, 280
Compression, uniaxial, 250, 264F
Compressive stress, 267
Concentration boundary
  condition, 22
Conductance
  equivalent, 43
    concentration dependence, 43
  measurement configuration, 216F
Conduction
  steady, 56
  nonsteady, 58
Conductivity, electrical, 42
  cells, 237
  compared with number density, 42
  effective for compound medium, 64
  measurement, 131F
  nonuniform, 67
  of heterogeneous media, 63
  surface conductivity, 237
  tissue and polyelectrolyte charge
    density, 131
Confined compression, 250, 250F,
  264F
Connective tissues, 240–244
  swelling behavior, 240
Conservation
  of charge, 36–37
    electrokinetic surface charge
      boundary condition, 220,
      230
  of mass
    fluid, 5, 173–174,
    poroelastic material, 262
  of momentum, 174, 182
    poroelastic material, 263
  of solute species, 3
Consolidating porous media, 260
Constitutive law
  electric current
    electroquasistatic, 42
    magnetoquasistatic, 54

Continuity, 50, 108
  binary electrolyte, 72
  for ith solute species, 2, 4
  for electrolytes, 4
  for species in an electrolyte
    solution, 5
  magnetoquasistatic, 51
  of co-ion, 117
Continuum approximation, 8, 9
Continuum Electromechanics, 186
Convection, 41
Convective diffusion, 189, 190F
Convective mass transport, 189
Convective momentum, 176
Convective surface current, 220
Convolution integral, 257
Cornea, 243
Corneal stroma, 240, 241F, 243, 275
  regular dense connective
    tissue, 243
Couette flow, 182
Couette viscometer, 198
Coulter counter, 64
Counter-ion, 81, 209F
Coupled diffusion
  binary electrolyte, 74, 76, 80
  charged porous media, 108–111
  neutral membrane, 123
Creep flow, 186, 209, 219
Creep
  experiment, 253F
  viscoelastic, 253, 254
Cross-linkages, collagen, 17, 244
Current density
  constitutive law
    electroquasistatic, 42
    magnetoquasistatic, 54
Current dipole, 66, 232
Cytoskeleton, 284

Damköhler number, 24
Darcy's law, 262
  with electroosmotic fluid flow, 269
Darcy hydraulic permability, 260
Dashpot element, 254
Debye length, 66, 71, 93, 104
  definition, 152
Debye–Hückel
  linear approximation, 66, 152
  correction, 96
  ionic atmosphere, 204
Debye length, 66, 73
Decorin, 19, 275
Detailed balance, principle of, 11, 90
Dielectric constant, 40
  values of, 46
Del (gradient) operator, 2
Dielectric suction, 145
Dielectrophoresis, 146–150, 147F
Differential operators, 291–295
Diffusion, 5, 41, 71
  ambipolar, 110
  binary electrolyte, 126
  magnetic, 194
  of mobile ions, 245
  of vorticity, 194

Diffusion (*Continued*)
 time constant, 108
 poroelastic wave, penetration
  depth, 271
Diffusion coefficient
 ambipolar, 110
 effective
  binary electrolyte, 74
  interdiffusion, 114
 of vorticity, 191
 values of, 41
Diffusion equation, 3
 poroelastic, 263
 poroelastic–electrokinetic, 270
Diffusion lag time, 30
Diffusion-limited chemical reaction,
  24, 25, 27
Diffusion potential, 77, 79
 charged membrane, 99–103
 multi-ionic, 87
Diffusion-limited binding, 245
Diffusion–reaction
 in charged membranes, 30
Diffusivity
 coupled, 80
 porous membrane, 84
 solute, 2
 Stokes–Einstein relation, 7
 viscous, 195
Dilatational (longitudinal) waves,
  elastic, 252
Dipole
 adsorption, 204
 centric dipole, electrocardiography,
  66
 current, 66, 232, 233
 moment, electric, 38, 148
Dipole field
 electric, 50, 50F, 57
Dirichlet condition, 21
Disaccharide, 20F, 21
 charge groups, 20F, 20–21, 229
Dispersion parameter, 200
Displacement current, 35
Displacement flux density, 39
DLVO theory, 159
Donnan
 equilibrium, 89, 92, 96, 129, 130
 exclusion, 106, 107, 108
 intra-tissue charge density, 96–99,
  99F
 intra-tissue pH, 95
 nonuniformly charged tissues, 96
 partitioning, 94, 96
 potential, 89, 92, 94, 102
Double layer, 150–156, 159–165
 capacitor, 66
 charge distribution, 204F
 diffuse, 150, 204
 electrical, 65
 Helmholtz layer, 204, 211
 mobile charge, 220
 planar, 150F
 plane parallel interacting, 159–165,
  161F
 potential distribution, 154F, 155F,
  161F, 168F, 204F
 repulsion, 159–165
Drug release, 24F

Einstein random walk
 diffusion time, 55
Einstein relation, 29, 41, 74
Einstein summation convention, 140
Elasticity, 246–252

Electric displacement flux density, 34
Electric field, 34–37, 48–54, 56–61
Electric fish, 62
Electric permittivity, 40
Electric potential, 49, 87
Electric susceptibility, 40
Electrocapillarity, 205
Electrochemical potential, 42, 87
Electrochemical reaction, 44
Electrode, 84
 Ag/AgCl, 86, 88
 calomel, 85, 86
 membrane, 124
 metal, 85
 nonpolarizable, 85
 pH, 85, 132, 133F
 platinum black, 86
 reversible, 85
 standard potential, 86
 working, 85
Electrode–electrolyte interface, 43
Electrodiffusion, 107, 245
 applied field, 115
 steady state, 134
 transient, 118, 135
Electrokinetics, 203–238
 intracellular, 284, 284F
Electrokinetic backflow, 269
Electrokinetic boundary condition,
  220, 220F
Electrokinetic coupling coefficients,
  212–215, 269
 cylindrical rod model, 224
 macroscopic model, 224
 dynamic measurement, 226, 227F
Electrokinetic slip plane, 210, 220
Electrokinetic transduction
 pore model, 222
 in poroelastic media, 268
Electrolysis, 43, 43F, 214
Electromagnetic blood flow meter, 121
Electromagnetic induction, 35
Electromagnetic spectrum, 48
Electromagnetic waves, 47
 plane wave, 47
Electroneutrality, 73, 89, 104, 125
Electroosmosis, 206F, 207, 208–213,
  219F
 planar, 221, 221F
Electrophoresis, 206F, 207, 217–222
 free, 217
 including surface conductivity, 237
 of microtubules, 219F
 polyacrylamide gel disk, 218F
 Tiselius moving-boundary
  apparatus, 218F
Electrophoretic velocity
 insulating charged particle, 222
 liquid metal droplet, 231
Electroquasistatic, 49
Electrosurgery, 68
Energy method, 145
Epiphycan, 19
Equivalent conductance, 43
Error function, 194
Excluded volume effects,
  polymer, 240
Extension, uniaxial, 248
Exterior flow, 188, 230
Extracellular matrix, 15, 26
Euler's equation, 48, 257

Faraday's constant, 41, 151
Faraday's law, 35
Fibromodulin, 19

Fick's first law, 2
Fick's second law, 3
Finite Fourier transform method, 23
Fixed charge membrane models,
  99–107
Fluid equation of state, 175, 197
Fluid flow
 creep, 186
 fully developed, 182
 in blood vessel, 72F
 incompressible, 175
 inviscid, 175
 irrotational, 177
 laminar, 198
 low-Reynolds-number, 182, 189
 Plane Couette, 182
 Poiseuille, 182, 212
Fluid isotropy, 179
Fluid mechanics
 Newtonian, 173
Fluid pendulum, 177
Fluid velocity, relative to solid, 262
Fluid volume flux, 262
Fluid viscous stress tensor, 181
Flux
 binary electrolyte, 72
 diffusive, 1
 electric field, 108
 molar, 2
 of ions, 40, 108
 through porous membrane, 83
Flux boundary condition, 22, 22F
Flux density, 2
Force, 139
 density, 141
  Kelvin, 145, 169
  Korteweg–Helmholtz, 146
  polarization, electric, 144, 146
  and stress tensor, 276
 London–van der Waals, 161
 total, 140
Fourier series expansion, 124
Fully developed flow, 182, 184F
 in electrokinetics, 210

GAG, 18
Gauss' integral theorem, 37
Gauss' law
 for electric field, 34
 for magnetic field, 35
 electroosmosis, 210
Gauss' theorem, 2, 36, 140
Gels, 242
Gel diffusivity, 242, 260
Gel swelling, 242, 260, 261F
Gibbs free energy, 87
Glutamic acid, 19F
Glycosaminoglycan, 18, 20, 20F,
  99, 240
Goldman equation, 128
Good solvent, 240
Gouy–Chapman diffuse double layer,
  204, 213F
Golgi apparatus, 19
Grahame's model, 155
Gravitational force density, 175

Helmholtz coil, 68
Helmholtz double layer, 211, 213F,
  220, 221
 Helmholtz capacitor, 155, 207
Henderson equation (junction
  potential), 101
Henderson–Hasselbalch equation, 12
High-Reynolds-number flow, 183

Histidine, 19F
Hodgkin–Katz model, 128
Hooke's law, generalized, 249, 250
Hyaluronan, 20, 20F, 242
Hydraulic permeability, Darcy, 262
    open circuit, 269
    short circuit, 269
Hydrodynamic filtration coefficient, 214, 216F
Hydrogels, 242
Hydrostatic pressure, 163

Ideal polarization, 203, 205, 230
Ideal semipermeability, 81
Incompressible flow, 175
Incompressible viscous fluids, 182
Inductive coupling, 59, 61
Insulin-like growth factor 1, 13, 14F, 26, 26F
Insulin-like growth factor binding proteins, 13, 14F, 26, 26F
Insulin-like growth factor receptor, 13, 14F
Interfacial matching conditions, 21
Interior flow, 188, 230
Intervertebral disk, 240, 241F, 273, 273F
Inviscid flow, 175
Ion exchange membranes, 113
Ionic transport, 40
Ionizable charge groups, 15–21
Irrotational flow, 177, 196
Isoelectric point, 131

Junction potential, 80

Kelvin force density, 145
Keratan sulfate, 20
Keratocan, 275
Kinematic viscosity, 191
Korteweg–Helmholtz force density, 146
Kronecker delta function, 143, 249

Lag time, diffusion, 30
Lamé constants, elastic, 249, 267
Laminar flow, 198
Laplace's equation, 23, 51, 53
    separation of variables, 56
    solution to, 60, 297
    Laplacian operator, 3
Law of detailed balance, 90
Legendre polynomial, 57
Leibnitz rule, 173, 287
Lenz's law, 35
Levich model, electrophoresis, 222
Ligament, 244
Link protein, 20, 20F
London–van der Waals force, 161
Longitudinal modulus, 251, 266F, 267
Lorentz force law, 37
Loss modulus, 257
Low-Reynolds-number flow, 182
Lumican, 275
Lumped damping term, 177
Lumped element models
    fluid, 177
    viscoelastic, 254, 282, 282F
Lysine, 19F

Magnetic diffusion, 54, 194
Magnetic field, 33, 35, 39, 54
Magnetic flux, 36

Magnetic flux density, 39
Magnetic Reynolds number, 194
Magnetic susceptibility, 40
Magnetization, 39
Magnetization density, 39
Magnetoquasistatic, 51
Mass action, principle of, 10
Mass density, 173
Maxwell model, viscoelastic, 255, 256F
Maxwell stress tensor, 143, 210
Maxwell's equations
    differential form, 36
    electroquasistatic, 48
    in media, 38–44
    integral forms, 33–36
    magnetoquasistatic, 51
    summary, 63
Mean normal stress, 252
Melcher, James R, 186
MEMS, 203
Membrane, charged, 95, 95F
    ion exchange, 127
    electrokinetic transduction phenomena, 215, 216
    transductive coupling, 227–228
Membrane potentials, 84, 86
Membrane thickness
    effective, 126
    mercury–electrolyte interface, 205F
Microfabricated channel, 198, 199F, 210F, 219F
    current within, 56
Migration, 41, 71
Migration currents, 109
Minority carrier, 111
Mobility
    ambipolar, 111
    electric, 40
        values of, 41
    liquid metal drop, 229
Mucin, 242
    gastrointestinal, 243F

Nanoporous membrane, 209F
Navier–Stokes equation, 178–182, 210
NEMS, 203
Nernst film, 190F
Nernst–Planck equation, 42
Nernstian relation, 87, 167
Neumann condition, 21
Neurocan, 18
Neutral membrane, 77
Newtonian fluid, 179
Newtonian fluid mechanics, 173–196
No-slip condition, 189, 211
Nucleus pulposus, 241
Number density
    compared with conductivity, 42

Ohm's law
    electrolyte solutions, 42
    with pressure-induced current density, 269
Ohmic conduction, 40
Ohmic model, 120
Ohmic power dissipation, 68
Ohmic surface current, 220
Oligosaccharides, 242
Osmotic pressure, 8, 163
Osmotic swelling pressure, 161–165
Osteoarthritis, 244
Osteogenesis, 68
Osteoglycin, 275

pH
    definition, 13
    intra-tissue, 94
pK
    definition, 13
    proteoglycan charge groups, 95F
Parallel shear flow, 179F
Particle image velocimetry, 199
Partition coefficient, 22, 28, 129
Péclet number, 191
Penetration depth, poroelastic wave, 271
Peptide backbone, 19F
Pericellular matrix, 20
Perlecan, 19
Permeability
    Darcy, hydraulic, 262
    magnetic, 35, 40, 51
    membrane, 82
    solute, 82, 217
Permittivity, 34
Permselective membrane, 81
Physical constants, 303
Pipe flow, 182
Plane Couette flow, 182
Plane fully developed flow, 182
Poiseuille flow, 182, 183F, 199F, 212
Poisson's equation, 51
Poisson's ratio, 249, 251, 272
Poisson–Boltzmann equation, 66, 93, 151, 152, 159
Polarization, electrical, 38
Polarization density, 38
Polarization force density, 144
Polyelectrolyte charge, 103
Polyelectrolyte hydrogels, 242, 243F
Polyelectrolyte macromolecules, 157
Polyelectrolyte membranes, 158, 223
Polyelectrolyte/electrolyte interface, 156
Poroelasticity, 239, 259–268
    intracellular, 284, 284F
Poroelastic diffusion equation, 263, 281
Poroelastic electrokinetics, 268
    diffusion equation, 270
Poroelastic media
    linear, isotropic, homogeneous, 261, 280
    inhomogeneous, two-layer, 282, 282F
    pressure-induced consolidation, 283, 283F
Poro-viscoelastic material, 285
Porosity, 262
Potential, 49
    chemical, 87
    Donnan, 92, 94
    electrical, 87
    electrochemical, 87, 90
    measured, 87
    sedimentation, 206, 207
    streaming, 206, 213
    vs. pH, 105
    zeta, 211
Potential difference
    charged membrane, 102
Potential-determining ion, 166
Pressure
    osmotic swelling, 164
Proteoglycans, 18–21

Quasineutrality, 73
    condition, 108
    validity test, 125

Quasistatic approximations
electroquasistatic, 48
magnetoquasistatic, 51

Rate of strain, fluid, 178
Reaction
equilibrium constant, 11
reversible, 10
mechanism, 10
molecularity, 10
order, 10
rate, 10
Reference frame, 28
Relaxation time
charge, 53
dielectric, 53
Repulsive electrostatic
interactions, 240
Resistivity, 45
Reynolds number, 183, 191
magnetic, 194
Rheology of tissues, 23, 239–272
Rigid body
rotation, translation, 178, 248
Rough endoplasmic reticulum, 19

Salt bridges, 86
Scaling, 25
Schmidt number, 191
Sedimentation potential, 206F, 207,
232, 232F
Self-assembling peptide hydrogel,
209F
Semipermeability, 81
Separation of variables, 23
electric potential solutions, 57, 297
Shear, elastic
modulus, 249
electrical and ionic interactions,
267
simple, 248F, 250
pure, 248F
shear (transverse) waves, 252, 277,
277F
Shear stress, 140, 274
SI units, 301
Skin depth, 55
Slip plane, 211
Solenoidal
flux, 23
velocity, 174
Space charge, 34
Spring, viscoelastic element, 254
Standard electrode potential, 86
Standard linear solid, 256, 256F, 278
Stenosis, 197
Stern layer, 213F
Stokes drag, 7, 189
Stokes equation, 185, 207
Stokes stream function, 186
Stokes theorem, 36
Stokes–Einstein relation, 7
Storage modulus, 257
Steady conduction, 56, 57
Strain, elastic, 247, 248
Cauchy strain tensor, 248
Strain rate, fluid, 178
Stream functions, 185
Streaming potential, 206F, 208–213
dynamic measurements, 225
measurement, flow-induced, 269F

oscillatory, compression-induced,
270, 270F, 271F
Streamline, 176
Stress, 139
chemical, 267
compressive, 267
elastic, 247
function, 139
normal, 140
shear, 140
Stress relaxation
experiment, 253F
poroelastic, 264
viscoelastic, 253, 255, 256F
cellular, 279, 279F
Stress tensor, 140
elastic materials, 247
Maxwell (electrical) stress tensor,
142–143
viscous, 181
Stress–strain relations
elastic, 246–252
poroelastic, 261–262
electrical and ionic interactions,
265
Stress–strain rate relations
viscous, 178–182
Sulfate groups, ionizable, 21
Summation convention, 140
Surface charge density, 54
Surface integral, time rate of
change, 287
Surface tension, 177
effects of surface charge, 231
electrocapillarity, 231
Swelling, 240, 246
chemical, 246
electromechanical, 246
kinetics, 245
mechanical, 246
nonequilibrium, 245
of tissues, 239–246
osmotic, 246
time constant, 260

Taylor dispersion, 200
Taylor expansion
of solute concentration, 5
of fluid velocity, 178
of elastic displacement, 248
Tectorial membrane, 277, 277F
Tendon, 240, 241F
rabbit flexor tendon, 281, 281F
regular oriented connective
tissue, 244
Teorell–Meyer–Sievers membrane
model (TMS), 99, 103
Theorem of vorticity, 196
Thermal equilibrium, 89
Thermal voltage, 75
Thermodynamic equilibrium, 82
Time constant, characteristic
charge relaxation, 53, 108, 245
coupled diffusion, 123
creep, viscoelastic, 255, 255F
creep poroelastic, 265
diffusion, 25, 108, 245
Einstein random walk, 55
diffusion–reaction, 28, 245
electrodiffusion, 118, 245
electrodiffusion–reaction, 245
fluid transport, 184

magnetic diffusion, 55
migration, 43
reaction, 25
stress relaxation, viscoelastic, 256,
256F
stress relaxation, poroelastic, 245,
264
viscous diffusion, 184
Tissue swelling kinetics, 245–246
Titration
acetic acid, 12, 12F
collagen, 18F
polyelectrolyte, 96
ribonuclease, 131F
tectorial membrane constituents,
278
Torsional strain, 253
sinusoidal experiment, 253F
Tortuosity factor, 83–84
Traction, 139, 139F
Transmembrane potential, 105, 105F
Transport cell configuration, 104F,
136F
Transport number, 79

Umbilical cord, 241, 242
Unconfined compression, 250, 250F
Uniaxial tension, 251F
Uniqueness, 22, 52, 67
Units, 301–302
Universal gas constant, 151

Valence, 40, 151
van't Hoff's law, 8, 163, 170
Vector identities, 299
Vector Laplacian, 186
Vector potential, fluid, 185
Vector stream function, 185
Velocity
mass-averaged, 4
molar-averaged, 4
Versican, 18
Viscoelasticity, 239, 253–259
quasilinear, 259
solid phase, 258
Viscosity
dynamic, 179
kinematic, 191
second coefficient, 181
Viscous diffusion coefficient, 184, 195
Viscous forces, 178–182
Viscous stress–strain rate relations,
178–182
Voigt model, 254, 255F
Volume rate of flow, 186
Volume integral, time rate of change,
288
Vorticity, 176, 196

Wave equation, 47
Wave propagation speed, 252
Weak electrolyte, 91
Wharton's jelly, 241
loose irregular connective tissue,
242

Young–Laplace equation, 231
Young's modulus, 249, 251

Zefal pump, 197
Zeta potential, 204F, 211

Printed and bound by CPI Group (UK) Ltd, Croydon, CR0 4YY

01/11/2024

01782605-0017